The Handbook of
PRIVATE TELEVISION

The Handbook of
PRIVATE

A Complete Guide for Video Facilities and Networks within Corporations, Nonprofit Institutions, and Government Agencies

TELEVISION

Nathan J. Sambul, Editor in Chief

McGraw-Hill Book Company

New York St. Louis San Francisco Auckland
Bogotá Hamburg Johannesburg London Madrid
Mexico Montreal New Delhi Panama Paris
São Paulo Singapore Sydney Tokyo Toronto

To
Murray Pollner
Special Thanks

Library of Congress Cataloging in Publication Data
Main entry under title:
The Handbook of private television.

   Includes index.
   1. Closed-circuit television—Addresses, essays,
lectures.   2. Industrial television—Addresses,
essays, lectures.   I. Sambul, Nathan J.   II. Title:
Private television.
HE8700.6.H36   384.55'5   81-20852
ISBN   0-07-054516-2   AACR2

Copyright © 1982 by McGraw-Hill, Inc. All rights reserved. Printed in the United States of America. Except as permitted under the United States Copyright Act of 1976, no part of this publication may be reproduced or distributed in any form or by any means, or stored in a data base or retrieval system, without the prior written permission of the publisher.

1234567890   DODO   898765432

ISBN 0-07-054516-2

The editors for this book were William A. Sabin and Charles P. Ray; the designer was Mark E. Safran, and the production supervisor was Sara Fliess. It was set in Vermilion by The Kingsport Press.

Printed and bound by R. R. Donnelley & Sons Company.

Contents

Contributors **xi**

Preface **xix**

PART 1 — An Overview of Private Television **1-1**

1. The Rise of Private Television **1-3**
PETER CARANICAS

2. An Organizational Perspective of Corporate Television **2-1**
ROBERT PASSARO

3. The Pitfalls of Private Television **3-1**
RON WHITTAKER

PART 2 — A Manager's Guide to Private Television **4-1**

4. Implementing a Private Television Operation **4-3**
PAMELA J. NEWMAN

vi CONTENTS

Appendix:
A Case Study in Implementing
a Private Video Network **4-16**
NATHAN J. SAMBUL

5 Operating a Video Facility
with a Limited Staff **5-1**
SHERRY LIEBOWITZ

6 Operating a Major Private
Television Operation **6-1**
BILLY BOWLES

7 Using the Outside
Producer **7-1**
BRUCE FINLEY AND BETSY FINLEY

8 Funding and Budgeting **8-1**
MEG GOTTEMOELLER

9 Contracts **9-1**
NANCY F. GURWITZ

10 Guidelines for Success **10-1**
NATHAN J. SAMBUL

PART **3** Production
within Private
Television **11-1**

11 Production Values **11-3**
JEANETTE P. LERMAN

12 Program
Development **12-1**
ROBERT T. HIDER

CONTENTS

13 Scripting **13-1**
DIANE M. SHARON

14 Set Design **14-1**
MARILYN REED

15 Graphics **15-1**
JAN THIEDE

16 Lighting **16-1**
JOEL WILLIS

17 Makeup **17-1**
LESLIE SHREVE

18 Audio in Television **18-1**
MARK DICHTER

19 Directing and Rehearsing the Nonprofessional Talent **19-1**
JUDY ANDERSON

20 Directing **20-1**
LON MCQUILLIN

21 Shooting on Location **21-1**
JEFFREY J. SILVERSTEIN

22 Editing **22-1**
PATRICK B. HOWLEY

viii CONTENTS

Appendix: Guidelines for
Purchasing Equipment   **22-12**
PATRICK B. HOWLEY

23 Videocassette Duplication and Distribution   **23-1**
WILLIAM B. FOLLETT

24 The Copyright Law and How to Use It   **24-1**
THOMAS VALENTINO, JR.

PART 4 International Video Networking   **25-1**

25 International Video Equipment Standards   **25-3**
GERALD CITRON

26 International Video Operations   **26-1**
PATRICIA TIERNEY WILSON

PART 5 Staffing   **27-1**

27 Staffing a Private Television Operation   **27-3**
THOMAS WM. RICHTER

28 Getting the Job You Want   **28-1**
NATHAN J. SAMBUL

PART 6
Tomorrow's Private Video Communications   29-1

29 The Future of Private Television   29-3
JOHN P. WILISZOWSKI, JR.

Index (follows Chapter 29)

Editor in Chief

Nathan J. Sambul is president of his own consulting firm NJSambul and Company Inc., serving the media, corporate, and financial communities. In addition, he is an instructor at The New School for Social Research (New York City) for an advanced media management course, has taught various courses at the International Television Association and Video Expo, and has authored articles in the fields of media and finance.

Prior to forming his own company, Mr. Sambul was assistant to the president at Video Corporation of America where his responsibilities included contract negotiations, budgeting, and administrative management of the consumer video division.

From 1970 through 1979, Mr. Sambul was employed by Merrill Lynch, Pierce, Fenner and Smith Inc. He was an officer of the firm and manager of the Video Network, one of the nation's most sophisticated private television networks. He has won awards from the International Television Association, Chicago Film Festival, and International Film Producers Association. During his tenure at Merrill Lynch, Mr. Sambul became a registered representative and completed M.B.A. courses as part of his Ph.D. candidacy at New York University. He holds an M.A. from the University of Michigan and a B.A. from Queens College. (*Chapter 4 Appendix; Chapter 10; Chapter 28*)

Contributors

Judy Anderson is director of audiovisual services at Peat, Marwick, Mitchell & Co., an international Big Eight public accounting firm. In this position, Ms. Anderson is responsible for the firm's audiovisual activities, including both training and communications. A communications arts graduate of Michigan State University, Ms. Anderson came to PMM & Co. from the Metropolitan Life Insurance Company, where she was manager of corporate audiovisual media services. Ms. Anderson has written a regular column in *Videography* on corporate television, and has served as editor of the *International Television News*, the publication of the International Television Association. She has also previously been published in *Public Relations Journal* and several other magazines and newspapers. (*Chapter 19*)

Billy Bowles is director of media resources for General Telephone Company of Florida. He has held this position since 1972. He manages a major private television facility which utilizes the medium for both training and informational video communications. His expertise has been recognized internationally, having served as president and chairman of the board of the International Television Association (ITVA). His background in television since 1960 is unique, having worked in public, commercial, and educational television. His articles have been published in popular magazines and professional journals, and he is asked to speak frequently at video conferences both here and abroad regarding private television. His B.A. degree was awarded in broadcasting and speech in 1960 from Florida State University. (*Chapter 6*)

Peter Caranicas is a consultant to the cable television industry and a former editor of *View*, which covers cable programming. From 1976 to 1980, he was editor of *Videography*, a leading trade publication of the private television field. Other publications he has been associated with include *Home Video*, a magazine for consumers and *Home Video Report*, a newsletter for video industry executives. He is a graduate of Yale, the London School of Economics, and frequently speaks at conferences and appears on panels relating to the video and cable television industries. (*Chapter 1*)

Gerald Citron, president of Intercontinental Televideo Inc., N.Y., received an economics degree from the Wharton School of the University of Pennsylvania. In 1974, after eight years with North American Philips Corp., Mr. Citron established Intercontinental Televideo Inc., a multinational video services and facilities company specializing in industrial and broadcast applications of video in all world formats and standards. In 1977, he formed Overseas News & Information Service Inc. (ONISI) to distribute video programs throughout the world. ONISI's principal programs are the *CBS Evening News* and *60 Minutes* which are now distributed to over seventy-five locations worldwide. Mr. Citron has written numerous articles and been a frequent speaker and lecturer on the subject of video hardware and software at industry conventions and gatherings throughout the United States, Europe, and the Far East. (*Chapter 25*)

Mark Dichter is an independent audio engineer, working in both film and television. He holds a B.S.E.E. from M.I.T. and an M.F.A. from Columbia University. Presently, he spends a majority of his time as a recordist for television films. He also is on the faculty of Columbia University's film department and guest lectures at other schools in the vicinity. He has written articles on recording and consults with various private television operations, devising ways to improve their use of audio. At this time he is designing one- and two-day seminars for private video. (*Chapter 18*)

Betsy Finley is a producer at Finley Communications Inc. She was previously employed by Abraham & Straus department store as an in-house producer of television commercials. She also worked at Flickers Inc., a production company, as casting director. Betsy Finley received a B.F.A. from Pratt Institute and for five years after graduating was an assistant at Irv Bahrt Photography Studio, a New York advertising photography firm. (*Chapter 7*)

Bruce Finley is president of Finley Communications Inc., a New York production and consulting firm specializing in corporate television programs. In the past four years, Finley Communications has produced various industrial videotape programs for Mobil Oil Corporation, Chemical Bank, AT&T, Bristol-Myers Company, Morgan Stanley, Coopers & Lybrand, Young & Rubicam, Merck, Sharp & Dohme, and others. Before starting Finley Communications, Mr. Finley worked for Towers, Perrin, Forster & Crosby, an international management consulting firm. Previously, Mr. Finley was audiovisual director for Coopers & Lybrand, and before that, a staff producer for Western Electric. Mr. Finley has an A.B. from Boston College and an M.S. from the Newhouse School of Communications at Syracuse University. (*Chapter 7*)

William B. Follett is president of VCA Duplication Corporation, a division of Video Corporation of America. During the past ten years, he has built and managed S/T Videocassette Duplicating Corp. and has assisted many major corporations in planning and making operational their videocassette communications network. He is a frequent speaker at industry meetings and active in all aspects of the videocassette duplicating industry. (*Chapter 23*)

Meg Gottemoeller is vice president and manager of the Chase Trade Information Corporation, a subsidiary of The Chase Manhattan Bank. Under her management, CTIC produces a wide range of information products on international trade and investment for corporate customers around the world. The corporation earns yearly revenues in excess of $1 million. Before moving to Chase Trade Information Corporation, Ms. Gottemoeller was the manager of Chase's media resource center and international video network. Before joining Chase, she held the position of senior producer at Merrill Lynch, Pierce, Fenner and Smith's Audio Visual Center. Ms. Gottemoeller received her A.B. degree from the University of Michigan and an M.S. in communications from Temple University. She is a member of the faculty of Fairfield University, Graduate School of Corporate and Political Communications, Fairfield, Connecticut, and is a member of the Board of Directors of the International Television Association. (*Chapter 8*)

Nancy F. Gurwitz is an attorney who provides legal, financial, and management services to small start-up organizations. Prior to the formation of her consulting business, Ms. Gurwitz was for three years an associate in the corporate department at Wachtell, Lipton, Rosen and Katz, a New York law firm. She received a B.S. in mathematics from Jackson College, Tufts University and obtained a J.D. degree from Rutgers School of Law, Newark, New Jersey. Upon graduation from law school, Ms. Gurwitz clerked for Justice Robert L. Clifford of the New Jersey Supreme Court. (*Chapter 9*)

Robert T. Hider is director of AV services for Arthur Young & Company, a large public accounting firm. Under his guidance, the company has produced over 400 programs in the past ten years, primarily for use in internal training and continuing professional education. Prior to that, Mr. Hider had several years in corporate education with Arthur Young and with IBM Corporation. He holds a bachelor of music degree from Westminster Choir College. (*Chapter 12*)

Patrick B. Howley has a strong knowledge of production and postproduction. He has been chief engineer of Merrill Lynch's video facility, a member of the SMPTE's Committee on Standardization of Digital Control of Television Equipment, and is currently supervisor of engineering maintenance at Teletronics (New York). He holds a B.S.E.E. degree from the Polytechnics Institute of New York and has taught electronics. He has worked as an editor, technical director, remote crew supervisor, and television camera operator. Prior to his move into video, he was a systems engineer at Grumman Aerospace, working on the Apollo lunar landing project. (*Chapter 22; Chapter 22 Appendix*)

Jeanette P. Lerman has specialized in media production since 1969. She is currently vice president and manager of Merrill Lynch's video network. The network produces weekly programs of financial topics which are distributed to over 600 Merrill Lynch locations around the world. An early career as a graphic artist evolved into film making when the social realities of the 1960s made the combination of art and social conciousness a paramount concern. She spent five years editing, writing, producing, and directing films for the National Film Board of Canada. The economic issues of the 1970s ignited an interest in finance and with it a career move into the corporate sector of private television. As supervisor of Canadian Industries Ltd. TV studio, she adapted documentary film making techniques to electronic field production. Ms. Lerman received a B.A. from Brandeis University and an M.A. from Stanford University. She has published several media magazine articles and is frequently a speaker at various media conventions. (*Chapter 11*)

Sherry Liebowitz is an assistant vice president and manager of the Audio Visual Media Services Department at Morgan Guaranty Trust Company. She began this media department, which includes a ¾-inch television studio and international video distribution network, in 1973. The main objective of the studio is corporate communications; specifically in the areas of product knowledge and Morgan's business activities. Programs are produced both in New York and overseas. Particular care is given to testing and evaluation. In addition to video, the department produces slides for sales presentations and some print materials. Sherry has been involved in the field of corporate communications since 1970. She is presently an adjunct faculty member of the Media Studies Program of the New School for Social Research, teaching two courses on the masters level in corporate communications. Sherry has a M.A. in communications from New York University. (*Chapter 5*)

Lon McQuillin has more than fourteen years experience in television production, and has his own company, McQ Productions, just south of San Francisco. As a producer, director, and editor, his background includes programming for entertainment, education, and corporate communications, as well as a large number of commercials. His productions have won the Golden Reel of Excellence from the ITVA, and Silver and Bronze Awards at the International Film & Television Festival of New York. He edited ITVA's *International Television News* for four years, has contributed to numerous industry magazines, and is the author of *The Video Production Guide*. He teaches a class in television production at the College of San Mateo, whose UHF station is a PBS affiliate, and is a frequent speaker and instructor at television industry conferences and seminars. (*Chapter 20*)

Pamela J. Newman heads professional development for Marsh & McLennan, Inc. She also serves on the Marsh & McLennan Management Council. Prior to joining Marsh & McLennan in 1979, Dr. Newman was director of management development and a manager in the Management Consulting Department of Peat, Marwick, Mitchell and Co. Dr. Newman's business and professional writing includes articles that have appeared in *Management Controls, Human Resources Magazine,* and *Training*. She has coauthored two books: *Meeting Time* and *Organizational Communication*. Dr. Newman serves on the Board of Directors of the Financial Women's Association. Dr. Newman received her Ph.D., M.A., and B.A. from the University of Michigan. (*Chapter 4*)

Robert Passaro is director, audiovisual services for Fisher Scientific Company, one of the world's largest manufacturers and distributors of laboratory equipment and supplies. A pioneer in corporate television, Robert Passaro started his video career at the Westinghouse Defense Center in the mid-1960s. Since joining Fisher in 1966, he has built a fully equipped 2-inch broadcast teleproduction facility and staffed it with an award-winning team of video specialists. Robert Passaro is a long-time member of the International Television Association (ITVA) and has served as international president and board chairman. Recently, he has become active in the Audio-Visual Management Association (AVMA). He has a B.S. in industrial management from Fairleigh Dickinson University, and has attended the RCA Institutes, Inc. (*Chapter 2*)

Marilyn Reed is a producer and director at the Merrill Lynch Video Network in New York City. She is responsible for all phases of her productions from developing the script through the final editing process. Ms. Reed joined the Merrill Lynch Video Network as a production designer and she still enjoys designing sets for her own productions. Among her experience, Ms. Reed worked as associate designer and costumer at Hiram College in Ohio. She designed sets for one season at the Utah Shakespeare Festival, and designed many Off Broadway sets. Ms. Reed received her B.A. in theatrical design at the University of California at Davis and she received her M.F.A. from Brandeis University. Ms. Reed is a member of the ITVA and she serves on the Board of Directors of the New York chapter of the United States Institute of Theater Technology. (*Chapter 14*)

Thomas Wm. Richter is manager of audiovisual service at Standard Oil Co. (Ind.). Since 1967, he has been involved with corporate television. Over the years, he has designed two media-equipped training centers; he was part of the organizational team that designed and built one of the most sophisticated audiovisual departments. He has been a teacher, data processing programmer, system analyst, industrial media director, media consultant, education consultant, and a manager. As a manager, he has established job descriptions for numerous creative positions. Outside of his business life, he is very active in the International Television Association and the American Society for Training and Development. He was one of the founding fathers of the ITVA and the ASTD media division, and held numerous offices in both organizations. Presently, he is associate relations chairman for ITVA. (*Chapter 27*)

Diane M. Sharon was president of DMS Services from 1977 to 1981. Her company developed and produced film, video, multimedia, and print projects for major corporations in the areas of sales training and motivation, corporation identity, marketing support, and employee information. She worked as a writer and producer for broadcast TV news and documentary programming, and was nominated for an Emmy in 1976. In 1981, Ms. Sharon moved from corporate video to cable television as director of sales support services for Home Box Office, Inc. She holds an M.B.A. in finance from the N.Y.U. Graduate School of Business Administration, and a B.A. in English literature from the State University of New York at Stony Brook. She has lectured, taught, and written on a variety of subjects including video production, self-employment, journalism, and career planning. (*Chapter 13*)

Leslie Shreve was recruited into the work of makeup for politicians and corporate executives several years ago. Since that time she has been responsible for the makeup design for countless international corporate heavyweights, "ordinary" people, sports figures, world leaders, and rock groups. Ms. Shreve also pursues a very active career as a performer. She has received notices in the *New York Times* for her stage work and awards for her nationally syndicated children's show, *Leslie the Shreve*. She has acted in dozens of television and radio commercials. She received a B.F.A. in Drama from Ithaca College. Leslie Shreve serves on both local and national boards of AFTRA and SAG, and on TIP-EAST (*To Increase Production in the East*) in conjunction with the Mayors Office of Motion Pictures and Television, and represents performers on the Board of Governors for the New York chapter of the National Academy of Television Arts and Sciences. (*Chapter 17*)

Jeffrey J. Silverstein is chief executive officer of Fusion Media, Inc. Fusion Media is a design and production firm that specializes in new technology video projects and corporate communications. As videodisc consultant to the IBM-MCA coventure, DiscoVision Associates, he has been instrumental in developing videodisc networks and interactive design. Mr. Silverstein previously managed the video department for ConEdison, New York's electric utility, where he produced award-winning video programs almost exclusively on location. He holds a master's degree in television and film from Syracuse University's Newhouse School of Public Communication. (*Chapter 21*)

Jan Thiede has operated an audiovisual design firm since 1974, specializing in slide presentations, filmstrips, and corporate television graphics. During this time, her graphics have been included in many videotapes for Merrill Lynch, Chase Manhattan Bank, and Union Labor Life Insurance Company. Prior to the formation of her own company, Ms. Thiede was for three years art director for Invision Association, and prior to that helped design and establish an audiovisual sales promotion department for TWA. A Pratt Institute graduate, she received her B.F.A. in design and visual communication. (*Chapter 15*)

Thomas Valentino, Jr., is president of Thomas J. Valentino, Inc., in New York City. The company is a total communications firm with an emphasis on music publishing and production. The firm is also in the commercial record field with such productions to its credit as coproduction of *Saturday Night Fever* and other hit records. The company primarily produces music for television, films, radio, and nonbroadcast productions. Tom Valentino received a B.S. and master's degree in economics from Fordham University in New York and also teaches classes at New York University. He speaks frequently to professional groups about music in all its aspects and also writes on the subject. (*Chapter 24*)

Ron Whittaker is director of broadcasting at Pepperdine University, Malibu, California. For over twenty years he has worked with corporations, educational institutions, and commercial and noncommercial television stations in developing documentaries, dramatic productions, music entertainment and public affairs programming, and daily newscasts. Dr. Whittaker has authored over eighty articles on state-of-the-art production techniques which have appeared in over a dozen broadcast magazines and journals. Prior to his position at Pepperdine, Malibu, he was a professor of film and television at the University of Florida. (*Chapter 3*)

John P. Wiliszowski, Jr., is currently the director of marketing and public relations for DPI. DPI produces motion pictures, plays, and the nationally renowned *Tiffany's Attic* and *Waldorf-Astoria* dinner theaters. In addition to producing for stage and screen, DPI arranges financing for Broadway and Off-Broadway productions. John Wiliszowski is responsible for all market analysis, advertising, and sales functions for the company. Prior to joining DPI, Wiliszowski was editor of *Video Systems* magazine, a national trade publication serving the television production industry. His career includes experience as a producer and director. He has taught college courses in radio and television and has supervised or served as a consultant during the development of several production facilities and communication departments. Wiliszowski is an experienced photographer, has worked as a freelance video producer and holds a master's degree in communications. (*Chapter 29*)

Joel Willis is president of Scarab Communications, Inc. Scarab is a media cooperative providing production, programming, and talent for private, public, and commercial television. Prior to forming his company, he was director of marketing for Rombex Productions, a New York City production facility. Preceding his five years at Rombex he was director of programming at Teleprompter Cable Television for seven years. Mr. Willis has served as director, producer, and lighting director on hundreds of programs in a wide variety of fields, from sports to drama, as well as children's programs, news, documentaries, and commercials. (*Chapter 16*)

Patricia Tierney Wilson is manager of Chase Manhattan Bank's Audio Visual Communications Presentations and Meetings, a unit of Internal Communications. Under her aegis are the TV studio, the senior management computerized meeting facility, and the auditorium. With her staff of six, Ms. Wilson is responsible for producing sound slide presentations, and film and videotaped productions in support of the Bank's management, training and marketing objectives. Ms. Wilson came to Chase from Merrill Lynch, where she was production coordinator and associate producer, primarily responsible for programming and distribution to Merrill Lynch's 400-unit video network. Patricia Wilson's background also includes work with several film production companies in New York City. She graduated summa cum laude from the City University of New York and received the President's Award as Highest Honor Student from Queensborough Community College. For the Joseph Jefferson Theater Company, Ms. Wilson produced, cowrote, and directed a film, *How Many Ways* which traces the development of Kathryn Forbes' best seller, *Mama's Bank Account*. Following a screening of that work, Alexander Cohen, the Broadway producer, bought the film as a promotional piece for Richard Rogers' last musical play, *I Remember Mama*, starring Liv Ullman. Patricia Wilson also has produced and directed several multimedia shows. (*Chapter 26*)

Preface

Upon graduating from the University of Michigan in 1969 with an M.A. degree in mass communications, I assumed that I would obtain a position either in the areas of broadcast television or independent production. The whole concept of "private television" was virtually unknown to graduates in that year.

After searching for a stimulating job for a number of months, I came across the Audio Visual Center at Merrill Lynch, Pierce, Fenner & Smith Inc. I was both surprised and delighted to find a small but growing full-color industrial-grade video operation in place. I was hired as the center's first production assistant.

During the next few years, the center grew and I received a number of promotions, to the point where I became an officer of Merrill Lynch and manager of the Audio Visual Center. I also met and joined associations with other managers of corporate in-house video facilities. The development of private video operations during these years was not limited to just profit-oriented corporations. Private television facilities existed in hospitals, museums, educational institutions, governmental agencies, and religious organizations.

The primary purpose of these facilities is to communicate messages which are indigenous to the parent organization and aimed at a specific audience, usually the employees. These messages may encompass: training, research, management reports, operations, morale pep talks, new product introduction, etc.

Corporations and not-for-profit organizations have embraced private television as an effective method to supplement other forms of internal communications. Those that have, quickly realize that new skills are required if a video facility is to produce quality and purposeful programming. In addition, because of the high capital investment and trained staff requirements, top management and managers of AV operations cannot afford to allow video operations to grow in an unorganized manner. Therefore, the following three items are needed to create a successful video facility: strong audiovisual management skills, knowledge of the medium, and organized plans for growth. This handbook addresses these key issues.

Recognizing that no one person has all the answers, I asked leading audiovisual communicators to focus on their area of expertise and to present pragmatic and constructive suggestions on how to operate a private television facility and produce it's programming.

This handbook places greater emphasis on the process of management and production than on technical hardware. Because of rapid changes in technological developments, this handbook would be quickly outdated if it focused on specific recorders, video switchers, and character generators. While equipment may change, sound management techniques and quality production values are more universal in nature and in the long run provide a solid foundation for growth.

The handbook is divided into six sections. Part 1 presents an overview of private television and gives management and managers a perspective of the field. Part 2 is a manager's guide to implementing and operating a successful video operation. Part 3 deals with the day-to-day production of video programs. Part 4 looks at a rapidly growing area of private television—international video networks. Part 5 examines staffing and getting a job in the private television field. Part 6 looks to the future in this expanding area.

As editor in chief, I would like to take this opportunity to thank the many people involved in the production of this complex volume.

To Professor Gary Gumpert of Queens College for planting the seed for this book.

To Joan Scherzer, Shirley Rivera, Joan Shore, and Lisa Ellex who helped during the formative days.

To John Rawlings, Mia Amato, and Jay Newell for providing much of the needed research.

To my 1980 class in advanced media management, at the New School for Social Research (New York), for acting as a springboard for new ideas and approaches. In particular, I would like to thank Susan Svigoon who made contributions to the funding and budgeting chapter and program development chapter, and to Joe Lyle for addressing the area of free-lance pay scale in the staffing chapter.

To Carol A. Friedland, Lana Golub, Diane Dowling, and Reyna Listokin (who has been with this project since its outset) for providing the day-to-day support in getting the manuscript into it's final form.

To Nancy Gurwitz for her continued understanding throughout the development of this volume.

To each and every contributing author who withstood my "gentle" harassment, I am truly thankful, for they in fact produced the book.

And finally, to my mentor Lee Roselle who gave me my first job at Merrill Lynch, who taught me all the elements of this field, and who made certain that I did not deviate from anything but the highest standards. Any strengths I have are a direct result of his dedication, direction, and tutelage..

Nathan J. Sambul

An Overview of Private Television

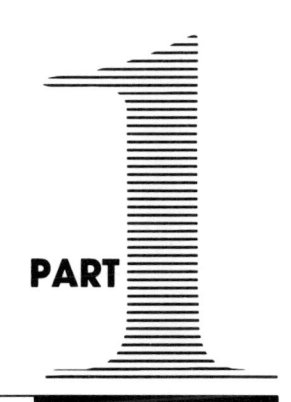

PART 1

To give shape and substance to the field of private television, this section looks at three major areas. Chapter 1 examines the various applications of video within corporations, not-for-profit organizations, and government agencies. Chapter 2 surveys the current users of private video and answers critical questions, such as "To which department should video report?" and "What is the typical production budget per program?" Chapter 3 examines the elements or pitfalls that reduce the effectiveness of private television (e.g., problems with goals, content, pacing, on-camera talent, etc.). Solutions to these and other problem areas are covered in later sections of the book.

The Rise of Private Television

PETER CARANICAS

WHY THE RISE OF PRIVATE TELEVISION?

Once television had emerged as the dominant medium bringing entertainment and information into the home, its expanded role in business communications became inevitable. If nothing else, television's explosive growth in the consumer market proved its superiority over print in holding people's attention. As television reached the proportions of a mass medium in the 1950s, it competed with—and contributed to the decline of—such industries as motion pictures, mass-appeal magazines, and newspapers. When a new generation, raised under the powerful spell of the TV image, entered the work force in the 1970s, all the necessary ingredients were present to facilitate television's rise as a communications tool of enormous power.

But it's a giant leap from consumer households to corporate boardrooms and training centers. Despite the ever-increasing complexity of corporations and their concomitant demands for better and more efficient internal communications, private television's early progress was agonizingly slow. Even as late as 1971, when television had penetrated into as many U.S. homes as had bathtubs, only a handful of businesses were using TV for communications. They preferred to rely, instead, on time-tested but outdated media such as film, slide, and filmstrip presentations, and, of course, print materials.

What blocked the advance of private television, and how were the obstacles overcome? Two factors stood in its way in the early days. The first was technical: In order to be effective, video communications required reliability and portability, and the early reel-to-reel black-and-white videotapes and videotape recorders (VTRs) were cumbersome and hard to use. The second factor was psychological: Television was untested as a private medium, and top management had to be persuaded of its effectiveness and its ability to save money in the long run before committing financial resources to it. The overcoming of these barriers contains countless case studies of how middle-level managers, attuned to the possibilities of the new medium, convinced their bosses to make the necessary expenditures to implement video communications networks as a way of boosting overall corporate efficiency.

TECHNICAL TRIUMPH

Private television's major technical breakthrough came in the late 1960s when Sony introduced its U-Matic videocassette recorder (VCR). Originally intended for consumer use, the U-Matic VCR entered a home market that wasn't yet ready for it. However, videocassette technology happened to be ideally suited to the needs of the budding private television network. Cassettes enable the easy duplication and transportation of instantly interchangeable videotapes. With the use of cassettes, video programs, produced at a central studio location (either inside or outside the company) or in the field, can be edited and duplicated onto any number of cassettes for distribution to an equal number of communication or training locations. Once used, the tapes can be erased and used again.

A video network has often been defined as a TV distribution system carrying user-originated video programs on a regular basis to several remote locations within the organization. The key word is *distribution*. It matters little whether the message on the tapes consists of sales training, employee news, management communications, or a Christmas message. The essential factor is that television is being used to distribute information on a regular basis. The videocassette—especially in its ¾-inch U-Matic embodiment—is the device that first enabled the quick and easy distribution of a television program within an organization.

Today, companies other than Sony market U-Matic format videocassettes and recorders. Also, the new ½-inch videocassette formats and the videodisc are making their appearance, relegating ¾-inch tape to an ever-narrowing share of the market. But cassette technology is what first broke the ice.

SELLING MANAGEMENT

Once cassettes and VCRs became readily available, the job of convincing top management of video's usefulness began. Naturally enough, the first companies to adopt video were the ones with heavy training and communication needs, especially high-technology, service-oriented firms with large sales staffs. IBM, AT&T, Xerox, Hewlett-Packard, and Texas Instruments were among the pioneers. All were early users of TV. Coincidentally, it also happens that each one either deals directly in data processing products and services, or relies extensively on data processing support services.

Historically, the adoption of video in large firms has followed the installation of major data processing divisions. The reason is that corporations, as they embraced computers, suddenly needed ways to train large numbers of people in a hurry—especially the computer programmers and the employees who work with them.

Video's advantages in data processing training were obvious from the beginning. The medium can endlessly repeat complex material; it reduces the boredom of learning intricate and tedious tasks; it can be used rapidly, with no need to darken a room, as with film; and a single video training program can be used by one employee, a small group, or a large assembly.

TYPES OF PROGRAMMING

Video's special abilities to convey complex information in a hurry, to convey it with ease to either a large or a small number of people in a uniform manner, and to do so less tediously than is possible with print, led to its rapid spread beyond data processing. It not only expanded to other departments, but also became adapted to other uses. Today, video has several major applications in the corporate environment. The main ones merit discussion here.

Training

No company activity better lends itself to video than employee training. Whether it be sales training, skills training, management training, or operations training, video gets the message out effectively.

Case Study 1. Bechtel, the large San Francisco–based engineering and consulting firm, uses video for employee training throughout the employee's tenure with the company. Many of the tapes are produced in several languages, reflecting the firm's multinational reach.

Another type of training using video takes place on the East Coast, and not in a private firm but in a public agency: the Suffolk County Police Department on Long Island, New York. One of the nation's most technically advanced, the department makes extensive use of individualized video training, which includes role play and instructor evaluation.

Also, training need not be confined to employees or members of the organization producing the tapes. Texas Instruments, of Dallas, has long been in the business of selling its training programs to outside users. Ball Memorial Hospital of Muncie, Indiana, distributes patient self-care programming to hospital rooms on a special channel, alongside training programs for physicians, nurses, and hospital staff. Eastern Airlines, of Miami, produces a basic TV workshop for personnel from other industries and government agencies.

Marketing

From reaching employees, it's only a short step to reaching dealers, retailers, and customers.

Case Study 2. An illustration of this form of marketing is provided by the AT&T Long Lines office in Oakton, Virginia, which produced an employee information program on energy conservation that also came to be shown to consumers.

Texas Instruments (TI) distributes some of its training programs to users of TI products, thus helping preserve and expand its share of the marketplace. The firm is now looking to use video to aid its marketing efforts in the burgeoning home computer field.

One of the most ambitious video marketing projects to date is that of General Motors (GM). The giant automaker has purchased and installed optical videodisc players at over 8000 dealerships. The versatile machines have two quite different purposes. They can be used to show new-car demonstration programs to prospective customers in the showroom. And they are used to play instructional programs on car maintenance and repair for personnel in the service shop.

American Express is also contemplating optical videodisc players in marketing its services, albeit on a smaller scale than GM. The travel, leisure, and financial services firm is experimenting with a computerized videodisc and travel reservations system that pulls programs of travel locations out of a videodisc data bank and plays whatever segments the customer wishes to view. If the decision is made to travel to that location, the system automatically makes the airline and hotel reservations.

Management Communications

Video is a very personal way for managers to communicate with one another and with their subordinates. The so-called face-behind-the-memo school of selling video to management involves emphasis of the medium's ego-gratification aspects. After all, practically everyone wants to see him- or herself on television. But more important, the personal approach has the magic of also improving corporate relations. And of course, most large organizations can benefit from personalized management.

Employee News

The TV version of the print house organ is more than a morale booster. In addition to providing employees with information about their company and the activities of their fellow employees, company news programs can carry such material as a specially edited tape of the company's annual meeting. Among innovators in the field of employee news tapes are General Telephone of Florida and GTE Data Services, both in Tampa; Norton Co. in Worcester, Massachusetts; Bechtel; American Motors; and Bendix in Kansas City, Missouri.

Meeting Regulatory Requirements

Video can help beleaguered companies meet the compliance requirements of the alphabet soup of federal regulations and agencies legislated in recent years: Equal Employment Opportunity (EEO); Employee Retirement Income Security Act (ERISA); Occupational Safety and Health Administration (OSHA); Environmental Protection Agency (EPA); and so forth. Companies are required by law to communicate appropriate regulations to their employees, and video is a simple and uniform way to do so.

In addition to outside regulations, internal rules are also frequent subjects of company-produced videotapes. For example, AT&T Long Lines in northern New Jersey recently produced a program stressing the importance of the company's "code of conduct." What could have been an ordinary, rather boring tape on rules and regulations evolved as a humorous, yet highly communicative, program that won an industry award. Furthermore, it probably went a far longer way toward accomplishing its purpose than any print brochure or handbook could have.

Advertising and Public Relations

Video is television, and corporate video departments can be enlisted to help out with the company's overall advertising efforts.

> **Case Study 3.** Proctor & Gamble, the giant package-goods manufacturer, uses its own internal TV setup to lend support to advertising and product development, in addition to putting it to the more traditional use of employee training.
>
> Airwick Industries, of Carlstadt, New Jersey, gets double use out of its video department by having it make test commercials for the company's products.
>
> In department stores like Bloomingdale's (New York), video is being used more and more for point-of-purchase displays that tout the store's products and entice passersby into making purchases.

Safety Instruction

Nothing can convey danger and emergency better than the intimate TV image. That's why safety instructions on tape work so much better than those in print. This point is made graphically by a video program recently produced by Union Pacific Railroad of Lincoln, Nebraska. The tape subtly and skillfully stresses that employees who don't follow safety directions may soon be dead.

Other Uses

There's no end to video's uses in a company. Other areas in which some companies use video include shareholder information, research, sales promotion, and investor relations. For example, programs can be produced to communicate information about the company to security analysts. Dealer and distributor meetings can be taped and the edited tapes shown to those unable to attend. And annual meetings can be edited and reedited in many combinations for presentations to any number of groups interested in the company's fortunes.

THE NEXT DECADE

Just as consumer TV arose in the 1950s and was followed by private TV, so the booming home video technology of the 1980s will be followed by innovations in company communications incorporating advances in cable TV, ½-inch videocassettes, and videodiscs. The home video revolution will further advance the cause of private television by allowing it to be consumed at home. Today, an employee can bring home a book or training manual for further study, but the videocassette stays in the office. Tomorrow, employee homes equipped with Video Home Systems (VHS) or Beta format videocassette recorders (VCRs), or with videodisc players, will be ideal after-hours training or communication centers.

As cable television spreads to a majority of U.S. households, and as most cable systems increase their capacity to fifty-four channels or more, it makes sense for a large corporation even to lease time on a special cable channel to transmit information to employees. And those trainees or executives who happen to be away from home or otherwise unable to watch TV at the time of the cablecast can, of course, use their home VCRs to record the program for later screening.

As the technology advances, home video and private television will become more and more intricately linked. The term *video revolution* will apply not just to changes in home entertainment, but also to the way in which television will enhance communications in our professional lives.

2 An Organizational Perspective of Corporate Television

ROBERT PASSARO

It is a sad, but truthful, statement that private television is oftentimes a stepchild department within the larger parent organization. As opposed to the accounting and quality control departments, which appear on the organizational charts in predictable places, video centers crop up in every conceivable area: in the personnel, training, administration, publications, sales, research, advertising, photographic, data processing, or even the maintenance department. As one personnel director stated: "Our AV Department evolved under Personnel and it survived. We later developed a rationale as to why it should remain here."

The reasons why audio visual (AV) departments start where they do are as varied as the areas themselves. For example, training departments traditionally have been major users and creators of audiovisual support material [overhead transparencies, public address systems, 16-millimeter (16mm) films]. So it is a logical extension for them to continue to produce and develop video tapes. Other organizations have felt that video centers should be housed and staffed by the people who will most often use their services; ergo, an AV department within a data processing center. There is also the rationale that video centers should report to those people who know how to repair the equipment or at least know how to use it. That is why photographic and maintenance departments are not uncommonly found with video production facilities—even though these departments may not be the most effective ones to encourage the use of video within the organization.

To present a more accurate view of the organizational structure of private television operations, I conducted a survey of thirty companies. They ranged from those in the Fortune 500 (with sales in excess of $63 billion) to companies with gross revenues of less than $50 million. By industry group, they were in automotive, petroleum, electronics, science, tobacco, utilities, finance, and medical.

The survey covered the following items:

- To which department or division does private television report?
- How many programs are produced a year?
- How many distribution points are there in the network?
- What is the typical production budget per program?

Although surveys of this nature provide useful mileposts for those video operations that are already in existence, they are of greatest benefit to

2-1

those corporations considering the establishment of a video department. They provide the needed background information that can be used in a presentation to management or in the development of an annual budget for staff and production.

MANAGEMENT SUPPORT

While conducting the survey, I uncovered a repetitive pattern: *two* key people within the organization are apparently needed to launch a video venture—the *visionary* and the *backer*. It is difficult to give their job functions, considering that their job titles are never the same even within corporations within the same industry. Yet, if this chapter is to provide any assistance for those contemplating the installation of a video operation, we must define who these people are and how to seek them out.

The visionary is generally the pioneer, the believer, the idealist who is convinced that the use of video can contribute to improving a firm's training, sales, marketing, employee relations, and personnel programs. The visionary is generally a line manager who has been with the firm a number of years, who understands the problems the firm faces, and who sees potential solutions. While this individual may have the confidence of senior management, he or she may not have budget authority to spend the capital funds needed to start a video operation. You, in fact, may be your firm's private television visionary.

The greatest strength of a visionary is determination. Senior management requires constant reminders that video can and should be playing an important role in a company's communication process. The visionary may feel that each presentation or reminder to management is an uphill battle, but, for any one of several reasons, he or she will also believe that winning the war is important. First, there is the belief that the firm will be a stronger, more productive, more profitable company if video is used on a regular basis. Second, the visionary may have the selfish motive of wishing to have the AV department report to him or her. Third, the visionary may want to vent some creative energies and either produce or appear on video tapes.

If you don't see yourself as a visionary but would rather report to one, or if you know of one in your organization, then lend your support to this individual and, if necessary, feed his or her ego. In the long run, the firm, the visionary, and you will benefit.

The backer, on the other hand, is much higher on the corporate ladder and has the authority to release funds for a video operation. He or she may be a division director, a vice president, or even chairman of the board. There are three questions that influence a backer's decision: (1) Will video truly benefit the firm? (2) Will video be cost-justifiable? (3) Does the backer really believe in the visionary's ability to execute the installation of the video operation?

The backer has far less to lose than the visionary. If the video operation is successful, everyone will share in the accolades. If the video operation goes over budget, alienates other departments, or fails to produce the type or quality programs that were anticipated, the visionary will probably be held accountable. The best approach in convincing the backer to support the video operation is to stress the cost-effectiveness of video and the ways it can contribute to the overall profitability of the firm.

TO WHICH DIVISION OR DEPARTMENT DOES PRIVATE TELEVISION REPORT?

There appears to be no direct rhyme or reason as to where video operations are placed or to which department has organizational authority over them. In my survey, I found that certain departments were most frequently cited as the parent department by the following percentages of respondents:

Training (23.6 percent)

Corporate services or administration (21.1 percent)

Photographic or film (18.4 percent)

Public relations or public affairs (13.2 percent)

Marketing or sales promotion (10.5 percent)

Personnel or human resources (7.9 percent)

Other or no answer (5.3 percent)

Examples within the corporate world are as follows:

1. Training
   - R. J. Reynolds Tobacco
   - Deere & Company
2. Corporate services administration
   - Texaco
   - Prudential Insurance Company
3. Photographic, film
   - Ford Motor Company
   - General Motors
4. Public relations, public affairs
   - General Telephone Company of Florida
   - Phillips Petroleum Company
5. Marketing, sales promotion
   - Merrill Lynch, Pierce, Fenner & Smith, Inc.
   - The Foxboro Company

6. Personnel, human resources
- Fisher Scientific Company
- Hewlett-Packard

As noted, the survey looked at a small section of the private television industry, so the above percentages are only an approximation of the location of parent departments. Accurate and up-to-date figures are kept by D/J Brush Associates (Highland Road, P.O. Box 210B, Cold Spring, NY 10516; telephone: (914) 265-3098) and Knowledge Industry Publications, Inc. (701 Westchester Ave., White Plains, NY 10604; telephone (914) 328-9157).

To Whom Should the Video Operation Report?

It is fairly simple to report the statistical results of a survey question such as the one regarding a parent department. A more beneficial finding for a new operation is a rationale for where a private television operation should be placed. I mean no offense to any audiovisual or video manager or to any existing departments. Rather, I would like to set forth a model for future departments.

It is not surprising that training departments are often the first to acquire and use television equipment. Most training material could be improved with strong visual aids and dramatic illustrations. Their acquisition usually poses no problem until management decides to request programs that extend beyond the realm of traditional training. For example, the chairman of the board may request a special presentation for the annual stockholders meeting. By the very nature of the occasion where it will be shown, this presentation must be informative and entertaining. While I know capable training managers who could easily accomplish such a demanding task, this type of presentation is not within the realm of expertise of most training managers. Therefore, I would not normally recommend placing a video operation in the training department.

In a number of industries, video reports to the photographic or film department, which in turn is wholly dominated by marketing and advertising. These industries generally produce large consumer items whose features and benefits must be demonstrated. While I believe that selling the company's products is of primary priority, I'm left with the uneasy feeling that the lure of such "glamour" programming may divert interest from producing video for employee communications and training. Therefore, I cannot recommend that video report to the photographic department.

Personnel departments (or, as they are currently called, human resource departments) will also frequently house video operations. Since most videotapes are for internal viewing, what better department to handle employee communications than the department responsible for all employees—personnel. Here again, I have a problem with placing video under personnel. I question whether personnel can adequately respond to the needs of marketing or public relations. Personnel would not be my first choice.

Some progressive organizations have established a communications department. As a parent unit, it will house a group of smaller departments, such as public relations, employee communications, shareholder relations, advertising, sales promotion, and audiovisual. By grouping them all together, the organization can speak effectively with one voice. Communications departments have a greater overview of the needs of the organization and can draw upon the experiences and information of many different divisions. This is where the television operation *really* belongs.

Communications should report to a top governing body so it can be immune (even if only partially) from that organization's mid-level political forces. Furthermore, there are ancillary benefits in the combination of these various departments within one communications division. For example, the division will be capable of producing accompanying print pieces or press releases. An effective communications department or division can provide stronger support to a wide variety of other departments within the firm and can usually do so in a cost-savings manner.

HOW MANY PROGRAMS ARE PRODUCED A YEAR?

Obviously, the number of programs produced by a video department in a single year will depend on many factors, including permanent staff size, skill levels of permanent staff, budget restrictions, length of typical program, other departmental responsibilities, methods of production, quality of production, and number of approvals required for each step.

In conducting my survey, I found Fortune 500 companies producing as few as 6 programs a year all the way up to 150. Two-thirds of the respondents in the Fortune 500 group reported typical annual production of 40 to 80 programs. That is somewhat higher than I have personally observed in my travels as an International Television Association (ITVA) officer. A more realistic range, in my judgment, would be 35 to 60 programs, typically of 10 to 20 minutes in length, produced by a typical permanent video staff of ten reasonably skilled people. Simplistically (if a rule of thumb is needed), this equates to an annual expectation of 3.5 to 6 programs per permanent video staff member. However, use *any* rule of thumb with caution.

Clearly, there is no single, absolute formula for top management to apply. If 100 programs per year are desired, then the goal of 100 should be realistically weighed according to available resources (including free-lance help and commercial production facilities) and adjusted according to sound business practices.

Another editorial point: While it is important for a video operation to produce a reasonable number of programs in any given year, first be certain that your programming truly fulfills your organization's needs. Grinding out great numbers of meaningless programs has been the downfall of many video departments.

HOW MANY DISTRIBUTION POINTS ARE THERE IN THE NETWORK?

The survey indicated that networks had as few as 14 locations and as many as 8000. As mentioned earlier, my population group was relatively small. (For more detailed information, contact D/J Brush Associates or Knowledge Industry Publications. Their addresses are given following the list of corporate examples in a previous section.)

Perhaps the important question is, How many distribution points should be in the network? That, of course, depends upon the goals and objectives of the video operation. If you want to improve employee communications, then the answer to the question is rather straightforward. You have to have a video distribution outlet in each branch office and plant. If your objective is to improve sales, perhaps you need playback units only at points of purchase. The important factor to remember is that you do not want to exclude important segments of your audience either deliberately or unintentionally. Information should flow freely to all employees who have a need to know. Key questions to ask are:

- Who are the intended audiences?
- Where are they?
- How large a group are they?
- When will the programs be viewed?
- How many times will they be viewed?

WHAT IS THE TYPICAL PRODUCTION BUDGET PER PROGRAM?

If you were to ask a dozen private video producers how much they would budget for a typical 15-minute video production, you'd unleash a veritable flood of questions like "How many shooting locations are involved?" "What type of professional or free-lance talent is required—and how many people of each type?" and "Will original art be required?"

There are many other factors that will affect the program budget as well. For example, if an organization uses a full charge-back system (all labor, materials, and overhead are charged back to the requestor), the budget figure can be high—on the order of $10,000 to $25,000, or $667 to $1,667 per finished program minute. This may be close to rates charged by commercial production houses.

Some video operations are considered strictly as overhead by their organization. Here, there is little or no charge-back to the requestor, so we're liable to get any kind of answer regarding program costs. It may be as low as $50 to $100 per finished program minute.

Partial charge-back systems (all costs except labor and overhead are charged back to the requestor) are by far the most widely found. In this instance, we find the rates to be half as much as those quoted above for full charge-back operations, or $330 to $830 per finished program minute.

I've attempted to give a broad-brush overview of how video departments are structured and how they function. Perhaps the most important lessons to be learned here are: 1) a department should take as broad an overview as possible for the needs of the firm and should attempt to service those needs, 2) a video department should distribute programs to as wide an audience as possible within the corporation or organization, 3) a video department should be as cost-conscious as possible, and 4) the video department should work within the organization to support the needs of *both* the visionary and the backer.

The Pitfalls of Private Television

RON WHITTAKER

In the past decade, countless private television users have seen their medium move from a "helpful embellishment" status to that of an "indispensable tool." Almost daily, we hear of additional corporations, schools, medical facilities, and government agencies moving to meet an ever-accelerating need to communicate ideas and information more rapidly and effectively.

This trend does not mean that the whole concept of private television has been totally successful or is not questioned in many sectors. The reasons for its less-than-wholehearted acceptance range from resistance to new technology and presentation techniques to disappointments in finding that costly investments in television hardware and facilities have fallen short of original expectations.

So, because some private television operations have not come up to expectations and because many others are searching for ways to improve their service, it may be well to examine some of the factors that have been shown to be related to successful (and unsuccessful) television operations. These factors fall logically under seven headings:

Process versus product

Time management

Pacing

Nonprofessional talent

Organizational needs

Professionalism

Foresight

PROCESS VERSUS PRODUCT

There are a number of "errors of emphasis" which can make the difference between an operation that is generally conceded to be worthwhile and successful and one that is viewed as something of an expensive toy that is in danger of being phased out at the first sign of economic difficulty. The first of these errors is a rather subtle but basic frame-of-reference problem: an emphasis on the *process* rather than on the *product*.

Admittedly, the very rapid developments in audio and video hardware are fascinating, especially when one regularly sees new generations of microprocessor equipment and digital technology. The numerous trade publications are largely responsible for keeping the emphasis on hardware. This stress is understandable, of course, when one considers that most of the revenue of these publications comes from equipment manufacturers.

In private television, the placement of emphasis on process rather than product is a rather easy error to make, not only because of the emphasis by trade publications and conventions, but because the skills, concerns, and problems of production personnel typically center on the operation of television equipment.

The Importance of the Bottom Line

Outside the production facility, however, where an understanding of hardware and what it can do is typically limited, only two "bottom lines" typically emerge: What results can be expected? and, What will they cost? Certainly, the private television producer who is honest will have to admit that it is not practical to try to outdazzle what can be seen on commercial television every night. In that league, the competition is so strong that millions of dollars are spent annually on the latest and most elaborate special effects and postproduction capabilities, produced by the brightest and most highly paid technical and artistic talent available. Not only is it virtually impossible for private television producers to compete with the razzle-dazzle of Hollywood productions; it is generally undesirable.

It is interesting to note in this regard that most viewers understand or appreciate little of the electronic wizardry which they regularly see on TV. Generally speaking, it is the content and how it affects them, first emotionally and then intellectually, which determines whether the production is successful or not. It may not be irrelevant to recall at this point that the most highly rated TV show in history was a simple, one-camera production: the moon walk!

TIME MANAGEMENT

Under the heading of time management is perhaps the quintessential question for the viewer: Is this program going to be worth the personal time invested in watching it? There are many variations on this basic question: Is there anything contained in this program that can directly affect me or what I'm doing? Would I be better off investing my time in another area (reading a book or magazine article covering the material)? Will the time I invest viewing this program provide benefits of at least equal value?

A producer must be realistic about the reason a viewer will be watching a tape. Is it because it holds the promise of important information (which may make the viewer's job or future more interesting, rewarding, or easier), or is the individual watching only because it is expected or required? If the reason is the latter, you would be well advised at the outset to sell your audience on the importance of the production. Numerous studies indicate that motivation is crucial in positively affecting viewer attitude toward the time invested in watching the production and, even more important, toward the amount of information communicated and retained. The whole principle of internal versus external motivation is much more important than is commonly realized.

PACING

Each viewer has a kind of internal time clock which results in a keen sense of subjective time. The adage "If you are having a good time, time will go quickly" is important here. By keeping the production's content and pace carefully tuned to the viewer, the subjective movement of time will work to your advantage.

The fact that many private television productions have been guilty of quickly boring their audiences (or worse, *slowly* boring them) attests to the lack of understanding of subjective pacing. One of the most commonly asked questions by a novice in this field is, How long should a video presentation run? The answer is simple: It should run a few moments shorter than the viewer expected it to run, thus creating a desire to see more. There is no advantage in showing the audience too much. Too much of anything is boring.

Because the viewing audience has become rather sophisticated in grasping ideas and concepts without the need of complete and detailed verbal and visual development, the boredom which has been associated with many private television productions can often be traced back to overwritten scripts and slow, inefficient presentation techniques. Once you have given a viewer the opportunity to grasp a point, don't dwell on it. Sophisticated viewers generally like to draw their own conclusions rather than to be led painstakingly through the obvious.

In this regard, keep in mind that after about 15 minutes of unfamiliar, factual information, learning and attention levels start dropping off significantly with most people. Most television viewers have been conditioned to 2 or more minutes of commercials regularly interrupting program content, giving concentration a rest (assuming, of course that the basic program material required any real concentration in the first place).

THE PROBLEM OF NONPROFESSIONAL TALENT

A certain lack of professional expertise behind the scenes can often be covered up in the production or postproduction process. Such disguise is much more difficult with on-camera personnel. (On-camera personnel are referred to as *talent*, whether the literal meaning of the term seems appropriate to the particular people in question or not.) With this issue, we come to probably the most difficult problem of all for private television producers. And, in fact, it's a twofold problem.

First, the talent for on-camera presentations are typically nonprofessional television personalities. As a result, their method of delivery or style often falls short of most viewers' expectations. If the director can exercise total control over content, presentation time, approach, etc., then many of the shortcomings can be corrected. This leads to the second problem. In private television, on-camera talent consists of executives and top management personnel. The resultant problem in rank and authority leaves the television director with less control than is generally needed to correct inherent problems.

Too often, the talent is motivated in varying degrees by ego. Although this contingency may present some significant positive aspects for the private television producer, it can also arouse rapid resentment with a captive viewership.

ORGANIZATIONAL NEEDS

Up until now, I have discussed the pitfalls that relate to producing programs. Now let's turn to some organizational pitfalls.

Documenting Your Worth

If you cannot constantly justify your existence in terms of the space, capital investment, and the monetary resources you require, you are in a precarious position. Even though your services may be much harder to quantify or measure than the output of an assembly line, it behooves private television producers to carefully think through the strongest justification possible for their existence and products. This information may be invaluable during times of budget requests (or threatened cuts) as well as in presenting proposals for new personnel, added space, capital expenditure, etc.

Ask yourself some questions:

- How have your productions made a beneficial difference?
- In what way would the corporation or institution suffer or be diminished without your services?
- Are before-and-after statistics available for the whole production unit, or for specific productions?
- How have productions affected sales, productivity, quality, competitive position, safety, efficiency, or job satisfaction?
- Is there a way of measuring the effect either objectively or subjectively?
- How much can you quantify and get down on paper?

Just as focusing on process rather than product can cause problems in production, focusing on goals and operations totally within your division, without fully taking into consideration the corporate or institutional goals and purposes, will present major problems for your future. Private television is a service, and as a service it must be totally responsive to the needs of the organization as expressed by its management.

Once you completely accept this fact, it becomes a matter of seeing needs from the viewpoint of top management executives; putting yourself in their position; understanding their directions, needs, and goals. If there is a single most important concept that you should remember above all else in this chapter, it is this. Seeing things from the viewpoint of your superiors breaks down "them" and "us" barriers and puts you and your division directly into the flow of things. The more you can do this, the more secure your position and services will become. (And, by the way, developing this understanding is one of the very best ways for you yourself to move up in your organization!)

PROFESSIONALISM

Acting in a nonprofessional manner is a dangerous pitfall. Rather than list a series of negative statements, I prefer to suggest seven guidelines to professional private television behavior.

1. Professionals take full responsibility for their work. They make their own decisions—sometimes after seeking advice from experts or consultants—and they stand behind what they do. They do not attempt to transfer responsibility for mistakes.
2. Professionals typically supervise their own activities and direct their own work. Their goals and decisions are based on an understanding of and sensitivity to the organization's activities, goals, and limitations.
3. Professionals have a responsibility to themselves and others to keep abreast of information in their field. They participate in professional conferences and set aside regular time for reading literature in their specialty. Not only do they openly share information with others in the field; they also realize their responsibility to add new information whenever possible.
4. Professionals are sensitive to their responsibilities to fellow workers. Their standards include respecting confidences, maintaining a positive, productive approach to problems and tasks, looking out for the mental and physical welfare of workers, and treating grievances objectively rather than responding in ego-centered ways.
5. In trying to get and hold attention within productions, the private video professional does not resort to cheap gimmicks, to satire or debasement of select groups of people, or to fabrication or misrepresentation of facts.

3-6 OVERVIEW OF PRIVATE TELEVISION

6. Professionals in private television respect the work of other professionals by staying within copyright and authorship restrictions.
7. Professionals realize that their chief function is to render service. They believe in, value, and are proud of the service they render.

FORESIGHT

A private television facility cannot operate solely for the moment and expect to meet emerging needs effectively. Five-year plans didn't start with the Russians. Every professional should try to project as realistically as possible where the function, work load, and technology of the facility are heading. "The future belongs to those who plan for it" is especially true for a rapidly changing area like television.

In such a volatile medium as television, it is necessary to see the pattern of change and move into the future fully prepared. This future can even represent the phaseout of a particular facility. Remember, however, that a professional operates on a level above a single job. A phaseout or retrenchment, if anticipated, may represent just a jumping-off point to newly emerging possibilities elsewhere. Change is inevitable. Your being prepared to meet the challenge and opportunity inherent in that change is something to which you must constantly be alert. You must gear your operation and personal approach to desirable change, but, at the same time, still maintain a solid economic, technical, and administrative base within your private television structure.

A Manager's Guide to Private Television

PART 2

Before an ongoing video operation can produce video programs, it needs sound management. Part 2 presents management-oriented chapters designed to assist those individuals who currently supervise a video facility or who will shortly assume that responsibility.

In Part 2, Chapter 4 looks at steps required to implement a video operation, including preparing a proposal to management. Chapter 5 presents methods of producing quality programs with a limited staff. Closely aligned is Chapter 6, which discusses the operation of a major facility. Next, Chapter 7 talks about using outside producers to increase or supplement output. Chapters 8 and 9 address critical issues: how to fund and budget a department, and how to prepare letters of agreement and contracts with vendors. Finally, Chapter 10 presents a number of guidelines to strengthen an existing video operation.

Implementing a Private Television Operation

PAMELA J. NEWMAN

INTRODUCTION

The purpose of this chapter is twofold. It will attempt:

1. To reassure you that there is a logical way to convince top management of the practicality of corporate television.
2. To remind you that, regardless of how logical the presentation to top management is, you will probably have a rough battle on your hands to succeed at getting private TV accepted.

Please keep the following ground rules in mind:

1. Do not accept any discouraging reactions or negative responses with gracious resignation. Instead, keep in mind Sinclair Lewis's observation: "Persistency is the one great element in success."
2. Keep in mind that private television is one of the highest visibility management decisions possible. Expect people to move cautiously.
3. Proceed with a plan. Try the one outlined in this chapter.

HOW TO GET YOUR COMPANY TO IMPLEMENT PRIVATE TELEVISION: AN OVERVIEW

There are many ways to go about implementing private television. Some methods are more successful than others. You naturally are looking for an approach that will allow you to achieve your mission (having a private TV structure) without making very many people unhappy. Your method of implementation must be extraordinarily sensitive to people's fears of change, *any* change—much less, anything as dramatic and visible as television. The approach described in this chapter is organized to this purpose. The steps are:

Step One. Write a memo (or a series of memos) about the possibility of private TV. (Thus, you get people to focus their thinking.)

Step Two. Conduct a needs analysis. (Take a step backward to verify what you already know.)

Step Three. Form a task force. (Committees increase commitment and reduce resistance.)

Step Four. Submit a proposal to the board of directors.

STEP ONE: WRITE A MEMO

The most important reason for writing a memo for the record is to clarify your own thinking. Since implementing a private TV network is an activity analogous to preparing and carrying out the Allied Forces campaigns of World War II, it is imperative that you thoroughly *analyze the opposition.*

The second reason you write such a memo is to inform its readers you are not personally biased toward a private TV network; rather, you understand the pros and cons, and you believe the pros outweigh the cons. The third reason you will want to write a memo is because you will be able to use it as a vehicle for action. A good memo requires a response from the reader: a yes, a no, a request for more detail, a formation of a task force. Whatever the response, a memo gets an answer.

Some pointers regarding a memo are these:

1. Distribute this memorandum to the highest level of management possible.
2. Do not accuse anyone in the memo of being old-fashioned because of resisting private TV. Do not ever make the resistance to television into a personal attack.
3. Suggest in your memorandum why people are usually reluctant to consider private television:
   - People have difficulty seeing how television solves communication problems within the organization.
   - They do not know how to buy television equipment.
   - They are confused about where the television equipment should be located (physically) or to whom it should be assigned (departmentally).
   - They are concerned that they will not be able to operate television equipment.
   - They are reluctant to put large sums of money into capital expenditures.
4. It is useful to draw comparisons between private television and other state-of-the-art technologies, such as:
   - Using minicomputers instead of manual labor for processing information
   - Using photocopies instead of carbon copies for distribution of memoranda
   - Using word-processing equipment instead of electric typewriters for producing major reports or contracts

Figure 4-1 is a sample topical outline for a memo. In all our examples, we will use as our company International Horizons, Inc. (IHI), a full-service financial brokerage corporation.

International Horizons, Inc.

```
Date: _____, 19__
From: _____, Manager, Corporate Communications
To: My Immediate Superior
Subject: Viability of Corporate Television Network for
         International Horizons, Inc.

cc: Chairman
    Vice Chairman
    Executive Vice President
```

I. Brief Overview of Purpose and Format of Memo

II. Need for Sophisticated Communication Network
    * fast-changing technology of the business
    * highly decentralized organization
    * wide geographic spread of units
    * need for uniform application of procedures, standards
    * need to improve standard of professionalism within organization

III. Specific Reasons to Consider Television
    * our competition has decided to implement TV
    * corporate TV will make us look like an innovative, modern company
    * corporate TV will replace national and regional sales, training, and management meetings

IV. Reasons Not to Consider Television
    * requires capital expenditure
    * demands increased specialized staffing (perhaps at a premium)
    * uncertain who will determine priorities

V. Steps for Analysis
    * needs analysis
    * task force
    * formal proposal

VI. Summary
    * thorough review of need
    * possible alternatives
    * next action step
        - request for comments
        - informal discussion
        - placement of topic as agenda item at next Operating Committee meeting

Figure 4-1 Initial memo for a needs analysis study.

STEP TWO: THE NEEDS ANALYSIS

A key strategy to remember when trying to implement a private television network is that you are not trying to encourage people to buy television equipment. Instead, your mission is to set up a communications network that will help address the problems from which the organization is suffering. Your first step to accomplish this will be to conduct a needs analysis.

To conduct a needs analysis, you will want to:

1. Select an audience of fifteen to twenty people (generally on a senior management level) to interview regarding communication difficulties within the organization. (I interviewed all the members of the operating committee.)
2. Design an interview questionnaire to use when conducting these interviews.
3. Conduct the interviews in person rather than by mail or phone.

Outlined below are questions and areas you will probably want to address in your needs analysis questionnaire:

1. Questions about the potential audience for a private television network:
   a. What is our work force turnover rate?
   b. What does the work force know, and what does it not know, about how this organization works?
   c. What is the level of technological competency of the work force?
   d. How much do changes in legislation and government regulations impact the work force?
2. Questions about the organization:
   a. How often are there changes in the company's organization chart?
   b. How often should news about recent events in the organization be reported?
   c. How does information provide an opportunity to motivate new clients, customers, and employees?
   d. How can communications, as a methodology, improve outstanding performance?
   e. How do problems with geographic spread of the organization impact quality and consistency?
3. General questions:
   a. What are the pressing communications problems of this organization?
   b. Would a better communications vehicle help train individuals?
   c. Are there examples that prove the necessity of quickly communicating changes?
   d. Are there arguments indicating where the TV operation should be located; i.e., the training department, the marketing department, or perhaps personnel?
   e. Who are the people who should be taped for broadcast in the organization?

f. What are the major issues people want to hear about?

g. In what ways can television improve revenues?

It is also important to ask the managers in various geographic units how they would incorporate private television. Figure 4-2 is a sample user questionnaire.

The purpose of conducting a needs analysis is to allow people throughout the organization a chance to get involved in the decision-making process for private television. This technique allows you to achieve the three C's of commitment, cooperation, and conflict reduction. Don't stop with merely conducting the interview. Afterward, be sure to:

1. Send a personal thank-you letter to the interviewee.
2. Summarize interview responses and send to all interviewees within three weeks after your final interview.
3. Acknowledge the time contribution the interviewees have made to your project by submitting a brief summary of this research to top management.
4. Select several of the most enthusiastic interviewers to serve on your task force (see below).

STEP THREE: FORM A TASK FORCE

Your needs analysis has allowed you to identify the communication problems within the organization. From your findings, you will be able to write a report to top management outlining the various solutions to the problems that should be studied. Your memorandum will suggest the formation of a task force to examine solutions to the communication problems you have identified and to present their recommendations to top management. The advantages to using this approach are these:

1. Instead of being seen as someone campaigning for your own cause (TV), you will be appreciated as somebody who is trying to attack true problems in the organization.
2. It is always better to stay detached from any one solution for fear of seeming fanatical.
3. Forming a committee relaxes top management. It allows the executives to think any decision will be carefully considered, slowly implemented, and delayed until the next financial quarter.
4. Forming a committee will allow you a second effort at involving the organization in the ultimate decision you wish to achieve. Therefore, when recommending your committee participants, carefully consider who should serve on it. As much as possible, you should draw on senior operating people in the organization. This approach will exclude staff (they are never really accepted by the organization); top management (they don't have time for this committee), and the training and communica-

International Horizons, Inc.

Corporate Television Network
Needs Assessment Profile
User Questionnaire

Date: _____
Individual: _____
Title: _____
Area: _____
Address: _____
Phone: _____

1. What key topics are important to your staff?
2. Who, in your office, would most benefit from corporate television programs?
3. Why are you in favor (not in favor) of corporate television?
4. What possible applications for video could you envision if we installed a TV in your office today?
5. Where would you put a television in your office?
6. When would you arrange for people to watch corporate TV programs?

Figure 4-2 A needs analysis user questionnaire.

tions departments (they are usually perceived as ineffectual or biased, or both).

Once you've received approval for a formation of a task force to study the communication problems you've identified in your needs analysis, you will be able to send committee members a memorandum announcing the following:

1. The formation of a blue-ribbon committee to study corporate communication

2. The time schedule for the committee (e.g., two one-day meetings over a six-month interval of time)

3. The goal of the committee—a formal written recommendation to top management on how best to meet the communication needs of the organization.

Once you have sent this announcement to other members of management (a million-dollar hint: get the chairman of the company to send the memo out under his or her signature), you are prepared to organize your agenda for the first three meetings. Figures 4-3, 4-4, and 4-5 show sample agendas that you may wish to borrow for the meetings.

International Horizons, Inc.

Task Force on Communication
A G E N D A

First Meeting

I. Welcome

II. Review of Purpose

III. Review of Today's Agenda
    A. Summary of needs analysis
    B. Discussion/Concurrence
    C. Alteration to conclusions reached from needs analysis
    D. Possible techniques to solve communication problems
       * newsletters
       * training programs
       * corporate television
       * status quo
       * management gatherings
    E. Research needed to determine size of staff
    F. Review of what competition does
    G. What cost factors need to be considered
    H. Assignments to investigate various alternatives and costs prior to next meeting

Figure 4-3 Agenda for the first meeting of the task force.

STEP FOUR: SUBMIT A PROPOSAL TO THE BOARD OF DIRECTORS

Your task force will have helped you develop a formal proposal to submit to the board of directors. Before completing the proposal, remember to send to the individuals, one month ahead of time, the following:

- An agenda for the meeting (see Figure 4-6)
- An outline of the formal proposal (see Figure 4-7)
- A covering letter asking for a critique of the proposal before the meeting

Make sure that your formal proposal to the board of directors of your organization requires the group to make a decision, take action, and reach a conclusion at the meeting. Let people know what decision you want reached; otherwise, they will never reach it. Make sure you continually bring the discussion back to this end.

Try to keep this presentation to approximately one and no more than two hours. Longer presentations tend to get into political contests and participants often tire. Keep the meeting crisp, formal, and to the point.

Watch overdoing your visuals and handouts for this meeting. Stick with

International Horizons, Inc.

Task Force on Communication
A G E N D A

Second Meeting

I. Reports of Efficacy of Different Solutions to Problems
   * newsletters
   * training programs
   * corporate television
   * status quo
   * management gatherings

II. Discussions: Which Solution Do We Recommend
    (Note: Provided a corporate TV network is the solution, proceed as follows:)
    A. Why will corporate TV solve our priority problem?
    B. What kinds of programs could be presented on corporate TV?
    C. How often should programs be distributed?
    D. To whom should programs be distributed?
    E. What will it cost? Cost considerations should include:
       * estimates from three vendors for equipment
       * shipping costs
       * estimated cost for software (tapes)
       * average cost for 10 minute and 20 minute programs
       * cost schedule for one year of program production and program scheduling
       * cost of ancillary material to be used with programs
    F. Who should be involved in developing the programs?

((Note: If the committee decides that corporate television is not the answer, then I suggest proceeding with the answer suggested by the committee and tabling corporate television as a future implementation device after the committee's effort has succeeded or failed.))

Figure 4-4 Agenda for the second meeting of the task force.

understatement. Do not devote much money to slides, print materials, or multimedia, as they will probably tend to put off participants who see this suggestion for private TV as a money gobbler.

When delivering the formal proposal, keep in mind the necessity of responding positively to any criticisms. Make sure, at the end of the formal presentation, that you thank every participant. Follow up with a summary memo of what happened to all members of the board. If the group agreed to go ahead with the expense, congratulations. If not, then consider the following backup plans:

- Regroup your task force.
- Look for a pilot project of any other corporate endeavor that could be best served by private TV.

International Horizons, Inc.

Task Force on Communication
A G E N D A

Third Meeting

  I. Designing the proposal for top management
 II. Organization of proposal
III. Background of problem
 IV. Solutions studied
  V. Recommended solution
 VI. How to implement the solution

((Note: At the end of the day a detailed outline will be completed to submit to top management))

Figure 4-5 Agenda for the third meeting of the task force.

- Do something different from private TV (such as a training and development program or a communications program) which brings you high visibility and a great deal of recognition and success. Success begets success. If you're successful in something else, people will be more inclined to let you handle a new project (such as private TV).
- Whatever else, do not get discouraged. Private TV requires remembering the classic observation of the famous fighter James J. Corbett: "Fight one more round."

International Horizons, Inc.

Presentation and Proposal
to the
Board of Directors

July 24, 1981

AGENDA FOR PRESENTATION

Time	Topic	Speaker
9:00	The Purpose of a Corporate Television Program	John King
9:10	An Overview of the Corporate Television Program	Mary Jones

    A. Why a Corporate Television Program?

    B. Why develop Our Own Corporate Television Operation?

    C. Who the Programs Will Address:

    D. What Kind of Programs Will be Developed:

    E. What Sequence of Programs Will be Developed?

    F. How Will the Programs be Developed:

        * Phases of Development

        * Program Procedures

        * Program Budget

    G. Benefits

    H. Recommendations

    I. Summary

10:00   Discussion by Board Members

Figure 4-6   Presentation to the board of directors.

International Horizons, Inc.

Proposal to Management
Corporate Communication Program

A. WHY A CORPORATE TELEVISION PROGRAM?

   * Increased technical sophistication of our business
   * Increased growth rate of professionals
   * Industry regulations -- quick changes
   * Competition
   * Increase production potential

B. WHY DEVELOP OUR OWN CORPORATE TELEVISION OPERATION?

   * Motivates
   * Encourages professional interchange
   * Communicates current information

C. WHO THE PROGRAM WILL ADDRESS?

   * Line professionals serving in a technical, sales or managerial responsibility
   * Both present employees and future employees

D. WHAT KINDS OF PROGRAMS WILL BE DEVELOPED?

   * Brokerage Operations
   * Client Industry Specialization
   * Sales Data
   * Management Training

E. WHAT SEQUENCE OF PROGRAMS WILL BE DEVELOPED?

Audience	Probable Age	Training Audience: Responsibility/Experience	Program Sequence
Technical Professional	21 - 23	New Hire: No insurance or sales knowledge	* Orientation
	22 - 24	One year experience -- has some technical knowledge	* I -- The Principles of Brokerage
	24 - 26	Two years experience -- has more technical knowledge but little brokerage knowledge	* II -- The Role of the Broker
	25 - 65	Three years experience -- is prepared to understand sophisticated nature of brokerage business and/or is being groomed to make sales calls	*III -- Lines of Brokerage * Fundamentals of Sales

Figure 4-7 Proposal to management for a corporate television operation.

Sales and Account Executive	25 - 65	Increased Account Representative responsibilities or new hire with insurance background or technical consultant	* Managing a Client Engagement
		Account Representative responsibilities demanding frequent oral and written client presentations	* Written Presentations * Oral Presentations
		Account Executive responsibilties requiring thorough knowledge of the clients' financial and operational problems	* Client Industry Training * Business development Strategies
Managerial Administrative	25 - 65	Managerial responsibility for other account representatives and technicians	* Management Skills
		Ongoing client and managerial responsibilties with a need for quick updating on technical issues and new firm products	* New Product Development * Account Executive Update
		Coaching and counselling responsibilities	* Coaching and Counselling skills
		Recruiting responsibilties	* Recruitment Skills
		Retirement Planning	* Retirement Focus

F. HOW WILL THE PROGRAMS BE IMPLEMENTED?

Phases of Development

The training plan will be carried out in three phases over a three year period:

        Phase 1 (August, 1980 - July, 1981)
        * Orientation
        * Fundamentals of Sales
        * Managing a Client Engagement
        * Oral Presentation Skills
        * Management Skills
        * New Product Developments

        Phase 2 (August, 1981 - July, 1982)
        *   I - Principles of Brokerage
        * II - The Role of our Professional
        * III - Lines of Brokerage
        * Written Presentation Skills
        * Client Industry Training
        * Business Development Strategies

Figure 4-7 (*Continued*)

Phase 3 (August, 1982 - July, 1983)
* Account Executive Update
* Coaching and Counselling Skills
* Recruitment Skills
* Retirement Focus

Instructors and course developers will be drawn primarily from the firm's professionals. Outside consultants will be used for specialized and advanced training programs.

Program Locations

Video playback units and television sets will be installed in all offices housing 10 or more professionals (89 offices).

COURSE COSTS

BUDGET

Summarized below is the cost of the Corporate Communication Program.

Phase	Date	Direct Costs
1	July/1980 - July/1981	$ 500,000
2	August/1981 - July/1982	1,000,000
3	August/1982 - July/1983	1,500,000
	TOTAL	$3,000,000

Percentage of 1980 Operating Revenues of $280,000,000	1%

The costs for the Corporate Communication Program will be an addition to Executive Office expenses. The costs will include time charges from areas and other operations providing individuals to help develop programs. Areas will pay for corporate communication through usual overhead allocation processes based on revenues.

BENEFITS

* Better communication = better services = better results = better reputation

* Achieves up-to-the minute knowledge in a timely fashion

* Develops professional colleagueship

RECOMMENDATIONS

1. Will develop a corporate communications (television medium)

2. Will implement the program in three phases. Each phase will be allocated a budget.

Figure 4-7 (*Continued*)

Appendix:
A Case Study in Implementing a Private Video Network

NATHAN J. SAMBUL

In 1975, during my tenure as manager of Merrill Lynch's Audio Visual Center, the firm authorized the implementation of the Video Network. That meant that we would be able to install over 200 videocassette players throughout the United States, expand our staff from seven to fourteen, and switch from an industrial-grade video studio to a full-blown broadcast-quality quadruplex operation.

The acceptance by management was reported in a number of video trade publications, and we became at that time one of the ten largest private video networks. I have been asked often since then how we accomplished that task. That is, how did we get management to approve the expansion of the video network?

Using a game plan that my immediate management and I had developed, we proceeded on a course of action similar to the one Dr. Newman outlines in Chapter 4. Specifically, to bring her comments into sharper focus, let me present my case study.

What we did not release to the press in 1975 was that this was the fourth proposal to management regarding the video network. We had submitted a major proposal year after year for the preceding three years. In each case we were rejected. The reasons for our proposal being turned down varied: In one year, it was due to an insufficient level of corporate profits. In another year, management was in a period of transition and had not fully developed its own approach as to how to improve employee communications and training in the field.

All management's reasons for rejection were quite acceptable and understandable. Despite the rejections, we continued to perform, as a department, in the most productive manner we could. Management had indicated to us that it supported the concept of the video network and was just waiting for the right time. With that stated commitment on management's part, we knew that patience would be its own reward and that we would just have to wait our turn.

A point Dr. Newman stresses is the importance of conducting a needs analysis. This is most assuredly true. Let me suggest that one should conduct the needs analysis on a face-to-face basis with division heads or senior management. We had a questionnaire quite similar to the one reproduced earlier in Chapter 4. But we started our interviews with one standard question: "If you [the division director] could address employees throughout the firm at one time, which problems or topics that are under your auspices would you place highest on your list?"

Obviously, we were asking the directors to identify the "jugular" topics within a division or department. Once we had compiled their list of topics, we mapped out our production schedule for the remainder of the year. We knew that if we were to be an effective department within the corporation, we would have to produce programming that would satisfy the true objectives of the firm. We got those objectives from that standard question. In our proposal to management, we included the list of programs we planned to produce and each one of them came from our needs analysis survey.

Another suggestion that Dr. Newman advocates is the formation of the task force committee. Here too, I wholeheartedly agree, although the composition of our committee was different from the one she recommends. Ours consisted of a representative of the office of the president, the four executive vice presidents, the head of training, the head of employee communications, the head of public relations, a major division director, and the audiovisual center manager. We would meet on a quarterly basis and decide which programs we would produce in the next quarter.

The committee served two functions. The first was that, as AV manager, I never assumed that I knew all the critical objectives of the firm. The committee had a far better picture of where the firm was going and therefore of which programs should be produced. Second, no department, be it audiovisual, marketing, advertising, or data processing, can exist without political support from other areas of the firm. There was no better way for us to develop that political support than by having key officers of the firm on our committee.

Naturally, there were times when the committee felt that some of our programming was not on target (e.g., that we should be changing our approach or creative style). At face value, these comments may appear to be criticisms, but in fact they were constructive evaluations of our performance. Furthermore, it was far better for our department to receive those comments from committee members who supported the AV department, rather than to have those comments generated without our knowledge and thus cause, perhaps, irrefutable damage to the existence of our unit.

After reading Chapter 4 and this Appendix, one may walk away with the impression that starting a network is easy. That is, if one performs A, B, and C, certainly D and E will follow. Of course, that is not the case. The AV manager must be aware of the importance of time and patience. If management is not ready to accept a full-blown video operation, perhaps an alternative proposal would be a test program. In 1974, we got management to approve ten programs for ten offices. This approval helped us lay the groundwork for our full network expansion. The name of the game is to get one's foot in the door. After that, it is up to you to make certain that the department fulfills the needs and objectives of the parent organization.

Using the firm that Dr. Newman created, International Horizons, Inc., let me present a proposal to management (Figure 4-1A), which you may find useful when preparing your own version within your own organization. In preparing your own proposal, you should include a description of the equipment, a list of programs to be produced, and a budget for the first three years.

PLAN TITLE:
INTERNATIONAL HORIZONS VIDEO NETWORK

I. Reason for Plan (why this plan is needed)

A serious communications problem exists between home office staff and branch office personnel. This problem covers all areas—new services, new procedures, updates on existing services, management communications, new operations procedures, and so on.

II. Objective (desired result)

To contribute to improved understanding and acceptance by the branch office system of information disseminated by home office divisional and top management by providing an effective communications system involving videotape TV programs designed to effectively convey the desired messages to the desired audience within an established time frame.

III. Measurement Standards (objective will be achieved when these standards are met)

    A. The International Horizons Video Network is approved and implemented with an operational target date of _____ , 19___ .

    B. Cooperative relationships with appropriate division directors and other senior management, developed as a part of the implementation of the network, are employed to maintain a continuous flow of effective programming for use in branch offices and elsewhere.

    C. Branch office staff enthusiastically accept the network as a valuable communication tool and make effective use of it on a continuing basis.

    D. Measurable increases in business and reductions in errors attributable to the network occur within two months of distribution of appropriate programs.

IV. Planned Actions to Achieve Objective

- Prepare comprehensive program and schedule for implementation of the network.
- Establish cooperative relationships with International Horizons' divisional and subsidiary management.
- Prepare programming outlines for users.
- Establish advisory board.
- Produce procedure handbook. Distribute to users.
- Order equipment.
- Deliver equipment to branch offices.
- Verify that Video Network is test-operational.
- Verify that Video Network is fully operational.

(Note: The above steps are highlights from the detailed action plan. There are many other steps which have been excluded for simplicity.)

Figure 4-1A Proposal for a video network.

Background

International Horizons has been involved with the use of TV as a communication tool since 1973. Various attempts have been made to get the full network into being, but they have all failed, owing chiefly to the expense involved. However, it has been the habit in the past to view the Video Network as primarily a single problem solver, for example, as a great training aid. In such terms, it has been hard to justify the expense of its operation.

But the network is now being viewed realistically as a major communication tool, to be used by many home office departments. Thus, its cost can be better justified. In fact, the leverage potential is so great that we believe that we will be able to produce measurable, attributable increases in business within the first six months of operation.

Part of the work which has gone into producing this recommendation has been a comprehensive series of meetings or conversations with divisional and regional management, as well as with key branch office managers and the Sales Force Advisory Council to Management. From all these come agreement that we do have a serious communications problem and that this can be effectively reduced by means of a well-handled video network.

Two tangible results from this work are (1) the answers to a questionnaire in which twenty-five branch office managers indicated their feelings about the ways we communicate now and the deficiencies which now exist in the understanding and acceptance by their staffs of the firm's policies and services, and (2) the willingness of divisional management members to associate themselves with the implementation and ongoing use of the network.

National versus Regional Operations

The network will be installed throughout the domestic branch office system. Phased regional introduction has been considered and rejected for the following reasons.

    1. Management pressures and requests indicate total involvement is required at this time.
    2. Pending introduction of many new services and promotions demand total branch office participation.
    3. Regional tests of the network have already been successfully carried out.
    4. There would be no significant cost savings, since a large part of the capital investment and the expense budget would have to be spent anyway.
    5. A morale problem would be created among the "have-not" offices during the phased introduction.
    6. It would be difficult to use the network intelligently if duplicate, non-TV methods would have to be created to communicate with non–network offices.

How a Program Gets Aired

The following is the sequence of events which will have to take place to enable a program to be produced and distributed:

Figure 4-1A (*Continued*)

1. Need for program is determined. Originator has informal discussion with the manager of the Audio Visual Center. The program is deemed feasible.
2. Originator completes Video Network Production Request Form and obtains departmental and divisional approvals.
3. Preproduction meeting is held between originator and Audio Visual Center staff.
4. Program is scripted. Script is approved by originator's division director.
5. Program is produced and edited. Answer print is approved by originator's division director.
6. Program is duplicated and distributed.

Steps 1 through 4 will take varying amounts of time depending on complexity, need for specialized scripting, sets, and so forth. Steps 5 and 6 should normally be completed in four to six business days.

Benefits of the International Horizons Video Network

The International Horizons Video Network, accompanied by appropriate backup printed matter, is the only effective way to meet our many complex communications objectives. Once the system is operative, many benefits will arise.

Credibility

The credibility and authority of video communications are rated second only to face-to-face communications, according to surveys carried out by large-scale users of such systems, including the Columbia Broadcasting System and IBM, in addition to *Broadcasting* magazine and others.

Effectiveness

By viewing the same program, all involved employees will better understand and accept the information shown. Communications dilution caused by the usual chain will be minimized. Variances caused by individual interpretations will be minimized.

Branch managers will be better able to control their requirement that all appropriate personnel be informed of a subject. "Command performance" viewing of a program is more thorough than memos or bulletins, which are often thrown out unread or misunderstood.

The International Horizons Video Network will provide a total ability to disseminate a high volume of transient information on a wide variety of subjects. Since human identification is involved, the motivation of our sales force may be constantly expanded and improved.

Speed

Speed of execution and distribution of programming is vital. A program can be taped, duplicated, and distributed to offices in four to six days. Fast follow-up amendments, or changes can be similarly swiftly carried out.

Figure 4-1A (*Continued*)

Familiarity and Acceptance

Many people have grown up with television as a learning medium, particularly in our educational system in recent years. They are used to getting information this way, and are often more willing to view and learn from a TV program than from a bulletin or memo.

Control

By using the International Horizons Video Network to disseminate information, a highly effective control filter comes into existence. In the past, anyone could send anything whatsoever to the branch office system, often causing confusion, irritation, and unnecessary expense. With the Video Network, however, controls provided by the advisory board will eliminate the random issuance of unnecessary or inaccurate communications.

Reliability of Equipment

The videocassette equipment we recommend has proven to be the industry standard. Over 250,000 units are in use. It is highly reliable and very easy to operate.

Figure 4-1A (*Continued*)

5

Operating a Video Facility with a Limited Staff

SHERRY LIEBOWITZ

This chapter will explore the view that there is no direct correlation between studio size and its importance to the organization. Running a well-planned studio that is responsive *to* and innovative *with* its organization's needs is not a "quantity" issue, but rather a "quality" issue. What is important is that the studio be incorporated into the organization and its way of doing business. As a member of the management staff responsible for a video operation, the manager of a video facility must understand a number of key elements and how they interrelate.

As a manager of a video facility, one knows a variety of ways to produce programs. One, you can contract with outside production houses. Two, you can purchase moderately priced equipment, expecting to run your operation with a couple of staff people and to supplement your facility with freelance people and more elaborate outside production capabilities when needed. Or three, you can decide that you want a top-quality, high-volume operation and that you don't want to waste money on an interim stage.

For the purposes of this chapter, let's explore the second option—a modest (small) facility. This chapter will address the key elements that a manager must coordinate and control if he or she is interested in producing quality programs. These elements are: managing the video design process, staffing the video facility, and planning the department's budgets.

VIDEO DESIGN PROCESS

The impact that the video communications effort will have on its organization is a function of how relevant the video effort is to the overall business objectives of the corporation. This is particularly true in a modest television operation where, because you will be producing only a limited number of programs per year, each may come under close scrutiny. Your success ratio has to be high. Identifying which corporate issues need communications support is a large question and needs attention from the senior management group involved in the overall planning and decision making in the organization. It is important that the manager stay informed about what is going on in the firm both from formal corporate communication channels and from an informal network of personal contacts.

Program Selection

If the studio is intended to serve the entire organization and assist with a mix of corporate planning and policy, product knowledge, and training tapes, the business of setting priorities is complex. A senior management program review committee is one effective way to approach this process. The committee should include a group of five or six key people who represent the major interests of the organization.

Their time is critical and the media manager is responsible for organizing the meetings so that decisions can be made quickly. A review every three or four months is most effective. This scheduling allows for ample program development time without sacrificing the responsiveness to the changing issues in the organization.

A list of possible programs should be made and distributed to committee members prior to each meeting. This list would include program requests from the field (collected from audience program-evaluation forms) and topics that the video manager gathers from other department heads in major activity areas. The list items can then be discussed and ranked for priority or replaced with more important topics. You should always agree on several more programs than your budget allows, in case of cancellations.

Because program decisions are being made by senior people who have an overview of the organization, your programming will probably be serving the overall needs of the organization as opposed to narrower interests. This process makes it more likely that you are assigning your production dollars and time to the most critical corporate issues. It also makes the video medium truly available to senior management as a problem-solving tool.

Every program must have clear objectives and a defined audience. Hider and Sharon (Chapters 12 and 13) discuss this aspect in greater detail. I have personally conducted two activities that have helped maintain high-quality video productions.

Approvals and Testing. A final and often neglected phase in program development is approval and testing. Too frequently a program is running tight on its deadline and no time is allotted for adjustments, improvements, and reedits. Many producers get final approval for the project based upon the scripts and thereafter deliver the first edit as a finished product. There is much to be gained by prescreening and testing. This phase, which can be planned into your production schedule, would include the following:

1. Rough-edit screening with content-approval experts
2. Rough-edit screening with sample audience
3. Collecting comments and planning changes
4. Screening of rough edit with management with report of intended changes
5. Final edit

This testing and reedit procedure can improve the effectiveness of your program. Considering the cost of your original production, oftentimes the cost of a final edit is insignificant. After the final edit, it is a good policy to have prescreenings for program participants and senior management.

Program Evaluation

In addition to testing the rough edit, it is also important to establish a feedback system for program evaluation of the final tape. An evaluation system involves your audience in the video effort. Utilizing both informal comments and formal questionnaires that are delivered with the tape throughout your network, you can evaluate whether the program has met its objectives. These questionnaires can also identify additional information needs in your organization.

The most organized way to monitor this system is to assign a communications coordinator in each location. They will distribute and return the questionnaires to you. The coordinators will also note general attitudes and comments about the tapes that may not appear on the questionnaire. It is important to have good relationships with the coordinators so you can hear what your audience is really saying. The coordinators will not report the informal comments unless they know you and understand how important negative as well as positive feedback is to program development.

Finally, reports should be written based on the questionnaire responses. They should be forwarded to senior management to inform them of the impact the programs had on your audience and to give them additional insight into your program development process.

STAFF INVESTMENT

Now that we have discussed the program design process in a modest facility, the next question is, What are the staffing requirements?

A small video facility can effectively operate with a minimal in-house staff supplemented and supported by free-lancers, if the facility is located in an area where these professionals are available. I would opt for a minimum team of three: a manager, a production coordinator, and a studio engineer.

The Manager

The manager occupies the first key staff position. Managers should have a strong background both in management and in media. In a modest facility, they will many times be functioning as producers of the programs. In a thumbnail sketch, they should be:

1. Able to develop good relationships with other managers
2. Initiating and responsive
3. Good promoters and good listeners

4. Capable of accepting the responsibility for bringing an entirely new form of communications into an organization
5. Innovative in meeting obstacles and problems as the medium is tested and tried

They will be hiring the free-lance people best qualified for each project, identifying the program's objectives with the client, choosing a format with the director, reviewing and editing the scripts, coordinating the production, testing the program, and obtaining approvals.

If the manager does not function as the producer and outside producers are needed, it is important to have a steady arrangement with these persons. It is their responsibility to create a media format that works for both the content and the particular style of the audience. Each producer can then lead the creative team in creating a customized program.

The Production Coordinator

A second key staff person is the production coordinator. This person is responsible for coordinating all the logistics and details involved in production. He or she schedules the members of the free-lance team and interfaces their needs with the organization. Often a company will have its own carpenters, painters, and maintenance staff who can be effectively used for set design. Additional rental equipment may be needed for specific shoots. And every organization has its own protection or security department and building personnel who need to be notified regarding internal location shooting. The production coordinator is the link between the media production needs and the organization's way of doing things, making maximum use of internal and external services without disrupting the systems and procedures of the organization.

Although the producer will choose the writer and the director, it is the production coordinator who usually keeps the free-lance book for the production crew (i.e., camera people, lighting directors, set designers, audio technicians, and production assistants). This individual also schedules castings and makes all the travel arrangements for location shoots. Furthermore, besides production responsibilities, the coordinator tracks the budget and keeps detailed records by project, as well as budget categories of all expenses for the media center. When a distribution network is part of the video effort, he or she may also be responsible for distributing and recalling tapes as well as for sending program announcements and evaluation forms to the coordinators at the locations served.

The Studio Engineer

The third critical staff position is that of the studio engineer. If an organization believes that its program volume justifies a facility, that facility must have a staff engineer who will be responsible for both the upkeep and the maintenance of the equipment and the technical quality of all programs. He or she will also be critical in planning the growth of the facility's capabilities with the manager.

During the development of the facility, the staff engineer can and should hire outside free-lance help for audio and editing. Thus, the department will have a period within which to discover its technical needs. There is no one ideal combination of in-house technical and free-lance staff. Each facility's needs will vary depending upon the program format and production schedule of the facility.

The arguments for keeping the internal staff small are that this policy offers flexibility in hiring specialists for particular program needs and the increase in the level of professionalism that can be hired. (For example, one can afford an experienced broadcast director for a limited time but not full-time.) The costs of staff employees include not just their salaries but also the fringe benefits and other payroll-related expenses paid by the company. Bringing free-lance staff in on a project-need basis can therefore be very cost-effective.

Free-Lance Support

A small television unit that has free-lance support provides the freedom and flexibility to put together a creative team that can most effectively handle the content and style requirements of each individual program. For example, a director who calls live shows on a regular basis will be more effective in calling a tight panel program than a director who is experienced with drama. Likewise, a documentary approach to material requires different creative skills from those needed in a scripted studio presentation. (See Chapter 7 for additional examples.)

Once you have found the appropriate director and agreed on style, you or your production coordinator will then hire the rest of the support crew. This group may include writers, camera people, a lighting director, a set designer, graphics and animation artists, a sound engineer, and actors. Often the directors will recommend their own book of people with whom they work. Or you may have a regular group who have been working in your facility and whom you prefer and trust. It is effective to build up a nucleus of free-lancers who produce the quality you want at the rates you can afford. If you hire people regularly, they will develop a sense of loyalty and commitment to you. This favorable relationship is important for scheduling tight deadline productions.

BUDGETS

One of the manager's responsibilities is the preparation of the budgets for the department. There are three budget classifications: capital expenditures, operating costs, and personnel. In certain organizations, personnel is a subunit of operating costs.

Capital Expenditures

In order to make capital expenditures (equipment-purchasing) decisions, it is critical to direct your thinking back to the fundamental questions: What are your organization's needs, and what media treatment is most appropriate? If most of your information is derived from field locations, you will need portable (remote) equipment and a good editing facility. If you plan to mix studio work with remote, you can either design a studio around portable cameras and recording equipment or purchase remote equipment and rent a studio when needed. In most cases, *it is judicious to remain flexible.*

In addition to the studio versus the remote issue, there is the quality question. There is a large difference in quality and therefore in cost between a ¾-inch videocassette facility and a 1-inch facility. Similarly, there is a wide price and capability gap between the simple electronic editor used with the ¾-inch system and the more sophisticated computer editors used with the 1-inch systems. Although this quality difference is noticeable to the audience and the more sophisticated systems are enjoyed by the creative people using them, good, effective programming can be created in either format.

Case Study 1. Rather than deal in the abstract, let's examine one company's facility in detail. It is a ¾-inch videocassette studio serving a Fortune 500 company. After a year of programming with outside production facilities, the media manager, with the support of upper management, decided to build a moderately priced TV studio for the following reasons:

1. The content of the company's programs would require a mix of remote and studio productions.
2. Continuing to rent remote equipment and studio time was not cost-effective because of the number of programs planned.
3. The programs required heavy editing. A rough edit was always used for prescreenings and approvals before the final edit was made. It was not cost-effective to use outside editing facilities.
4. Management wanted a studio facility within the company premises for the convenience factor. Furthermore, to secure the confidentiality of their information, a ¾-inch format studio and editing and control room were built. The studio was large enough for simple set design and held three camera positions. The cameras were actually remote units that had studio attachments. Each was priced in the low $20,000 range. The control room had the necessary support equipment needed to produce good-quality ¾-inch video. It also had an 8-track audio mixing board and a good switcher. For remote shoots, the cameras went out with portable ¾-inch decks. With this investment of approximately $250,000, any style program could be produced. When additional quality or special effects were needed for a program, an outside editing facility was used for the final edit and the footage was transferred to a 1-inch format on a computer editor.

Figure 5-1 Morgan video studio. This in-house facility allows management to control the confidentiality of material. In the foreground is a portable camera which is "tied to" the studio system. (*Courtesy of John Brumage.*)

In addition to the initial cost of setting up an in-house facility, the second critical expense will be purchasing the equipment needed to play the tapes. Figure 5-1 shows one way to set up a video studio.

If programs are to be viewed exclusively within one location, playback can be centralized and fed from the control room to monitors located throughout the premises. Obviously, the most cost-effective time to install the cable lines needed for this type of system is during the construction of a new building or the renovation of an existing one. If more flexibility is needed to accommodate the audience's time schedule, playback units can be coupled with the monitors to allow for user scheduling of the screenings.

If the organization is decentralized with many locations throughout the country or the world, it will need a distribution network to reach the employee audience. This network will require playback units and monitors at each location.

Operating Expense

In order to estimate your budget for a given period accurately, you will need to define, together with management, your production volume and the studio's clients during that period. Since video is meant to be responsive to the particular corporate needs and problems that arise, it can be difficult but possible to know fifteen months in advance (your budget time) what your specific program requests in the coming year will be.

As an example, if you have a network and decide to produce one program per month covering either product knowledge or corporate policy and planning, you can easily budget for twelve tapes at an average cost that will cover the varied styles used. You do not need to identify the specific topics but only unusual costs (say, for overseas production) at this time. In addition to the network programming, you may have corporate approval for shows that are used only by individual departments for training or sales promotion. Regardless of how the budget is structured (e.g., centrally or on a direct, charge-back basis for department users), you need to identify your costs and stay within the budget. If extra volume is required during the budget period, you should request a variance based on demand rather than reduce program quality to accommodate unexpected requests.

It is important for management to understand the costs that go into production: First, there are basic operating expenses that are needed to operate the video facility (e.g., maintenance, supplies, tape stock, electrical junction boxes). Second, there are specific expenses related to producing the style and number of programs agreed to with management. A small studio that is being supplemented by outside services and free-lance staff may have a very detailed budget in this area. As mentioned earlier, expenses in the free-lance category include the cost of hiring writers, directors, camera people, lighting directors, set designers, production assistants, casting directors, and actors. In the outside services area, costs include duplicating, equipment rental, graphics, special effects, set rental and construction, and transcription services.

Case Study 2. Let us refer to the small studio situation introduced earlier. This ¾-inch, three-camera studio with a good-quality remote rig is valued at about $250,000. It meets its production and network needs with a full-time staff of three (including the manager) at a staff cost of $100,000 (including salary and benefits). The staff has corporate budget approval to produce 11 programs a year for the network (2 are produced overseas) and 5 programs for training. A variety of low-cost role plays are also produced. Additional individual department requests are fitted into the production schedule and out-of-pocket expenses for these tapes are charged back to the user department.

Of the total 16 programs, 10 are projected to be fully scripted, fully produced, and edited tapes (with either a mix of studio and remote or all remote); they are budgeted at about $12,000 each. Six are projected as semiscripted and fully produced (mainly panel-discussion studio programs) and are budgeted at $5000 each. The role plays will generally be shot in office locations with corporate participants and little editing. They typically cost only about $300 each, and therefore the number produced may vary depending on the training department's needs.

The total cost for outside services and free-lance professionals for these 16 shows and role plays is $150,000. The additional expenses involved in running the facility, including staff overtime, outside typists, supplies, maintenance, organization dues and conventions, delivery services and cabfare, travel, and entertainment, total about $31,000. Therefore, for a total cost of $181,000 per year plus $100,000 for staff, the facility meets its corporate needs. It should be noted that this operating budget affords both top-quality free-lance professionals and the external services needed to produce maximum-quality ¾-inch format programs. However, duplicating costs in this particular example are low since the entire network includes only 28 locations. Many studios have to service large networks and their duplicating costs can be sizable.

EXPLORE AND REASSESS

Regardless of whether a television facility is large or small, it continually needs to reassess its programs in terms of both content and treatment. Your organization is constantly changing. Your audience is changing and therefore your programs should be changing. Your program planning therefore requires exploration and experimentation. To integrate the values of the organization into the video effort, there should be an ongoing interaction between the media manager, the audience, and senior management.

The growth of a studio should be measured in terms of increasing the quality and sophistication of the programming. Increases in volume demands will affect staff requirements, operating budgets, and equipment needs. They should not, however, in any way reduce the effort invested in program design or evaluation.

In summary, the size of your studio depends on the specific needs of your organization. But the contribution that your programs will make to the organization depends on how sensitive you are to its needs.

6 Operating a Major Private Television Operation

BILLY BOWLES

WHAT IS A MAJOR PRIVATE FACILITY?

What is a major facility? If you don't have a staff of twenty, a capital investment of $1 million plus, a distribution network exceeding 500 receivers or more, and an expense budget knocking on the door of, or exceeding, $1 million annually, you probably won't be classified as a major private television facility.

Big! Size and numbers, however, can be deceptive. Having a large physical plant, but producing only five programs a year, does not make a *major* facility, only a *big* one. A major facility must have depth and breadth. It must use its plant and personnel in a cost-effective manner to communicate programs that are meaningful and important to the company.

Operating a private television facility, whether it is a major or a modest one, requires clearly defined goals, a capable and motivated staff, and a high degree of dedication that is demonstrated by energy, flexibility, and foresight. Managing a major private television facility also requires handling many projects at the same time, defending the video operation from those that would like to see it pared, and expanding the scope of the center to meet the growing communications needs of the parent organization.

These are not easy tasks. They require more effort than producing any series of programs. But to be successful, a manager must have a mastery of the following topics:

Planning

Motivating staff

The production environment

Budgeting

Program distribution

PLANNING

Short- and long-range planning is a must. Ongoing needs studies and program evaluation are also necessary. Also of dire importance is the continuous promotion of television's capabilities and usefulness to a usually less than informed management. Your department may be considered a costly fifth wheel compared with other departments. As a manager, you cannot afford that evaluation.

You must from the outset become a knowledgeable manager regarding not only your television facility but also the needs, purposes, and goals of your company. Justifying a million-dollar investment is no easy task.

Shortsighted managers may make the mistake of using television for only one of its many bottom-line uses—training, for example. Company management seems more attuned to accepting television for training than for most other uses. Top-level people understand the need for training and most likely have a training department. If the television manager, however, puts the entire emphasis on training, then the investment in equipment and people is being underutilized and is not cost-effective.

The Executive Steering Committee

The formation of a television executive steering committee can help achieve full utilization. The importance of this committee, which is stressed often in this book, cannot be overemphasized.

The committee should be made up of department heads from at least five key departments within your company and the television department manager. The manager then has a forum of top management to keep informed on important programs that are being developed, capital needs, and problem areas that may arise. There are several basic factors in achieving good results:

- Direct dialogue with the committee, at least quarterly, to resolve major questions and point out the next quarter's activities is a necessity.

- These meetings should be held at the studio—"on your own turf." This location helps to make the committee more comfortable in the media environment and gives you the opportunity to demonstrate new techniques and equipment when applicable.

- The committee members should receive reports of important happenings on an ongoing basis.

- The members' opinions of the programs should be solicited. This documentation can be invaluable to you in promoting the program when comments are favorable. Unfavorable comments also give you immediate feedback as to how some top managers view your products and services.

In addressing your committee, it is important, if not imperative, to use management's vocabulary, not that of television. Expense reduction, investment tax credit on capital purchases, increased productivity, net savings, and the bottom line are words that top managers want to hear. They couldn't care less if a double reentry switcher and a time-base corrector are needed to achieve results.

The committee membership should be alternated. You will get people to accept membership more willingly if the terms are for only two years. However, to prevent a complete turnover every two years, terms should be staggered. For example, when the committee is created, two of the five members should serve only for one year. You will have an effective two-year rotation schedule at the beginning of the second year of the committee.

Don't underestimate the value of this committee. Take the members' counsel! If the meetings are properly handled, you should be able to come away from them with many meaningful program ideas and needs. Developing the media needs of a large company and determining priorities of such needs are no easy tasks. The committee can be an invaluable source of information and assistance.

The Production Schedule

A major facility is often under pressure to service many departments within the organization. Therefore, determining priorities is essential in establishing a productive production schedule. Each plant or department obviously thinks that its program deserves top priority. Priorities, however, are relative, ever-shifting. To cope with them, the production schedule must be well planned but not inflexible.

Evaluate your schedule, first for a year, then quarterly, monthly, and so on. Every company has annual events, stockholders' and sales meetings, a president's conference, etc. Using history as a guide, you should be able to get an accurate estimate of productions during these times. Since major programs take several months to complete, the quarterly outline becomes essential. As we drop down to the week at a glance, the real importance of good scheduling becomes more obvious. Does the schedule utilize your capital investment and staff abilities to the fullest?

The manager must have some say in the determination of program deadlines, or the scheduling will be totally fouled up. It will be feast or famine. Good managers develop good rapport with their staff and fellow managers, and they suggest when a program can be reasonably completed.

You must also know your staff and equipment. How many projects—major and minor—can your staff handle at a time? Was the equipment designed to do the job you expect it to do? The secret is knowing your people, their special skills, approaches, and overall capabilities. Such knowledge will make it easier for the manager to select the right person or persons for each project. A simple flowchart mapping out the different phases of each production is a must. Completion dates can then be listed for preproduction, scripting, talent auditions, remotes, etc. At a quick glance, you can then see the progress of all productions in-house.

Proper scheduling means *productivity. Productivity is the byword in business.*

Selling Your Facility

One effective but often unused method of selling your facility is to conduct a day-long, well-planned, and structured seminar. This seminar should be held at your studio and kept to approximately fifteen people. At day's end, you will have identified a plan for potential clients, met them face to face at your facility, extolled your virtues, and removed many of the misnomers and fears.

Begin the day with a general overview of the capabilities of television for training and informational programming. Program excerpts should be shown to augment the discussion. It is helpful to view your firm's programs as well as those from other companies. Procedures for using the television department's facilities are outlined in a booklet for participants. This booklet also suggests a list of television dos and don'ts.

After lunch, the group is divided into three smaller groups. Each group is assigned a producer from the studio, who will take the group members through all the steps from preproduction to completion of a 2-minute program. They will act as camera operators, floor directors, audio persons, and various other workers to complete their own show. Obviously, they are under the helpful hands of a professional in these jobs, but they do the work.

They leave the day's seminar excited. Without exception, the seminar generates understanding and program production requests far beyond normal expectations. It is a worthwhile endeavor.

The manager must have a well-developed five-year plan that will provide answers to these questions:

- Where is the department going and how will it get there?
- How many staff members are needed and how should they be incorporated into the department?
- How many programs will the department produce?
- How will such programs be distributed?
- What are the costs involved?

For the short term, it is important to have a one-year plan and to have it broken down into quarters. The quarterly plan must be tight and well structured for full utilization of the staff and the facility.

MOTIVATING STAFF

Part of a manager's job is hiring, preparing job descriptions, and monitoring salaries (see Part 5). But another very important part is motivating staff.

Motivation and Training of Staff

Motivation and training of your full-time production employees are imperative. This is, however, easier said than done. There is no way you can "find the time"; there is no "good" time. But the task must be done and the time must be scheduled.

There are four areas that should be considered. New employees should have a good indoctrination regarding the company. A series of tours through major areas is also helpful. Your employees will be working horizontally and vertically across all company lines, and the more familiar they are with the company and its activities, the better.

In most large firms, in-house management and supervisory training courses are offered. Make sure your staff is included. These classes develop the employees' capabilities and also give them more insight into the company's opportunities and problem areas.

Cross-training within your own department will help develop depth in television and other media skills. Exchanging jobs for a period of at least three months is a good method of increasing depth. A well-planned and monitored program must accompany this job switch.

External training, such as seminars and conferences, is also available. Before sending an employee to either, it is advisable to check with someone who has participated in a session similar to the one you are considering. Some are a waste of time and money. The International Television Association (ITVA) holds several regional conferences and an international one each year—they are excellent. Most are workshop-oriented, giving the participant a small-group approach, or in some cases, hands-on involvement.

The demonstration of interest in the employee's welfare by offering training is one excellent motivator. Stopping to give praise—another motivator—cannot be underestimated. Too often, employees hear from supervisors only when something is wrong.

THE PRODUCTION ENVIRONMENT

The complexity and types of program formats determine the needed production environment. For an ongoing major facility, these environments will be varied.

A studio will almost always be needed. On-location production is used more frequently now with the advent of miniaturized, mobile equipment. For special production needs, even major facilities must occasionally use outside production houses for their particular expertise and equipment.

The In-House Studio

The in-house studio is a must for any private system. (Figure 6-1.) The trend is for smaller *insert stages* (large enough for an average set) and not the large sound stage. This has been brought about by the miniaturization of production hardware and more location shoots. The larger studio may be required, however, by some companies because of the nature of their business or production needs. An example might be a company that manufactures large equipment or one that requires permanent sets.

The location of the studio is important. It should be situated within, or

Figure 6-1 A major video facility. This studio allows for sets of varying sizes. (*General Telephone Company of Florida, Media Resources Dept.*)

near, the company headquarters for maximum utilization. It must be convenient for the clients. This location also gives visibility to the television department, which thus appears as an obvious part of the business and not as a separate company.

Private television is still new in the eyes of management, and it is imperative at this point to do everything possible to promote company television in a positive light. Convenience is expected and is a must.

On-Location Production

The private television users were the originators of electronic field production (EFP). They struggled through the development of the first ½-inch portapacks and helped pave the way for minicamera and videotape recorders of today. Their suffering, however, has helped revolutionize the industry, both in private and broadcast television.

The credibility achieved by shooting on location has been the key. The stodgy studio productions of the past have become the exception and not the rule. Location production has, as much as anything, accelerated television's acceptance in industry.

The Outside Production House

You can be recognized as a major TV facility without owning a studio. Depending on the city or area of the country where you are located, your productions may be handled by an outside production house. If your firm's headquarters is not located in or near a large city, the outside production house becomes impractical.

Production houses can also offer specialized services that are not cost-effective for you to own. Computerized editing, digital special effects, or sound sweetening and film chains are only a few.

No facility can be so major that it can justify the cost of the needed hardware for all productions. The answer is: When needed, use outside facilities.

BUDGETING

There are two major budgets that a manager must administer: capital and operating. The capital budget includes hardware and plant. The operating budget includes day-to-day expenses, salaries, rents, and supplies. A manager should have a background in budgeting and accounting procedures. Budgeting and planning should be used synonymously. You plan for the future and must budget monies to accomplish the plan. I recommend that you refer to Chapter 8, where this topic is discussed in great detail.

PROGRAM DISTRIBUTION

A manager is responsible for more than just the creation of the programs. As the manager, you must make certain that they are properly distributed. You can produce the best programs in the world, designed specifically to do the job, but if no one sees them, they are worthless. It is a common error to build a magnificent production facility and ignore the distribution system.

Identify your audiences and provide the hardware necessary to deliver the programs. Set up a system to alert viewers of new programs. A program guide (see Figure 6-2) for each program produced, when sent to key managers, is very helpful. On this guide, show the program title, a short synopsis of its contents, the audience to which the program is aimed, and when and how it will be made available. A program catalog, updated on a monthly basis and available throughout the company, is also desirable. Care should also be taken to assure that programs are eliminated from the catalog when they become outdated. A three-ring loose-leaf binder with an appropriate cover works well and gives flexibility.

There are a variety of methods for distributing programs: videocassette,

6-8 A MANAGER'S GUIDE

```
GENERAL TELEPHONE COMPANY of FLORIDA COLOR TELEVISION CENTER
PROGRAM GUIDE
```

OPEN FLAME IN THE MANHOLE

This program describes the ONE and ONLY approved method for using an open flame in a manhole at General Telephone Company of Florida. The required PRIOR authorization and proper procedures necessary to do the work is emphasized to enhance the safety of the employee and the general public.

OPEN FLAME IN THE MANHOLE is 9 minutes in length and will be aired on our closed-circuit network in Tampa Main, Madison Street, Ashley Street, Tampa Street, Flagship Bank, First Florida Tower, and the EAX building on Channel 2. The program may also be viewed on Channel 2 in Clearwater, Sarasota, St. Petersburg, Lakeland and Winter Haven Mains.

This program is of interest to employees in the Engineering and Construction Department. It will be scheduled as follows:

Tuesday October 2	Wednesday October 3	Thursday October 4
	1:35 am	1:35 am
8:05 am	8:35 am	7:35 am
10:05 am	10:35 am	9:35 am
12:05 pm	12:35 pm	11:35 am
2:05 pm	2:35 pm	1:35 pm
4:05 pm	4:35 pm	3:35 pm
10:35 pm	10:35 pm	

If you have any questions regarding your nearest viewing area; would like to schedule additional airings on the network; or would like to order a videocassette, call Jackie Christianson at Tampa 224-4925.

GTE
GENERAL TELEPHONE COMPANY OF FLORIDA

Figure 6-2 A program guide. (*General Telephone Company of Florida, Media Resources Dept.*)

cable, and microwave. A closed-circuit network (cable) works well and is cost-effective if the firm's facilities are centrally located. Microwave can be added for more distant locations; however, it is very expensive. It does open up video conference capabilities if needed. Videocassette playback and receiver locations are the most common for private television distribution at this time. The most popular format is the ¾-inch U-Matic, with the half-inch Video Home System (VHS) and Beta second and third, in that order. The growth of ½-inch videocassette usage is due to the reduced cost of players, cassettes, and shipping as compared with ¾-inch U-Matic.

A successful distribution system also includes an effective and accurate logging and retrieval system. This facilitates the return of videocassettes for bicycling or reuse.

SUMMARY

A manager, whether for a large or small facility, must stay on top of new developments. In addition, he or she should be concerned about developing subordinates through training and cross-promotions. It is the manager, more than any other individual in the department, who must stay abreast of the company's needs and match them with the resources made possible with television.

Nothing new (and private television *is* new) is easy to sell to a conservative business community. Success stories must be documented and related to the bottom line. Calling a completed program a "show" should be avoided. As trite as this observation may seem, "show" has entertainment connotations, and you are in the business of producing training and information programs—not "shows."

The opportunities are enormous in this industry, but it requires a separate discipline. Visualizing complex management theories or technical materials takes persistence and ingenuity. The rewards, however, are worth the effort. The audience can be very appreciative and the professional and monetary rewards gratifying.

The most exciting thing about private television is that it is still open to the pioneer. All the rules have not been written, nor all the roads explored. You can take chances. You will make mistakes. You can write your own chapter in the book.

7 Using the Outside Producer

BRUCE FINLEY AND BETSY FINLEY

The television producer has a mysterious occupation. Nobody is quite sure exactly what producers do except that they smoke big cigars, sit in on complicated and often somewhat shady financial deals, and attend "interesting" casting sessions. Mothers and fathers of producers usually cannot pinpoint any specific activity that their progeny engage in during the day. But their pride for their lofty position is revealed by the title they use: "My-son-[daughter-]the-producer." Their pride is also revealed in the promises made to friends for cheap repair jobs for broken television sets.

In the old days, the task of hiring and firing producers was left up to moguls like Louis B. Mayer and Jack Warner. Now, with the expansion of private television, business executives, school teachers, and library administrators are getting into the act. And they often have to make these hiring decisions with little or no frame of reference.

After all, Louis B. Mayer smoked big cigars himself and knew how to handle himself in wild casting sessions. He could relate to producers. How is a manager for General Electric, whose job is developing packaging for toaster ovens, supposed to hire a producer? Or how is a hospital administrator, who spends most of her time with computer-controlled inventories, expected to price and supervise the television producer?

Fortunately, the task is not as difficult as it seems. This chapter will examine the step-by-step procedures that a member of management should follow in deciding which producer to use.

THE PRODUCTION

Before hiring a producer, you, as a member of management or the manager of the video operation, must have a clear understanding of the project. More specifically, you should have clear in your mind the *subject, object,* and *scope* of the presentation. Hiring a television producer is just one part of the continuum. Defining subject, object, and scope precedes the hiring decision. It is, in fact, the starting point in working with an independent producer, because it will be your responsibility to convey this information to the producer.

The Subject

What is the material under discussion? As an example, if the subject is the automobile, will gas, diesel, and electric cars all be presented, or only one or two? These are some of the important questions you must ask yourself. Defining your subject is fundamental. It sets your parameters and lets you know what you will or will not cover.

The Object

What do you want your audience to come away with after seeing your program? Using the example of the automobile, do you want the viewers to learn how to repair it or to be so impressed by its sleekness and beauty that they will buy it? When you are determining your objectives, the more specific you make them, the better the planning for your program will be. For example, "I want our audience to learn how to repair automobiles" is vastly improved by adding such specifics as "I want our entry-level employees with little mechanical training to learn the basics of repairing a one-barrel carburetor."

The Scope

Scope refers to the style, approach, breadth, depth, expanse, and view of the presentation. Determining the scope is really the next step after asking the basic questions of the subject and object of the program. This step can be time-consuming and jarring because here the creativity of the imagination meets the reality of the pocketbook. The style and approach you pick will greatly impact the cost, crew, locations, and talent used in your production. It will also determine which producer should or should not handle the project. Remember, no matter how much you might love the idea, shooting an eclipse of the sun as a backdrop for your car will be expensive. Here is where you'll be forced to be creative, but with an approach that works and is affordable. But, take heart, even programs with very modest budgets can look very interesting.

And, one final note, don't be too rigid about your format until you've locked it in with the producer you choose. You're going to pay the producer for his or her experience. Be ready to take advantage of the successful production techniques he or she has used on other projects.

THE PRODUCER DEFINED

This chapter started with innuendos about what producers are rumored to be and aren't. Let's examine the facts. We'll look at a producer's job description and the five reasons why you should consider hiring an outside producer.

The Role of the Producer

Basically, the producer is the boss on the project. Just like every job, a television program has to have one, and the producer is it.

Depending on the scope and budget of the project, a private television producer will perform most, if not all, of the following tasks:

Work with client to define subject, object, and scope

Write or supervise preparation of script

Prepare production schedule

Prepare budget

Hire director and crew

Hire production and editing equipment

Arrange for set design

Arrange for artwork preparation

Arrange for permits (for location shoots)

Cast for talent

Produce program (including all photography and audio elements needed for finished program)

Edit program

The producer has to have a good working knowledge of the elements that go into a production, because it's his or her responsibility to see that they all get done. In point of fact, the smaller the budget for a production, the more jobs each person has to assume. In many cases, the producer of private television programs takes on jobs that a specialist would be hired for in Hollywood. That's when you begin to see a name repeated often in credits or used with a lot of slashes (Producer/Director/Writer).

Reasons for Using an Outside Producer

Unless you personally have years of experience in television and have time to produce all your television programs, you're going to need a producer. You'll have to decide whether to draw upon the resources of your own organization's staff or to go outside. We're going to cover the five key reasons why you would choose the latter.

Often there is no internal production staff. It is safe to assume that the majority of private television programs produced today are done by internal staffs. However, many other organizations that do produce programs have no true staff to speak of. The staff may be limited to a single audiovisual (AV) manager who has neither the time nor the resources to handle the actual productions.

Obviously, in a case like this, one would use an outside producer. In fact, we would say that occasional users (i.e., AV managers or members of management) of television programs make up the biggest list of customers for outside producers. They certainly account for ours.

The internal staff is overburdened with work. Many organizations maintain a production staff to handle the average number of productions generated in a year. Given the nature of probability, they sometimes get swamped and can't handle all the work. They accommodate for this peak capacity by supplementing their internal staff with outside assistance.

Sometimes they hire individual directors or crew members at these times; sometimes they parcel out whole productions. What they do usually depends on whether they have made a significant investment in hardware. A company with a "studio" likes to hire people on a per diem basis to use its own equipment.

A particular skill or expertise is required. If your in-house staff people have a fine record of producing a certain type of program or of using a certain type of equipment, they may or may not transfer easily to another discipline. Also, there are people in the film and television business who are artists that few in-house professionals can duplicate. And sometimes a certain production calls for these special skills. For example, if your staff has used ¾-inch production equipment exclusively, you may want to hire an outside producer if you plan a higher-quality 2-inch production.

A particular programming skill is required. We have specialized in shadow-matte, chroma key productions for the past few years. This high-technology term simply means using television special effects to superimpose one scene over another. Not all producers have in-house experience with this technique because, even in the most expensive facilities, it's difficult to achieve. Many clients like AT&T, Young & Rubicam, and Chemical Bank have chosen our company for these kinds of productions.

Other producers have developed a great deal of proficiency in documentary techniques; others are expert in comedy, and still others in animation.

For certain kinds of production it is necessary to go outside to hire specialists. The outside resource is important in enabling you to expand the repertory of your productions and to maintain a high quality in the work that you undertake.

There are important cost considerations. Believe it or not, it is often cheaper to use an outside producer, especially if you don't have your own production equipment. Outside producers who are masterful at using expensive rental equipment efficiently can end up saving you a great deal of money. They can get services at a discounted rate if they produce a lot of programming, and they often know of services that can easily be dispensed.

WHAT SERVICES TO BUY

As we said before, pinning down the duties of a producer is not an easy task. To make the job easier, we've tried to break down the kinds of production services that are often dispensed by producers on the outside.

Packaged Productions. Here, the outside producer furnishes you with the entire package from soup to nuts. The producer handles everything from the initial client meeting and script preparation to the delivery of the finished master. This producer must be familiar not only with AV technology, but with business and other specialized subjects as well.

Many organizations like this approach. They don't wish to get into discussions of color correcting and lens apertures. Packaged productions are convenient for client and producer alike. But they are based in a solid understanding and trust built up between the two parties. The blanket authority given by the organization in the packaged production has to be controlled in some way, and it must be given only after the firm has proof that the producer has successfully completed similar projects. You must ask for evidence of his or her work.

Production and Postproduction Services. In these cases, the production is let out with the opposite philosophy from the packaged production. Here, the client says, "What I need is a technical producer-director. I'll handle the writing, ideas, and direct our executives on camera."

Here, a completed script is handed to a producer, who is then responsible for its execution. This is the way commercials are produced and some feature films are made.

The producer here is technically oriented, and usually not conversant in the subject matter of the program. Often, videotape facilities offer this kind of production service.

In deciding to use this kind of producer, you should try to determine how important it is for the producer to understand the subject of the program. If a standard format is required, producers with this technical orientation can work out well.

Parts of Productions. You'll find that as your productions become more complex, they can be broken down into parts. And each of these parts may end up being a production in itself.

The most obvious examples are introduction and title sequences. Think of a network sportshow. Whether it's the Masters Golf Tournament or the World Series, there is always a very jazzy opening title sequence.

These sequences may last only 30 seconds or even less. Still, they often contain the most complicated and expensive production values of the entire program. These specialized parts are best produced by producers who have experience in a given field. Computer-animated graphics, for example, is a production process requiring very specialized equipment not available in most studios. To produce a special computerized sequence, you would normally choose a production company that does that kind of work exclusively.

When choosing producers for specialized productions, you have to evaluate their past work. Producers should have available samples of their previous work (e.g., their *reel*). Before selecting someone for a specialized programming effort, see if anything the person has done before comes close to the effect you want to achieve.

Also, even though we haven't talked about money, be prepared to pay a premium for specialized productions. The extra cost may be due to a person's reputation, or to the complexity of the production, or to the cost of the equipment needed. Animation and computer work, for example, are the most expensive production formats, often costing 10 times more than an alternative production process.

RULES TO FOLLOW IN CHOOSING AN OUTSIDE PRODUCER

First, keep these important facts in mind when choosing an outside producer. Choose an outside producer who matches your needs. This, of course, means that you'll have to go through a bit of preliminary planning—of defining the subject, object, and scope that we talked about before. Know some general parameters in which your production must fall before calling in the outside producer.

Second, on the personal side, choose a producer to whom you can relate personally. Just because a producer may be the "best" this or the "most creative" that, and at the same time an obnoxious bore, the final product may well reflect your inability to communicate effectively with such a person.

PRICING THE PRODUCER

Perhaps the most crucial question in choosing a producer is whether the person can bring the project to conclusion within your budget.

Let's briefly touch on the various methods under which you might hire a producer. They relate very closely to the kinds of production we talked about earlier.

First, for the package production: If you hire a producer to handle an entire production from script to completed master, you will usually be quoted a package price for the job. This price will be based on the scope of the job, including shooting days required, studio or location, number of actors required, type of equipment needed, amount of editing, and a whole host of other variables.

The number of variables in a television production makes estimating production costs difficult. That's why the package price is best based upon a completed, final script that contains full shooting details.

If you go out for multiple bids, make sure all producers are bidding on the same specifications. Sometimes producers come in with high bids because they plan a complicated, high-quality production. Sometimes they bid high to make more money. With a full shooting script, you'll be most likely to have a sound bidding process.

It's almost impossible to estimate the costs for a typical production today. And a cost-per-minute figure is totally artificial. A 30-minute panel show with little editing can be done for $10,000, while a 30-second commercial with a half-dozen exterior locations might cost $50,000. And costs will vary around the country. Generally, costs are higher in Los Angeles and New York because of the influence of television, feature films, and commercials. Facilities and crew members get used to higher production budgets. On the other hand, productions in these cities can be very efficient. The per-day cost may be higher, but you may get more for your time and money in terms of quality.

To get back to package prices, once you've agreed on a price with a producer, he or she lives or dies with that price. This is so unless you change the specifications. But generally, if you get a quote based on a detailed treatment or script, and you plan no changes or cause no reshoots during production, that quote is the price. Changes should be negotiated on the spot, not left until the conclusion of the production.

Payments are usually made on a schedule with a third of the package price in advance, a third on conclusion of the shooting to your satisfaction, and a third upon delivery of the finished master. Sometimes, because of the speed of videotape production, the terms are a half in advance and a half on conclusion.

In other cases, producers are hired like other crew members, on a contract, job, or per diem basis. Fees range with experience and depend on the other jobs undertaken: producer, producer-director, producer-writer, etc. Prices range all the way from $150 to $1000 a day, depending on experience and the complexity of your job. There are no union rates, and fees are negotiable. Generally, you'll want to hire as much experience as the job requires, and be ready to pay for it if you require a lot.

SOME CONCLUDING THOUGHTS

A friend has pointed out that there can be a conflict in blending inside production staff with outside producers. We personally don't think that there should be a conflict if both parties are professionals. It's our opinion that outside producers should be able to work well with in-house people. However, everyone must have her or his duties explained carefully, and be held accountable in performing them.

Whether you are using in-house staff or an outside producer, your finished product should match the subject, object, and scope that you desire. If you use an outside producer, determine in advance the extent to which you plan to use the producer and the budget you can afford. Finally, determine whether this producer can meet your needs and work within the guidelines you set. Your intent is to have a quality program and the right producer will help you achieve this goal.

8 Funding and Budgeting

MEG GOTTEMOELLER

Video does not fit neatly into most corporate structures, physically or philosophically.

Existing office buildings were not designed for TV and the classic private television facility has a ceiling that's too low, inadequate air conditioning, minimal soundproofing, and a column in the middle of the studio floor. The company facilities department is used to designing office space, not TV studios.

Other departments also have difficulty understanding the needs of private video. The corporate controller often has a hard time fitting video into the firm's accounting structure. How does one measure the return on investment in an expensive facility that produces no income? How does one classify expenses that don't fit the standard budget categories? The purchasing department finds it difficult to bid intelligently on complex equipment packages that fit no description of office machines. Personnel is often puzzled by the job classifications of producer-director, set designer, and video engineer.

The situation has improved over the past few years, primarily because of the assimilation of data processing (DP), with its computer equipment and staff of systems professionals, into nearly every area of the corporation. DP has been sold to management as a way to reduce costs and increase productivity through automation. Where systems managers succeed, they have learned the accounting, purchasing, and personnel procedures of the corporation and make these procedures work for them.

A communications professional faces the same challenge: to act as an interpreter, translating the requirements and accomplishments of a video facility into language the corporation understands.

This chapter will cover the basic funding and budgeting issues involved in the operation of a private video facility. Since each corporation operates differently, I can't hope to offer standard solutions to all problems. I urge you, however, to do three things: (1) Read the chapters in this book on implementing a video operation (Chapter 4) and managing an operation (Chapters 5 and 6); (2) learn the basics of accounting and financial analysis either by taking academic courses or by studying one of the many books available for the nonfinancial manager; and (3) get to know your corporate controller and the budget analyst assigned to your area and go to them for information and advice.

FUNDING

Corporations fund and staff service groups in two basic ways: directly (as a self-funded cost center), or through a system of charge-backs to user departments (as a profit center). A manager may have no control over the way the organization is funded. The decision on how the facility is to operate is usually made during the initial proposal process and can be a matter of corporate policy. A manager should be aware of the advantages and disadvantages of each system when developing a proposal.

The Cost Center, Self-Funded Approach

In the case of a cost center, capital and operating funds are supplied directly to a manager's budget. The cost of the video operation (including programming) is absorbed by a department or the corporation as a whole.

This method has a great many advantages, especially during the start-up phase of an operation. The manager is not required to make a massive selling effort for each program. Since he or she has discretionary control over the production money, the "selling" job usually consists of approaching other departments with the following proposal: "You have a problem which I can help you solve, and it won't cost you anything but time and energy." It is a rare department manager who won't be willing to experiment with a new medium under those terms.

Under the self-funded approach, the video manager maintains control over the programming, sets priorities among competing projects, and allocates resources. Although perhaps subject to a management advisory committee of some kind that helps legitimize the resource allocation, the manager still has a great deal of leeway in managing the operation. Because the manager is paying the bills, decisions can be made quickly about whether a program warrants extra expenditure for remote footage, actors, and large sets, or whether it would be equally effective as a low-budget studio production. The manager can decide to refuse a project if, on the basis of professional judgment, another department's project will apparently make more effective use of the facility.

Having responsibility for setting priorities, the video manager can plan ahead and make the most efficient use of this corporate resource. Since the corporation as a whole is paying for productions, the manager can also insist that departments use the facility instead of outside production houses. This authority allows the video manager to maintain control over programming quality for the corporation and, not incidentally, ensures a steady flow of work for the facility.

There are, of course, disadvantages to a totally self-funded cost center. The first is illustrated by the adage, "If it is free, it has no value." Client departments that receive the videotapes at no charge may not "push the program through the system"; that is, they may not encourage wide viewership or provide support material.

Another disadvantage is that the video department may become too complacent. This attitude may result in adding too many people to staff or in not providing all the creative thinking that an outside facility may bring to bear.

Also, since the video manager has a great deal of responsibility for setting priorities for the facility, he or she must keep in touch with corporate objectives to make sure the available resources are used wisely.

Contrast this situation with that of a studio run on a complete charge-back system.

The Charge-Back Approach

The manager must charge user departments for every direct cost of a production (tape, set construction, free-lance help, equipment rentals, etc.) and will probably have to assess a daily or hourly charge for use of the studio. Charges may also be made for the time of the in-house staff.

This approach places the video facility on the same footing as an outside production house, and has advantages and disadvantages.

The major advantage from the corporation's point of view is that the true cost-effectiveness of the video operation is much easier to measure if it is forced to pay its own way. The rationale is this: If no one in the corporation is willing to pay the fully loaded cost of TV production, then video has no place in the business. Users have to be willing to pay for video, or it's just a corporate toy.

Many managers welcome the opportunity to run what is, in effect, a business of their own. They advertise within the corporation and spend a great deal of time selling their services to user departments. At the end of the year, they can show a profit or loss for the corporation.

This probably is the most realistic way to fund a video operation. In corporations with a long history of video use, with user departments that understand and need video, and with a mature viewer network in place, it can work well. It has some very serious disadvantages in most corporations, however.

First, if video is new to the corporation, the manager will spend most of each week in the often-discouraging effort of selling to user departments. Video is expensive and is an unknown quantity to most executives. Unless the senior management of the corporation is visibly and vocally behind it, most middle managers hesitate to spend money on something that costs a lot and smacks of "show biz."

Even when senior management is supportive, the first things to go when individual department budgets are cut are advertising, public relations, and video. It is unfair to say, as is always said in this case, "If they really wanted video, they wouldn't cut it out." Department heads have to make impossible choices at budget time, and when the choice is between firing three people or making five video programs, video goes. The video manager is left with plans and production schedules in a shambles and nothing to do for the next year.

Second, when the budget is not controlled by the video manager, he or she cannot control the content of programming. The maxim that "he who pays the piper, calls the tune" applies here. The facility will operate as a service organization, accepting all comers, with influence but no authority over the usefulness or technical quality of programs. So, as with the outside production house, the energies of the manager are expended in selling the service and running an efficient operation.

Again, in a start-up operation this can be fatal. When video is new to an organization, the programming has to be absolutely first rate in order to set standards for a skeptical audience. If the user departments are allowed to be the sole arbiters of content and costs, programs may not be the most desirable from an overall corporate point of view; budgets may be overspent when there is no need; or corners may be cut to the point where a program's quality is destroyed.

The video manager, the communications professional, has to be able to control content and budget to some extent if the operation is to succeed!

Third, it is difficult for in-house production facilities to compete with outside houses on a pure cost basis.

The private video manager, no matter how satisfying the thought might be, is not really operating an independent business. He or she is charged for expensive space in the headquarters building at fixed rates. If times get rough, the manager doesn't usually have the option of moving the studio to less expensive quarters or letting staff go temporarily. If all the user departments cut their programming, the manager seldom has the option of soliciting work from other corporations. If outside studios cut their prices, he or she can't follow, because rates were fixed by the corporate controller. It is difficult for a video facility to produce even a paper profit under these circumstances.

In a corporation using a charge-back system to fund audiovisual productions, there is rarely an executive committee that ranks projects according to project priority or corporate objectives, as is customary in a self-funded operation.

Priorities are set on a first-come, first-served basis, and the manager with the biggest budget gets the most attention.

Which Route to Go?

If I seem prejudiced against a total charge-back system, it's because I am. While it seems eminently fair to ask video to pay its own way in the corporation, it doesn't always turn out to be possible. Internal corporate accounting is not the open market and a total charge-back system forces video to labor under handicaps from which it may not recover.

It is my firm opinion that, in a start-up situation, the commitment and the budget money for video have to be there from the outset. Only after video has had a chance to prove itself in the corporation should a charge-back system be implemented.

An alternative, and perhaps a much fairer, system is a combination of direct funding and charge-back in which the corporation as a whole absorbs the capital and basic operating expenses (rent, salaries, utilities, etc.), and the actual programming costs are allocated to user departments.

This system requires the video operation to sell its services to departments, but does not expect users to bear the entire brunt of the studio overhead. Because the charges to the users are much lower, it is easier to sell video to them, and the video manager can maintain greater control over studio use and program content. At the same time, the corporation has a measure of the overall desirability of video, since user departments are required to bear some of the costs.

Whatever the system used for funding the video operation, it is the manager's job constantly to sell the capabilities and cost-effectiveness of the facility to the corporation. Without a yearly influx of cash—to buy and maintain equipment, to pay the best professional staff available, and to produce high-quality programming—the operation will die.

What to do with the funds under your control is covered in the next section.

BUDGETING

The difference between a capital and an operating budget was once illustrated to me very graphically. When asked why there seemed to be thousands of dollars available for equipment and very little for salaries, our manager explained, "Because equipment comes out of one pocket and salaries come out of another—and there's no money in the second pocket."

The *capital budget* is money set aside by management for investment in physical assets: building, equipment, etc. It is generally managed at the corporate level and the amount set aside each year will vary according to the profit and tax position of the company.

To receive a share of the capital budget, a manager has to make a proposal showing why this investment on the part of the corporation is worthwhile, and what sort of return on investment the corporation can expect.

The *operating budget* shows projected expenses and revenues, if any, for one year. It is subject to rigid corporate "guidelines" (i.e., a ceiling on the amount of increase over the previous year) that are exceeded only with great difficulty and that limit the manager's ability to grant salary increases of the size that are undoubtedly deserved.

Since a different accounting logic applies to capital and operating budgets (equipment is depreciable, salaries are not), it is often easier to buy a new camera than to hire an engineer to maintain it. This is a fact of corporate life and explains why there is sometimes money in the capital pocket and none at all in the operating pocket

The Capital Budget

Any request for capital funds should be accompanied by an analysis of the return on investment expected. The preparation of the capital budget request requires that you understand some basic financial concepts.

Benefit versus Cost

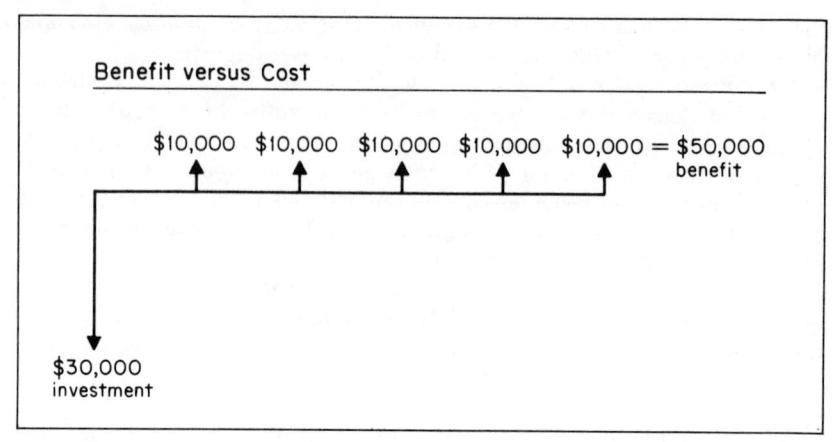

```
        $10,000  $10,000  $10,000  $10,000  $10,000 = $50,000
           ↑        ↑        ↑        ↑        ↑      benefit
     ┌─────┴────────┴────────┴────────┴────────┘
     ↓
  $30,000
  investment
```

Figure 8-1 Benefit versus cost.

First, you have to know what is the *useful life* of the proposed equipment. Three years? Ten years? This number will be used to help figure other values and should be an accurate indication of how long you intend to keep the equipment before replacing it.

You should know the *net present value* of the equipment. This figure will tell you whether the cost saving or revenue realized from the purchase of this equipment, spread over its useful life, is greater or less than its cost.

To give a very simplified example, you may buy a camera today for $30,000 and assume it has a useful life of five years. You figure it will provide a benefit worth $10,000 a year (for example, in rental savings). The benefits total $50,000 over the life of the camera (assuming, for the sake of the example, that the rental costs stay constant). It seems like a good investment. (See Figure 8-1.)

However, a dollar today is worth more than a dollar tomorrow. In our example, our dollar is reduced or discounted by 10 percent each year. As a result, our $10,000 benefit in year 1 is really worth only $9,100 (as measured in terms of today's dollars). Benefits in subsequent years are also discounted, so that our benefit in year two is worth $8,260, $7,500 the year after, $6,830 in year 4, and $6,210 in year 5. When the present or discounted values are added together, it appears that the total benefit of the purchase is actually $37,900, not the $50,000 we first assumed. (See Figure 8-2.)

The net present value is the difference between the benefit of the discounted dollars and the cost of the equipment (Figure 8-3) in this case, $37,900 minus $30,000, or $7,900. If the net present value is greater than $0, the investment is generally favorable; if less than $0 it is unfavorable.

Another way to look at the purchase is to figure the *internal rate of return (IRR)*. In our example, our IRR is 19.9 percent. If this percentage is better than the expected cost of money over the five years of the project

```
Present Value

        $9,100   $8,260   $7,500   $6,830   $6,210 = $37,900
          ↑        ↑        ↑        ↑        ↑          benefit
          |
          |
          ↓
    $30,000
    investment
```

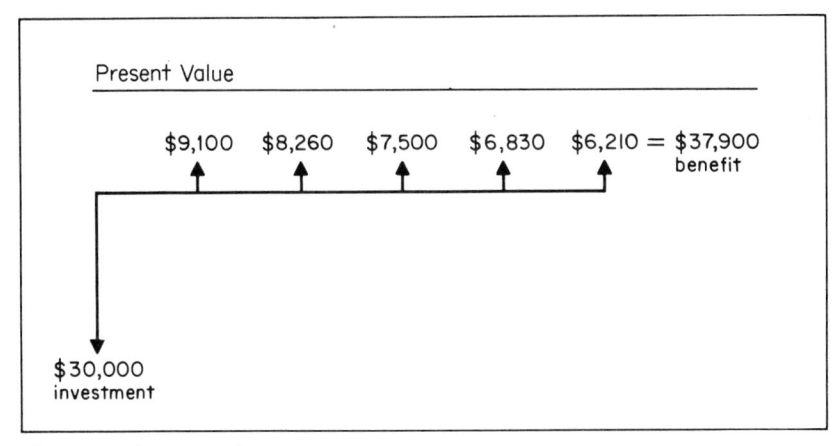

Figure 8-2 Present value.

life, then our camera is a good investment. In today's economy, an IRR of 5 percent would probably not be desirable.

You may also be asked for the *payback period*, or the time when the expected benefits will equal the cost of the equipment. In our example, that point would be reached in three years and nine months. This is derived by using our discounted dollar benefit ($9,100 from year 1; $8,260 from year 2; $7,500 from year 3; and $5,140, which is nine months of year 4). Management often looks at the payback period when comparing potential investments and may, in fact, routinely reject investments that take too long to show a benefit. Figure 8-4 illustrates the payback period.

The tax benefits are also weighed in the decision. The U.S. government will give an *investment tax credit* of 10 percent on all new production equipment that has a useful life of five years or more. The credit is taken in the year of purchase and can be an additional incentive to purchase equipment.

Capital equipment is depreciated on the basis of the equipment's tax life. That means that the cost of the equipment is charged over the life of the equipment rather than all at once when it's first purchased. Using *straight-line depreciation,* an equal amount is charged each year. *Accelerated depreciation* results in a higher write-off in the early years, and a lower one as the equipment ages.

For the first few times, you should consult someone from the controller's department when drawing up an equipment proposal. He or she will describe the corporation's procedures, explain some of the finer points of tax law, and assist you in doing the calculation. You can also learn a great deal about how and when to present your proposal and how likely it is to be favorably received. In addition, it never hurts to have an accountant's imprimatur when you send your request to senior management.

It is quite possible to submit a major proposal several times before it

```
Net Present Value
_____

            $37,900      Benefit
             30,000      Investment
            _____
            $ 7,900      NPV

            NPV > $0     Favorable
            NPV < $0     Unfavorable
```

Figure 8-3 Net present value.

is accepted. The reasons for rejection may have much more to do with the corporation's tax position in a given year than with the intrinsic merits of your proposal. You are competing with the proposers of other worthy projects for sometimes scarce resources, and you may simply have to wait your turn.

The Operating Budget

The *operating budget,* as its name implies, covers the expenses of running your unit for a year. It includes salaries and employee benefits, fixed costs such as rent and utilities, and variables such as supplies and travel.

Salaries and related benefit costs will always be one of the largest expenses. You will probably be allowed a fixed amount for raises, to be allocated to the staff on the basis of merit. The average percentage will probably bear no relation to the increase in the cost of living. To exceed your limit will require convincing a string of doubtful personnel officers that everyone on your staff is a star performer. Your success is highly unlikely. Therefore you are left with the option of giving everyone the same unsatisfactory raise or rewarding the truly excellent achievers and neglecting the rest. In either case, your staff will be convinced that you could have given them all 20 percent raises if only you would.

Compensation can be handled only if you have a plan for the year worked out by budget time. You then have a fighting chance to get the money you need, if you can justify it. The salary survey of corporate video professionals, published by the International Television Association (ITVA), can be a useful reference when writing job descriptions and working with the personnel department to set equitable salary levels for your staff. For further information, see Chapter 27.

You may or may not be charged for rent and electricity. If you are, you

Payback Period

$9,100 $8,260 $7,500 $6,830 $6,210 = $37,900 benefit

Payback
$30,000 = 3 years, 9 months

$30,000 investment

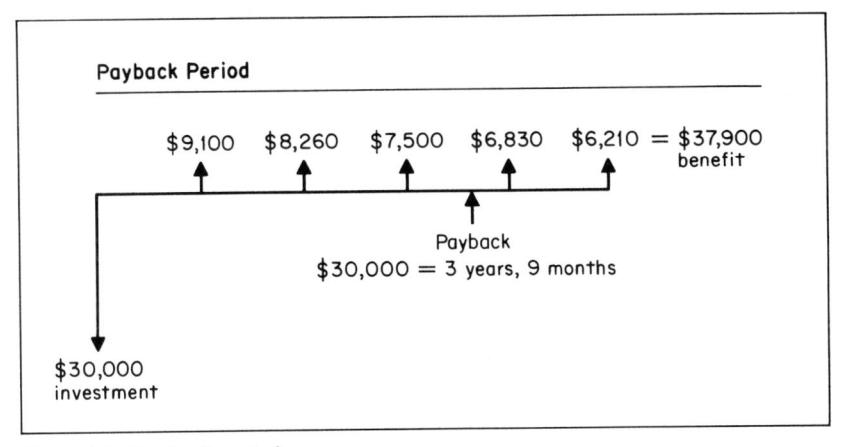

Figure 8-4 Payback period.

will probably be assessed a fixed rate per square foot, which is nonnegotiable.

Telephone is charged per instrument, plus toll calls, with additional charges for installing, removing, or relocating instruments.

The remaining budget items are more directly controlled.

If you are operating on a direct cost basis, it is best to estimate the number and types of programs to be produced in the coming year and to develop the budget based on your expected cost for the total number.

In the process, you may arouse attention for unusual budget amounts in unexpected places. Most corporate budget forms do not have a separate category for video programming costs. As a result, $50,000 under "Office Supplies" for "tape" may lead your budget analyst to suspect you of setting up a black-market operation in Magic Mending Tape.

Your "Supplies" category will seem swollen by normal corporate standards, since it will include such expensive items as replacement components and camera tubes. "Temporary Help" may include all free-lance technicians; "Consultants" may mean writers and directors; and "Furniture Rental" may be the only place to put set construction costs.

If possible, work with the budget people to set up a separate category called "Video Programming." This heading will allow you to place all your direct costs for programs in one place and to track them more efficiently.

This special category is particularly important if you operate on a charge-back system. In that case, you have to be concerned not only with your own budget, but with your client's as well. You have to spend several months of the year (probably June and July) meeting with clients, helping them to project the number and types of programs, advising them on the amount of money to be budgeted, and in what categories. The process is simplified if expenses can be consolidated in one video programming category.

After you've succeeded in getting clients to budget money, however, you have to keep going back to them through each successive level of budget

Media Resource
Work Requisition

CHASE

Please type or print clearly

Date Requisitioned	Organization Number to be Charged	Organization Name to be Charged	Requisitioned from ☐ Video Services ☐ Graphic Services
Date and Time Finished Work is Desired	Contact Individual's Name, Location, and Complete Telephone Number		
Type of Work Requisitioned	Purpose of Work		
Description of Work			
Name, Title, and Dept. of Requesting Officer (Type or print)	Signature of Requesting Officer		

Below for Media Resource use only

Work Order Number	Assigned To	Date Assigned	Time Assigned	Title Assigned
Remarks				

PHOT 8* REV. 6-78 PTG. 4-81

2. Requestor's File Copy (Remove and forward other parts intact)

Figure 8-5 Work requisition form. (*Chase Manhattan Media Resource Center.*)

call to make sure the program money isn't cut as the corporate pursestrings tighten. In a difficult budget year, this battle will require all your powers of persuasion as other managers see cherished plans axed and are sorely tempted to sacrifice video instead.

Operating on the charge-back system during the year requires rigid adherence to the agreed budget, accurate record keeping, and a good line of communication between the video manager and the client whose money you are spending.

Case Study 1. One very effective method is used in the Chase Manhattan Bank AV Communications Center.

First a budget for the program is prepared by the producer, working with the writer and the director. The total amount for the program is presented to the client on a requisition form, and the client is asked to approve the expenditure by filling in the appropriate charge number, signing, and returning all but one copy of the form.

The producer carefully monitors expenses during the production. As bills come in, they are approved by the video manager and sent directly to Expense Processing, with a copy of the client's requisition attached. In this way, vendors are paid promptly, an always important practice, but especially so with free-lance help. If it begins to appear that the cost of the program will exceed budget, the producer has to return to the client for further authorization.

A complete set of the forms used within the Chase Manhattan Bank AV Communications Center is included here. They start at the Work Requisition Form (Figure 8-5), proceed to a cover sheet for the entire production (Figure 8-6), specific forms for video (Figure 8-7), and support material (Figure 8-8), and conclude with the Script Approval Form (Figure 8-9).

FUNDING AND BUDGETING 8-11

PRODUCTION TITLE: _____	JOB NUMBER: _____	
	CODE NUMBER: _____	
CLIENT DEPARTMENT: _____	EXPENSE NUMBER: _____	
CLIENT REPRESENTATIVE: _____	EXTENSION: _____	
PRODUCTION DISTRIBUTION DATE (S): _____		

VIDEO PRODUCTION BUDGET	ESTIMATE	FINAL
Pre-production		
Production		
Post-production		
TOTAL		
SUPPORT MATERIALS BUDGET		
Slides		
Filmstrips		
Transcripts		
Print		
Audio cassettes		
TOTAL		
GRAND TOTAL: VIDEO AND SUPPORT MATERIALS		

Budget approval:

_____   _____   _____
(client signature)   (title)   (date)

Figure 8-6 Cover sheet for video production expense form.

These forms are a useful control mechanism and record, whether the expenses must be reallocated or are charged to the video manager's budget. The financial history of each program is available at a glance and the data are extremely useful when planning the facility's overall budget for the coming year.

CONCLUSION

No matter what funding and budgeting methods are used within the corporation, it is extremely important to maintain accurate records. Take it for granted that sooner or later your unit will be audited. Keep your files accordingly.

8-12 A MANAGER'S GUIDE

COLUMN 1	ESTIMATE	FINAL	COLUMN 2	ESTIMATE	FINAL
RESEARCH			SET DESIGNER		
SCRIPTWRITER			GRAPHIC ARTIST		
CASTING			NARRATOR/ ANNOUNCER		
DIRECTOR			TALENT		
ASSISTANT DIRECTOR			TALENT		
FLOOR MANAGER			TALENT		
CAMERA OPERATOR 1			TALENT		
CAMERA OPERATOR 2			PRODUCTION ASSISTANT		
CAMERA OPERATOR 3			PRODUCTION ASSISTANT		
ENGINEER, VIDEO			PRODUCTION ASSISTANT		
ENGINEER, AUDIO			TECHNICAL DIRECTOR		
LIGHTING DIRECTOR			OVERTIME		
LIGHTING ASSISTANT			MESSENGERS		
SUBTOTAL			SUBTOTAL		
			TOTAL PAGE ONE		

VIDEO PRODUCTION BUDGET PAGE ONE

JOB NUMBER: _____
PRODUCTION TITLE: _____   CODE NUMBER: _____
CLIENT DEPARTMENT: _____   EXPENSE NUMBER: _____
PRODUCTION DISTRIBUTION DATE(S): _____

Figure 8-7 Video production budget form.

For all equipment purchases, obtain competitive bids from several suppliers. When contracting for services from production houses, editing facilities, engineering firms—even free-lance personnel, such as writers, directors, and camera operators—ask for, and keep, a formal statement of their rates. Bring your files up to date on a regular schedule. An auditor once explained to me why this is so important: "We don't insist that you always go to the lowest bidder. We recognize that you may want to work with some

FUNDING AND BUDGETING 8-13

```
                    VIDEO PRODUCTION BUDGET PAGE TWO
    PRODUCTION TITLE: _____ CODE NUMBER: _____
```

COLUMN 1	ESTIMATE	FINAL	COLUMN 2	ESTIMATE	FINAL
TITLES, GRAPHICS, ARTWORK			EDITING		
MATERIALS, SET CONSTRUCTION			MEALS		
PROPERTIES: RENTAL PURCHASE			TRANSPORTATION		
LIGHTING: RENTAL PURCHASE			DUPLICATION		
TIME-CODE GENERATOR			DISTRIBUTION		
VTR RENTAL			MISCELLANEOUS		
TAPE STOCK			OTHER		
TELEPROMPTER SCRIPT			SUBTOTAL		
LOCATION FEES, EXPENSES			TOTAL PAGE TWO		
MUSIC			TOTAL PAGE ONE		
ANIMATION, SPECIAL EFFECTS			GRAND TOTAL		

Figure 8-7 *(continued)*

vendors rather than others for professional reasons that we can't judge. But we do want to make sure that the rates you are paying are within a normal range for the industry."

The records you keep will provide you with the information you need to manage your unit and will give you the support you need to prove the cost-effectiveness of your organization.

The role of the video manager often falls to those ill prepared to handle it. They have been attracted to the field and have succeeded because they are creative, adept at turning ideas into visual material that informs, teaches, and moves an audience to change. They find themselves with less and less time to create as they worry over personnel files and prepare budget estimates.

However, managing a creative unit can in itself be creative. It is the manager's efforts—vision, really—that drive the organization. The manager's choices, the allocation of resources, set the standards and the style. His or her boldness in seeking support from the corporation and success in obtaining it makes everything else—the staff, the equipment, the opportunities to be creative—possible. It can be a uniquely satisfying role.

SUPPORT MATERIALS BUDGET PAGE ONE								

PRODUCTION TITLE: _____ JOB NUMBER _____

CLIENT DEPARTMENT: _____ CODE NUMBER: _____

DISTRIBUTION DATE(S): _____ EXPENSE NUMBER: _____

COLUMN 1	ESTIMATE	FINAL	COLUMN 2	ESTIMATE	FINAL
FILMSTRIPS			PRINT		
RESEARCH			RESEARCH		
FILMSTRIP SHOOTING			WRITER		
GRAPHIC ARTIST			GRAPHIC ARTIST		
ARTWORK, GRAPHICS, TITLES			ARTWORK, GRAPHICS, TITLES		
PROCESSING			TYPESETTING		
ANSWER PRINTS			PRINTING		
DUPLICATION			DUPLICATION		
DISTRIBUTION			DISTRIBUTION		
MESSENGERS			MESSENGERS		
MISC. FEES & EXPENSES			MISC. FEES & EXPENSES		
OTHER: TAX, LOCATION FEES, ETC.			OTHER TAX, LOCATION FEES, ETC		
SUBTOTAL			SUBTOTAL		
			TOTAL PAGE ONE		

Figure 8-8 Support materials budget form.

SUPPORT MATERIALS BUDGET PAGE TWO

PRODUCTION TITLE: _____ CODE NUMBER: _____

COLUMN 1	ESTIMATE	FINAL	COLUMN 2	ESTIMATE	FINAL
SLIDES			TRANSCRIPTS		
RESEARCH			MISC. FEES & EXPENSES		
PHOTOGRAPHY SESSION(S)			OTHER: TAX, LOCATION FEES		
GRAPHIC ARTIST			SUBTOTAL		
			RECORDING		
ARTWORK, GRAPHICS TITLES			RESEARCH		
PROCESSING			STUDIO		
DUPLICATION			EDITING & MIXING		
DISTRIBUTION			NARRATOR/ ANNOUNCER		
MESSENGERS			MUSIC CLEARANCE FEES		
MISC. FEES & EXPENSES			RECORDING		
OTHER: TAX, LOCATION FEES, ETC.			DUP.: TAPES, CASSETTES, CARTRIDGES		
SUBTOTAL			DISTRIBUTION		
TRANSCRIPTS			MESSENGERS		
TRANSCRIPTION SERVICE			STOCK		
DUPLICATION			MISC. FEES & EXPENSES		
DISTRIBUTION			OTHER: TAX, LOCATION, FEES, ETC.		
MESSENGERS			TOTAL		

Figure 8-8 *(continued)*

PRODUCTION TITLE: _____	CODE NUMBER: _____
CLIENT DEPARTMENT: _____	EXPENSE NUMBER: _____
CLIENT REPRESENTATIVE: _____	EXTENSION: _____

PRODUCTION DISTRIBUTION DATE(S): _____

FIRST SCRIPT APPROVAL:

_____   _____   _____
(client signature)   (title)   (date)

SCRIPT REVISION APPROVAL:

_____   _____   _____
(client signature)   (title)   (date)

FINAL SCRIPT APPROVAL:

_____   _____   _____
(client signature)   (title)   (date)

STORYBOARD APPROVAL:

_____   _____   _____
(client signature)   (title)   (date)

ROUGH EDIT APPROVAL:

_____   _____   _____
(client signature)   (title)   (date)

FINAL EDIT APPROVAL:

_____   _____   _____
(client signature)   (title)   (date)

SUPPORT MATERIALS APPROVAL:

_____   _____   _____
(client signature)   (title)   (date)

Figure 8-9 Script approval form.

Contracts

NANCY F. GURWITZ

As the manager of a media department, you are, or will be, responsible for the purchase and leasing of equipment as well as the hiring of persons outside your department to perform various services. Whether or not you actually commit the arrangement to writing, you are "contracting" for services or equipment.

REASONS FOR A CONTRACT

Whether or not you should reduce the arrangement to writing and how complex this contract should be will depend on several factors.

Prior Approval. Your corporation may require that proposed expenditures involving large dollar amounts be submitted in writing for prior approval.

New Product or Service. If you are hiring a person with whom you have never worked or purchasing or leasing expensive equipment from an unknown vendor, you can protect yourself from later disappointment and problems by clearly stating in an informal letter of agreement, or a more formal contract, your expectations regarding the service or equipment.

Expected Exposure to Liability. A project may involve a deadline; for example, a special video program must be prepared for the annual meeting. Obviously, one wants to ensure that persons providing services and equipment as part of this project will deliver on time. By committing your requirements to writing, you have made quite clear the importance of any deadlines and of any quality specifications. In this contract particularly, you may want to make provision for damages in case of a failure to perform the contract (see below).

Psychological Impact. Even if none of the above factors exists, you may want to commit the arrangement to writing as a psychological hedge against next-level management. When you've spent large dollar amounts on a project, people in upper-level management may be more receptive if they can see in writing the extent and complexity of the services and equipment that were purchased for the project.

Accommodation. The parties providing the services may want, or may be required by their company, to commit the arrangement to writing. Obviously you will want to accommodate them, but not at your own expense. Therefore, make sure you review the contract or letter of agreement which they submit as carefully as though you had decided to submit a contract to them for one of the reasons enumerated earlier. They may say that it is a standard contract. It may even be similar to one which you have used with them before. Just remember that this is a new project and it may have different specifications and problems.

Standard Agreements. Finally, even if you have decided that a contract is not needed for your own protection, the arrangement may automatically include a written contract. For example, the vendor's invoice, issued in connection with the sale to you of a small equipment purchase, will probably include a standard contract of sale setting forth certain limitations on the liability of the vendor. These will be discussed in more detail later in this chapter.

As mentioned above, even without a written letter of agreement or a formal contract, an oral agreement for services or equipment is enforceable under law—if the services to be performed or the equipment to be delivered are to be performed or delivered within one year. There is nothing wrong with an oral contract—until something goes wrong with the equipment or the services. Then you may have to settle the dispute in court, and, without anything in writing, the outcome depends on your word against the supplier's regarding the terms of your agreement.

Obviously, if you have dealt with the contractor or vendor before, if the project is not unusually complex, and if it is without expected exposure to liability, you may very reasonably depend on an oral contract. If not, then get it in writing!

That brings us to the question of a letter of agreement versus a formal contract. There are no hard-and-fast rules for deciding how formal the contract should be. If the arrangement is complex or is going to cover a long period of time or a number of services, the letter of agreement may grow on its own accord into a formal contract during the negotiation period. Clearly, if time is of the essence, you may be lucky if you can produce a quick letter of agreement that lists the essential terms. A more formal contract may need to be reviewed by lawyers—either your in-house legal department or your corporation's outside counsel. This review can take considerable time and cause considerable delay. Fortunately, much of a long, formal contract is standard industry boiler plate which may reduce the review time. As with all boiler plate, it probably is not worth arguing about. In addition, you may have dealt with this contractor or vendor before and be familiar with his or her contracts.

At this point, I, as a lawyer, would like to put in a good word for your law department. If you have the luxury of time on your hands, then by all means send the document for review—whether it is a long, formal contract or a short letter of agreement. Try and develop a good relationship with your law department. If you include the lawyers when you are working on the early stages of a project, you may be the chief beneficiary. If the lawyers understand your goals and problems, they may notice language—even boiler-plate jargon—that can and should be changed. In addition, if you have developed a relationship with them, when you need quick turn-around on a formal contract, the willingness and comprehension will be there for you.

CONTENT OF CONTRACTS

Let's now discuss content. What are you going to include in your contracts? What are the points to be negotiated? Service contracts will be discussed separately from equipment contracts and equipment contracts will be categorized by the cost of the purchase or lease. But, first, let me stress the most important rule for effective contract negotiation: *Know your goals!* In each of the areas to be discussed, I will list terms that can or should appear in your contract. Theoretically, each one is negotiable. The point is: Know your immediate and long-term objectives. They will dictate the terms you can live with in the particular transaction. Just because you had a term in an earlier agreement doesn't mean it should be included in the latest contract. But it should be there if it is necessary to fulfill your objectives.

CONTENT OF A SERVICE CONTRACT

Generally, a service contract will cover the terms discussed in the following paragraphs. (Figure 9-1 shows such a contract.)

The Scope of the Project. The contract should briefly describe the item to be delivered and the expected form and format. For example, the item may be a script, videotape, or sound recording. If the videotape or sound recording will be based upon a completed script, the initial description in the contract should include the source of the script. It may describe, again briefly, the subject matter of the project. A comprehensive description of the item should be set forth later in the contract.

Ownership and Use of the Item. This term gives ownership of the item to the corporation. Generally, one provides for this assignment of ownership in the following, or similar, language:

> The corporation shall have unrestricted right to use this tape [script, whatever] for any and all purposes.

Cost and Payment Terms. This section of the contract should describe all items and services within the cost, e.g., photography and recording, and

AGREEMENT

Agreement dated November 1, 198_ between NFG, Inc. ("NFG") and World Traders Ltd. ("World Traders").

NFG will produce for World Traders five (5) scripts describing the export/import programs of Canada, Spain, Belgium, Ireland, and Australia. Thereafter, World Traders may commission additional scripts with respect to other countries. NFG may accept such commissions but is under no obligation to do so. The scripts are to be used in connection with World Traders' consulting division.

1. Each script will be for a presentation that will last approximately 15 minutes in duration. The script will be suitable for video or personal presentation.

2. The scripts will be delivered in legible hard copy or in soft copy compatible with World Traders' word processing equipment.

3. The scripts will be owned by World Traders. World Traders shall have unrestricted right to use the scripts for any or all purposes.

4. World Traders will pay NFG $8750 for the five scripts. Thereafter, the price per additional script shall be $1500. World Traders shall advance NFG one-half the total cost for the first five scripts on commission of the assignment and the remainder upon the delivery of the fifth script. Therefore, $750 will be paid upon commission and acceptance of each additional script and the remainder upon the delivery of such additional script.

5. World Traders shall be responsible for providing information concerning the export/import program of the countries commissioned and shall make available to NFG the services of employees and/or customers who are familiar with such countries' programs.

6. Before completion of a script, such script shall be submitted to World Traders for technical review by an accepted expert in the country's program. NFG shall make reasonable technical changes as World Traders, through such experts, may suggest without additional charges.

7. NFG agrees that any information gained about World Traders or its customers will remain strictly confidential. NFG, during the term of this agreement or thereafter, will not disclose to anyone information concerning the operations or business affairs of World Traders which NFG may have acquired in performing services under this agreement.

8. Notwithstanding any other provision of this agreement, nothing stated herein shall be deemed to prevent NFG from writing about export/import programs other than World Traders' even if such persons may be deemed to be competition.

9. NFG acknowledges that the scripts have been ordered and commissioned by World Traders and will be prepared by NFG as "work-made-for-hire." The content of the scripts will be original material except for such excerpts of copyrighted works as may be included with the written permission of the copyright owners, and that the scripts will not violate or infringe any copyright, trademark, patent, statutory, common law, or proprietary rights of others, or

Figure 9-1 Letter of agreement.

contain anything libelous. World Traders will secure and hold the copyrights to the scripts.

    10. NFG is solely responsible for its work described herein and may not delegate its responsibility to any other party.

    11. NFG warrants that all work will be done in a professional manner.

World Traders, Inc.　　　　　　NFG, Inc.

Figure 9-1 *(continued)*

including *reasonable* changes. If you expect that frequent changes will be made, you may wish to set the rate for the excess cost beforehand.

This section will also define when and how payment will be made. If partial payments are agreed upon, the expected dates and prior requirements to be met on such dates should be delineated. If the requirement depends upon partial completion of the project, you should include a description of what will suffice as reasonable completion.

Production Schedule. This term should describe the schedule expected to be followed and should include deadlines which have to be met by both parties. For example, if the corporation will be delivering certain data to the contractor or if the contractor will be meeting with corporate personnel, the dates which reflect such obligations by the corporation should be included. Caveat: Be very careful to live up to your deadlines if you want the contractor to live up to his or hers (see the paragraph "Expected Exposure to Liability" above).

Detailed Description. Having briefly described the project in the beginning of the agreement, you will now want to give a comprehensive description of the rights and obligations of the two parties. For example, this is the place where you will describe those areas for which the corporation (i.e., *you!*) will be responsible. This may include the provision of employees, contacts with customers, the script, location, studio, etc.

Copyright. If you are hiring a writer to prepare a script, you must include a proprietary clause which provides that the copyright belongs to the corporation. Without this clause, under the Copyright Law as revised in 1976, the copyright remains with the writer unless the contract provides otherwise. This is an exception to the general copyright law that work-for-hire belongs to the person who hired the contractor.

If you are hiring a producer, the contract may provide that the producer has the responsibility to secure the copyright in the name of the corporation.

More information about the copyright laws appears in Chapter 24.

Confidentiality. Generally the corporation will supply the data and sources needed by the writer or producer. This information may be confidential or proprietary financial or nonfinancial in nature. Remember to protect yourself and your corporation by expressly providing for the confidentiality of any information given to the contractor.

Credits. This is a negotiable point. It will depend upon your customary procedures, the contractor's customary practices, or a resolution of the two.

Provision of a Copy to the Writer or Producer. This provision is optional. In addition, you may have developed the practice of handling this informally. The reason to include the term in the agreement is to protect your copyright. You will want to limit the use of any copies of your program given to the writer or producer so that you do not give away your copyright. The following, or similar, language will protect you:

> [The contractor] has the right to make and exhibit prints for sales purposes in order to demonstrate its quality and capabilities, with *prior written consent* of the corporation. (Emphasis added.)

Of course, if this is extremely sensitive or confidential information, for the use of which written consent would never be given, you will do better to make this clear to the contractor from the start and to omit the clause entirely. The presence of this clause implies that consent will be given.

Nondelegation Clause. All work for which the contractor is being hired will be performed by the contractor except as expressly provided in the contract. Otherwise, why are you hiring the contractor?

Standards Clause. This clause refers to acceptable standards. Unfortunately, it does not really define what acceptable standards are. Nevertheless, the following boiler plate is generally found in all contracts, long or short:

> Work will be done in a professional manner and the material also used will be of highest quality.

Obviously, if you leave this clause out, it is probably not critical. You are hiring this contractor because you believe he or she will provide high-quality, professional service. Not so obviously, if the use of certain specific materials is important to you, include this requirement in the contract. Again, this recommendation comes back to the basic precept: Know your goals.

Obtaining Necessary Rights. This clause relates to the need to obtain permission from the holder of a copyright to use quotations from literary writings or sound recordings in your products. Therefore, this clause should be in any contract with a producer or writer.

Damages. A damage clause establishes beforehand the compensation by the contractor to you in the event of default. This clause is not necessary unless you foresee sufficient liabilities if the contract is not performed in a specific manner or is not completed by a specific deadline. In this situation, the clause should have a high priority on your list of required terms. The damages may be a certain dollar amount determined as a flat amount or as a percentage of monies already paid or still to be paid for the project.

Where performance is essential to meeting a deadline, the damage clause can provide that the contractor who fails to provide the object of the contract may be required to pay the *reasonable* costs of having the product provided by someone else. Remember, however, that this clause does not give you license to hire the most expensive producer, writer, or actor if the one with whom you have contracted defaults on the contract. On the other hand, you are not necessarily limited to the amount specified on the original contract; the replacement contractor will be working within a shorter period of time under considerably more pressure and *may* deserve a premium. Again, that wonderfully imprecise legal standard of *reasonableness* is determinative.

The question of damages is important for another reason. Contemplation of damages should trigger serious consideration of two other aspects of the contract: production scheduling and completion deadlines. When you are negotiating the production schedule of a product that is to be used on a particular date, it is essential to give yourself the maximum lead time possible. In addition, as part of the schedule and certainly before you start making partial payments, you must have progress checkpoints. The purpose of the checkpoints is to let you out of the contract if the contractor—without fault attributable to you—is not meeting the agreed-upon deadlines. The contract must permit careful monitoring as the deadlines draw nearer—but, remember, the contract is worthless if you haven't been meeting your obligations.

CONTENT OF AN EQUIPMENT PURCHASE OR LEASE (SMALL ITEMS) CONTRACT

Almost everything you need to know has been covered above; however, it will be reviewed quickly here.

Basically, a contract is not really necessary for the purchase or lease of a small item. However, most vendors will include a standard agreement at the end of the invoice. It is not worth arguing about the language or spending time trying to make it nonstandard. You would be wiser to spend your energy on careful research of reliable vendors.

A second point in connection with leasing small equipment is that the leasor may require a small bond. Again, don't argue. This request is standard. Call your insurance department (or whichever department handles insurance coverage) to arrange for a floater. If you are continually in need of bonds, your insurance department should be able to arrange a perpetual floater. In that event, be aware of the limits of coverage and keep track of the extent of your bond requirements at any given point to ensure that you are not exceeding that coverage.

CONTENT OF AN EQUIPMENT PURCHASE OR LEASE (MAJOR ITEMS) CONTRACT

You have decided that you need a major piece of equipment for the studio. Now you must decide whether to purchase or lease it.

Purchase

If you purchase equipment, you acquire outright ownership. Ownership entitles your corporation to investment tax credits in the year of the purchase and to depreciation deductions throughout the *tax life* of the equipment. If you sell or exchange the equipment prior to the end of its tax life, there will be residual value that you can use in bargaining. Even if you keep the equipment for its entire life, there may be some remaining salvage value.

However, in order to acquire ownership, you must use money that has been budgeted for capital expenditures or you must obtain financing. Unless you have an unlimited capital expenditure budget (lucky you!), you will need to be making difficult choices among major pieces of equipment.

Even if you obtain financing, it is unlikely that you will be able to finance 100 percent of the purchase price. The down payment comes out of the capital budget.

Lease

Leasing gives you the obvious benefit of using a major piece of equipment without committing yourself to spending large sums of money. If you are unhappy with the equipment, return it to the leasor and lease something different. Although there may be trade-in value in property that is owned, high-technology equipment can become outmoded quickly and thus have an adverse effect on that value.

If you lease a major piece of equipment, even with an option to purchase at the end of the lease contract, you have none of the benefits of ownership until you exercise the option. However, if and when ownership is acquired, it will have been achieved without the necessity of making a large down payment out of your capital expenditure budget. Lease payments come out of your operating budget.

Obviously, if you are tight on operating budget and have unused capital funds, the decision will be made for you. Remember, when you lease instead of purchase, the leasor gets the tax benefits which arise from the depreciation deductions. This should be reflected in a better lease price. Although you cannot take depreciation on leased equipment, the Internal Revenue Code permits the leasee of equipment to take the investment tax credit during the first year of the lease—but only if this arrangement is negotiated and included in the lease contract.

Another caveat involves the way you structure a lease with an option to purchase. If you are to pay most of the money during the term of the lease and can exercise the option for little extra cost and if the life of the lease approximates the life of the equipment, then the IRS may contest your "lease" and require you and the leasor to treat the arrangement as a leveraged buy-out, i.e., a financed purchase. The depreciation will belong to your corporation and you will have to restate your budgets to reflect that the equipment was obtained from the capital budget, not from operations. The rules and case law on leveraged buy-out are fairly complex. Make sure you have your legal and accounting departments check the financing and lease terms for compliance with Financial Accounting Standards Board (FASB) rules.

Specific Contract Terms

Particularly when purchasing a major piece of equipment, it is important to keep the basic rule in mind: Know your goals and needs. Why are you buying this equipment? What will be your dependency upon it? What will happen if it becomes nonoperational? When do you need it? And what happens if you don't get it on time? The answers to these questions and the priorities which you set on those answers will determine which of the following points will and should be covered by your contract. You must be willing to give away the less important points in order to get the ones you can't live without. But first you must know what you need.

Generally, an equipment contract will include many or all the items discussed below.

Service Contract. This portion of the contract not only establishes the annual rate for service but can be used to delineate your expectations and needs concerning turnaround time, provision for temporary replacement, or both. If you expect that the equipment will be in constant use and that quick servicing will be essential, you can attempt to negotiate for damages for excessive downtime. Although such a damage clause is rare, if the point is very important, try to include the clause to highlight your needs. In addition, it is easier to negotiate for *soft* dollars than *hard* dollars—i.e., excessive downtime can result in an extension of the service contract or the damage dollars can be awarded as credits toward future purchases from this company (if you still want to buy from it after this experience).

Payment Terms. Your goal should be to extend final payment (approximately the last 20 percent) beyond the date of installation of major equipment and until such time as the equipment is fully operational. However, you cannot be unreasonable in picking this date for final payment. If becoming fully operational will take an extremely long time or if it will depend upon delivery and installation of other parts from other vendors, you will probably have to agree to pay in full upon installation. However, that's when you negotiate for low-cost servicing during the period when all the equipment is coming together.

If, even after being extremely reasonable, you are unable to negotiate an extended final payment into the contract, do not worry. The practical solution is to withhold payment authorization of the last portion until you are satisfied—again, assuming a reasonable period of time. Otherwise, an important vendor might sue you unnecessarily (and very expensively, considering the high cost of legal services just to determine what the suit is all about) or might refuse to deal with you when you really need its product.

Delivery. You must consider not only the latest date at which you will be willing to accept delivery, but also the earliest date at which you can conveniently accept delivery. The dates may be unimportant to you. Nevertheless, it is advisable at the minimum to set some time frame within which you expect delivery.

On the other hand, if the delivery date is very important, you will want to elaborate. For example, it might be terribly inconvenient to accept the delivery before a certain date because your studio may not be ready. Or, budgeting considerations may prevent certain expenditures prior to a given date. Then, by all means include a clause in the contract which permits you to decline acceptance of the equipment prior to a specific date.

On the other hand, as is the usual case, late delivery might imperil your ability to meet certain deadlines. In that event, you may want to include within your contract the right to terminate the contract without payment of any money upon nondelivery by a specific time. This may go without saying, but I'll say it anyway: This right is useless if this vendor is the only supplier of equipment that is essential to your overall designs. Obviously, in that case, you will do better to include a variant of the damage clause that specifies a reduction of the purchase price directly related to the additional days of waiting beyond the specified delivery date.

Risk of Loss. Every contract should contain a risk-of-loss clause that clearly places transportation risk upon the vendor. It is unreasonable to make the purchaser responsible as soon as the equipment leaves the manufacturer. The manufacturer-vendor has the requisite knowledge and experience in shipping its products. The vendor chooses the carrier and packs the equipment. Let your vendor worry about insuring the shipment, just passing the charges on to you.

Full Specification Sheet. Particularly with state-of-the-art equipment, you should demand and get full specification sheets from the company so you have an objective basis against which to judge fulfillment of the contract. Note, however, that to get the benefit of the specification ("spec") sheet, you *must* incorporate it by reference within the contract, such as by this statement:

> The terms of the specification sheet set forth as Exhibit A hereto are incorporated by reference herein.

Without this incorporation, you have no legal right to rely upon the specs (or upon any other sales literature or oral commitment or written correspondence). Remember: Get it in writing and make it part of the agreement.

Necessary Operational Information. Make sure that you contract for all technical information and for any necessary updates. This is the time and place to negotiate a good agreement on instruction classes the vendor might offer to teach your present and *future* personnel who will be operating this equipment.

Finally, this is where you worry about what to do if this company, which has all the proprietary information about how to operate this newest and best equipment, goes out of business. Who's going to service you then? Remember the best service contract in the world isn't worth the paper it's written on if the company is no longer there. An acquaintance of mine was willing to negotiate on every point—including price—in order to protect his company's investment in a very sophisticated piece of equipment—a computerized editor. As he saw it, he would need all the computer software to operate the editor, if the company went under. As the company saw it, that was very valuable proprietary information. The resolution? The software was put into an escrow account with a lawyer (of course), with instructions to release that information to my friend's company if the manufacturer went under within a given number of years.

Standard Representations and Warranties. For the most part, this section is boiler plate. As said before, don't waste your energy arguing with boiler plate. (This, however, is one area where your law department's review can be advantageous. If you have carefully negotiated a point, the law department may pick up any language which appears contradictory. The other side should logically agree to excise the excess verbiage.)

Reasonable Substitution. The manufacturer will probably ask for, and should probably get, the right to make minor deviations from parts which have been specified. If having a specific type of equipment is very important to you, its provision should be articulated clearly in the contract.

CONCLUSION

Let me conclude by reminding you of the fact that you may not need to spend the time and energy to commit an arrangement to writing.

There are various factors to consider first: the need for prior approval; no previous relationship; exposure to liability; psychological factors; and the need to accommodate the other party. The existence of several factors will most likely point to a written contract, if only a short letter of agreement.

When committing the arrangement to writing, attempt to make the document a letter of agreement. The factors and the problems unique to this project or purchase will dictate the terms to be included. As the project and problems become more complex, your agreement letter will evolve naturally into a formal contract.

Always remember: Know your goals! This knowledge will determine whether to include any or all of the terms which have been discussed here under either service or equipment contracts.

10 Guidelines for Success

NATHAN J. SAMBUL

As mentioned oftentimes in other chapters of this book, the whole area of private television is growing at a phenomenal rate. Yet, despite the increase in overall numbers, some video operations are being trimmed out of company budgets, others are stagnating with little or no productive work, and still others fail to get management support. Not uncommonly, a company or institution authorizes the initial expenditure to start up an operation, but fails to provide for its continual existence. It is unfortunate when the operation folds, because people lose their jobs and organizations lose an important tool to communicate with their employees and the general public. They miss the opportunity of increasing their productivity, improving employee loyalty, capturing a larger share of the market, or providing better service to the community.

It is quite easy to point an accusing finger at management and blame its members for the lack of growth of a video operation. We have all heard colleagues bemoan the fact that management does not understand what video can accomplish or that management is too rigid and won't allow for new types of programming. In truth, management may be guilty of some of these accusations; but if any guilt is to be assigned, how much rests right within the video operation? Looking for guilty parties may be an interesting exercise, but it is not as constructive as working toward a successful video operation.

The following discussion presents ten guidelines for successful video operation management. I've limited them to management activities, for, although producing high-quality videotapes will also enhance a video operation, working within an organizational structure is often more difficult than producing a program. It is not uncommon for an audiovisual manager to be promoted through the ranks. Originally you may have been a producer, writer, or engineer, and now you find yourself at the level of manager, often without any formal managerial training.

First, let us understand what we mean by *managing*. Louis A. Allen, the noted management consultant, in his book *Making Managerial Planning More Effective* (McGraw-Hill Book Company, New York, 1982), describes four components of managing: planning, organizing, leading, and controlling. These are the necessary activities that one must perform regardless of whether one is managing a video operation, an aircraft assembly plant, or

SOURCE: Sections of this chapter were first published in Nathan J. Sambul, "Management: 10 Guidelines for Success," *Video Systems*, January 1979. Used by permission.

10-2 A MANAGER'S GUIDE

the Dallas Cowboys. Specifically, there are functions for each of the major activities.

1. Planning
   Forecasting
   Developing objectives
   Programming
   Scheduling
   Budgeting
   Developing procedures
   Developing policies
2. Organizing
   Developing organization structure
   Delegating
   Developing working relationships
3. Leading
   Decision making
   Communicating
   Motivating
   Selecting people
   Developing people
4. Controlling
   Developing performance standards
   Measuring performance
   Evaluating performance
   Correcting performance

One's first love is very hard to give up. So, when called upon to perform either managerial work (planning, leading, organizing, or controlling) or technical work (producing, directing, or writing) during the same period of time, a manager will tend to give first priority to technical work. Louis Allen calls this tendency the "principle of technical priority." Of course, it operates against a smooth-running, efficient audiovisual department.

There are a number of excellent books on classical management techniques. The ten guidelines presented in this chapter, however, are designed to go beyond the classical activities of planning, organizing, leading, and controlling. These guidelines have been prepared because, in many regards, video operations are unique departments within larger organizations. The equipment is different, measurement standards are different, and oftentimes staffing is different.

SELL THE VIDEO OPERATION

Video programs are products not unlike the products your company manufactures or the services your organization provides. Just as other departments vie for the attention of the sales force, or top management, or plant personnel, you and your video department have to be out of the studio and in the field selling your product—videotapes. The techniques for selling are the same: analyze the market (investigate which videotapes the audience needs the most); pretest ideas (touch base with opinion leaders and segments of the audience); position the product (before a videotape is sent to an intended audience, notify the recipients of its arrival date and suggest ways of using the tape effectively); and monitor results (conduct evaluations in person or through feedback forms).

Selling is not easy. It requires building a level of confidence and trust with your clients and prospects. It also requires that you deliver the merchandise on time and on budget. But perhaps the single most important factor in selling is filling the needs of your clients, that is, providing them with videotapes that truly satisfy their objectives. Unfortunately, clients are not always clear as to what their objectives are, and it behooves you to clarify and focus these objectives.

By selling on a needs basis you accomplish a number of goals. One, you demonstrate that you care about the product and the firm (a trait that top management and coworkers find admirable). Two, you create an environment of cooperation; that is, both management and staff feel that they are working toward the same larger goal. As a result, more give-and-take occurs and you find yourself with greater creative leeway. Finally, you are more apt to have repeat business. But recognize that selling requires an inordinate amount of time, so it has to be done continually—at annual meetings, staff functions, interdepartmental training sessions, everywhere.

KNOW THE PRODUCT

This can be a most challenging task, but an important one nonetheless. Regardless of how attractive a program is, it still has to convey a message. And if the message isn't accurate or at the proper level of sophistication, the video operation loses credibility. Some video personnel rely on the internal client to make certain that the content is correct. However, clients sometimes lose interest. Video operations cannot afford to have that happen, and therefore the audiovisual manager or producer should be as knowledgeable as the client about the product. Naturally, some products are more difficult to learn about than others—possibly because of specialized terminology. But that shouldn't prevent a video team from acquiring the language of the industry.

As an example, while the manager of Merrill Lynch's Video Network, I became a registered representative so that I could learn the language of the stock market. I then carried that knowledge one step further by conducting early morning training sessions in my department on the fundamentals of the stock market. By the eighth session, the staff understood the key buzz terms used throughout the firm: options, primary offerings, leasebacks, pass-throughs. Producers began to realize that they could not be bowled over by clients using industry jargon. They understood the vocabulary of the market as well as having a working knowledge of their own technical vocabulary (chroma key, matte shot, double reentry switchers). This knowledge gave the staff producers an upper hand.

STAY ON TOP OF ALL COMMUNICATIONS

Many corporations issue information packets that are sent to different departments. They may be daily, weekly, or monthly operating reports, new corporate policies, projections of sales, or management personnel changes. Government bodies issue white papers, while educational organizations report on new developments in the field. It is important for the audiovisual unit to receive as many of these reports as possible. They will provide leads for new videotapes and background information for other programs.

Equally important, a videotape operation should subscribe to trade publications covering both media (see list in Chapter 28) and the company's own industry. For example, an audiovisual manager in the financial community should read the *Wall Street Journal* or the financial section of the *New York Times* or *Washington Post*. Those in the fashion or retail clothing industry should read *Women's Wear Daily*, while those employed by a telephone company should read *Communication News*.

One last source of information which should not be overlooked is the company grapevine. This suggestion may strike some readers as amusing, since one of the major reasons cited for installing a video network is the elimination of the company grapevine. Of course, that notion is too simplistic for it ever to actually happen. Video, however, can be used quickly and effectively to minimize the erroneous rumors that travel through the grapevine, and it therefore can act as a stabilizing force. Nonetheless, the grapevine can often provide excellent leads on new projects, new personnel changes, or new corporate strategies.

KEEP THE VIDEO OPERATION SPECIAL

Back in the old days (in private television, "old days" can refer to ten years ago), any client who walked in and was willing to pay for a project was special. As a result, a number of programs were produced that were perhaps low on content and high on ego building. Today, a conscientious video manager is interested in producing programs that fit the category of "need to know" rather than "nice to know." An AV manager will try to dissuade clients who wish to produce video programs that have little or no function. If the concept does in fact need audiovisual support, the manager will suggest other media that may be more appropriate, such as audio, overheads, or print. Of course, keeping a video operation "special" does not mean producing only those programs that you want to produce. A successful video network produces programs on a regular and consistent basis, be it every week or twice a month.

GET MANAGEMENT INVOLVED

This guideline refers specifically to top management; the men and women of your company who approve operating budgets, sign personnel requisition forms, and authorize capital improvements. There are various ways you can get your management people involved. First, bring to their attention the dollars they have already invested in your operation (either through videocassette field units or studio costs). Second, explain to them why they should be involved (the importance of time, dollar, and personnel control; minimizing redundant communications; increasing sales potential, etc.). Third, keep management informed of all projects. Better still, have these executives get involved in the selection process of what programs will or will not be produced.

I have found, too often, that audiovisual personnel are intimidated by top management or feel that management does not want to bother with video. And yet, both parties are interested in improved communications. When open dialogue exists between top management and the audiovisual personnel, management will support their efforts in good times and in bad.

KEEP THE CLIENT INFORMED AND INVOLVED

Different video operations prepare programs differently. Some have the client write the script, others have writers prepare drafts for client approval. Regardless of the method in your operation, it is wise to keep the client informed of the status of the program at every stage of development and execution. By doing so, you demonstrate your energy and enthusiasm for the project. Furthermore, you educate the client as to the numerous steps required to get a videotape produced. As a result, he or she will be much more appreciative of your efforts. If a delay should arise (and it will), clients are usually more understanding if they have been kept abreast of developments. And with any sales or marketing effort, it is important to get repeat business from a satisfied client.

BUILD A PROPER STAFF

No matter how good you are as an AV manager, the success or failure of your operation often rests in the hands of your staff members. They are the ones who will normally execute the programs, have daily contact with your clients, and add or detract from the creative approach of the program.

While there is no problem finding a large quantity of people who want to work in television, finding quality is a different matter. When interviewing applicants, I look for the following traits: high technical expertise, good verbal skills, presentable manner, high degree of enthusiasm (without being overbearing), and team spirit. Of course, once you find such an individual, you have certain obligations in order to make this individual as productive and happy as possible. They will include the fair distribution of work (you can't always take the plums), a meaningful training program (as previously described under "Know Your Product"), and not keeping all the perks (travel or production credits) for yourself.

SET AND MAINTAIN STANDARDS

In the same manner that you turn to your clients for accurate information, they look to you for professional output. Working with cooperative clients, a video operation can develop new formats which then become the standards for future programs. Furthermore, once a standard has been set, it is best not to retract from it because of time or dollar pressures. Even simple "talking heads" can look polished, well paced, and well appointed.

Another method of improving the standards of the video operation is to let staff members set them. At least one operation has monthly review sessions of all the material the center has produced since the last meeting. They are always closed-door sessions with only staff members invited. Tapes of segments or whole programs are shown. There are the usual array of compliments, but then the hard work begins. The group offers very constructive criticisms and evaluations of the material. Outside examples from broadcast television are often cited and previous programs are shown again to clarify a point. Because these sessions are conducted on a regular basis, no one feels picked on or singled out. The end result: consistently high-quality programming.

KNOW WHEN TO COMPROMISE

There is no advantage in trying to win every point. There will always be instances when you've learned the topic, kept the client informed, and set a quality standard, and yet the client is still not satisfied.

First, determine the accuracy of the client's objections. If you find that they are valid, change the tape. If you feel that the tape has met its objectives, then refer back to your notes or production forms that outline the objectives of the tape. You can even suggest testing the tape with different sample audiences. However, as a team member (referring to the whole corporate or organizational team), you may find that a compromise is the best solution. After all, it is the client who will use the tape long after you have moved on to another project.

KEEP THE BOTTOM LINE IN MIND

This is probably the most important of the ten guidelines. It refers to both the operating budget of the video center and the organization as a whole. It is incumbent upon you to keep all costs at a reasonable level. That entails negotiating with suppliers, recycling of tape stock, purchasing on a volume basis, etc.

One important factor that must be included is the cost of personnel or labor. Sometimes going for the lowest bid may result in a greater toll on your personnel, which in turn may lead to greater turnover or employee dissatisfaction.

The key factor, however, is the company's bottom line. Corporations have objectives—be they building better products, increasing sales, launching new services, establishing effective training, improving profit margins, or developing new territories. A conscientious video operation makes certain that its programming supports the overall objectives of the firm.

These guidelines have been prepared to be of some assistance to any size of video operation—from a three-person shop to a thirteen-person production facility. As mentioned earlier, there are other management activities that deal with planning, organizing, leading, and controlling. Nonetheless, these ten guidelines should provide a basis for establishing and maintaining a viable, productive video operation.

Production within Private Television

PART 3

Part 3 covers all the key functions of video production. The chapters are designed to provide concrete suggestions and solutions for many of the typical problems that arise when preparing a video program. Private television often has to operate under severe limitations, such as lack of space, content that is abstract, tight time schedules, and limited funds. The chapters recognize these limitations and show how to eliminate or circumvent them.

Chapter 11 starts this part of the book with suggestions on developing a set of production standards. Chapters 12 and 13 examine the program development and scripting stage that is required before the first frame of video can be put on tape. It is not surprising that the preparation of a treatment and script can take four to six weeks for a one-day shoot.

Chapter 14 demonstrates how a well-executed set adds strength and even character to a production. Graphics are also vitally important, and examples are included in Chapter 15.

Lighting and makeup (Chapters 16 and 17) have to be applied carefully if the talent is to appear in the best possible light. If television has a stepchild, it is audio; Chapter 18 represents useful techniques to improve its status.

Chapter 19 pays particular attention to the skills required to rehearse and direct nonprofessional talent; that is to say, the company employees who appear on private television. Chapter 20 then examines other directorial skills that are required in order to produce quality programs. Shooting on location is often performed under time and physical constraints, and Chapter 21 discusses the various techniques that can simplify on-location production.

Editing, a critical element of video production, is covered in detail in Chapter 22. Videocassette duplication and distribution—getting the program to the audience—are analyzed in Chapter 23. Part 3 concludes with a discussion of copyright protection after the program has been produced and edited (Chapter 24).

11 Production Values

JEANETTE P. LERMAN

Few professional undertakings offer the freedom of private television. The technology which created the medium is approximately twenty-five years old. Private television is simply too new to have legions of cynical, tired old drones declaring what can or cannot be done. Private television was spawned into existence by the expectations of an electronically compulsive age. The qualities that characterize the industry are the same that mark the best of youth—raw energy, inventiveness, gusto, and bursts of activity unencumbered by years of dreary habit.

The resolution of most contemporary problems starts by acquiring the necessary information or expertise. One of the more commendable achievements of our civilization is the democratization of information. Has there ever before been a time when there were more individuals willing and eager to tell their counterparts how to accomplish the most essential, obscure, or complex tasks? how to perform brain surgery? how to market banking services? how to assemble dams and bridges millions of miles away from where their components were manufactured?

Information has become one of the chief commodities produced by the global village. The more that satellites and computers accelerate access to information, the more need there is to capture those fleeting moments of presentation for later reflection, assimilation, and ever-wider dispersal. Hence private television and the videocassette—the easiest way for an individual to share knowledge with an audience that can't purchase or cajole or schedule a live presentation at a time convenient to its own needs.

What a marvelous opportunity for the video professional! Great torrents of information begging to be harnessed and distilled into useful, comprehensible, entertaining packages! The ever-increasing supply of information seems to produce an ever-greater demand for tapes. Fortunately, an evaluative process is taking place simultaneously. Capturing real time on tape is not enough. Some video programs create the market for their successors while others alienate a once-enthusiastic audience. Both kinds of programs may do a decent job encapsulating the facts. What, then, is the critical difference? It consists of all the subtle nuances of presentation. The integrated effect of those elements is called production value.

WHAT IS PRODUCTION VALUE?

Production value is what makes the finished show different from the script. It's the look, the feel, the stylistic rendering of the show's essential ideas.

It's not icing, lathered onto a structurally defective cake. It is the cake, the rich combination of ingredients that induces the most satiated diner to take one more bite.

In video, the ingredients can include the style of language (parody as opposed to exegesis), lighting (atmospheric as opposed to a uniformly flat 250 footcandles), set design, music, editing rhythm, camera composition, directing pace, and all the other chapter headings in Part 3 of this book. There's nothing intrinsically mysterious about the separate elements. What is mysterious, or at least magical, are the results produced by combining all the separate parts—especially when each part has been chosen to support a central idea.

High production value can be injected into any program as long as the producer identifies and designs a specific concept for the show and insists that all elements used to execute that concept adhere to a unified standard of excellence. It's easy to use lack of budget as an excuse for lazy design. Elements don't *have* to be expensive to be elegant. A larger budget does offer more latitude for experimentation and mistakes. But discipline goes a long way in milking the most production value out of production dollars.

CREATING PRODUCTION VALUES

Identifying the specific concept for a program depends on answering questions about three elements: impact, audience, and content.

To figure out how a show is supposed to *look*, a producer must first understand what the show is supposed to *do*. What is the desired impact? The moment the program fades to black, how should the audience respond? Should the viewers be motivated to specific action? Should they be convinced of a particular point of view? Should they simply be stimulated to consider new possibilities?

The second step in identifying the concept of a show is knowing who your audience is. To create a specific impact, you must know what attitudinal preconceptions its members bring to the subject. Is there hostility to be overcome? Is there confusion to be cleared? Is consideration of the topic likely to be an aggravating or a pleasurable experience?

How else can the audience be defined? Are its members intellectually sophisticated? Will they respond to irreverent humor? Are there cultural or national sensibilities that must not be offended? The more comprehensive a portrait the producer has, the more accurately he or she will be able to choose the elements necessary for the tape to produce the desired impact.

Case Study 1. For the program *Commodity Funds*, the audience section on the proposal form stated that the targeted audience consisted of Merrill Lynch retail account executives and managed account prospects, both of which groups have little if any knowledge of commodities. The prospect is probably a male in his late forties to early fifties; he owns his own business or is a professional and very successful. He prides himself on making his own decisions, and he is loath to admit that he does not understand something in a seminar situation. The prospect should have a net worth of $250,000, annual income of $50,000 and trading capital of $25,000, and should be a risk-tolerant investor.

Finally, there is the hard information that the show must convey. What points must be made? Are they all of equal weight? What ideas or concepts, if any, must one clearly establish before additional ideas are introduced or defined?

Case Study 2. For the program *IRA/Keogh Sales Videotape*, the client listed the following key points for the program:

Account executives should perceive the substantial business opportunities IRA/Keogh provides. The essential points to stress about the product are:

1. Tax deferment
2. Investment flexibility
3. Asset growth potential
4. Group plans
5. Employer contribution flexibility
6. Low cost

Designing the specific concept for a program depends on evaluating the answers to these questions about impact, audience, and content in terms of one another.

Developing a concept (which is ultimately articulated in a script) is like drawing a blueprint. It is the basis on which every decision pertaining to execution is made. Once the overall plan is in place, a producer can begin to examine separate parts to decide how to allocate resources to yield maximum effect.

INJECTING PRODUCTION VALUE INTO THE SCRIPTWRITING PROCESS

Some people see the blank page as intimidating. In fact, it's freedom. The scriptwriting process is the only step in production where you have unlimited license to draw upon whatever resources you need to implement your concept.

For content, you look to the experts in the subject area. Make them partners in the scriptwriting process. Tap their enthusiasm as much as their expertise. The initial tendancy is for an expert to "enlighten" the producer with an avalanche of detail. Don't let the assault overwhelm you. Your job, whether you hire a scriptwriter or write the script yourself, is to temper the information with media techniques so that the expert's interest in the subject becomes palpable to the viewer. Production value means delivering the excitement, the passion, the conviction along with the content.

There are some guidelines to keep in mind. First: variety is the key to stimulating the imagination. Use as many characters, situations, and graphic devices as you need to keep the story line flowing at a lively pace. Experiment with styles. Does the narrator have to be standing behind a podium? Can he or she be the protagonist of a drama in which his or her information provides the resolution to a crisis? Can the narrator be an explorer? A surrogate for the viewer investigating dangerous or provocative situations?

Furthermore, if the subject is so abstract that the only way to explain it is via "talking heads," it's better to have more than one person talking. Find experts with dissimilar points of view and let each state his or her case. Get each person involved in a particular achievement to tell the entire story, so that, ultimately, the most vibrant moments can be intercut. The sparkle in the eye and a twist of phrase go further in telling the tale than a list of dates and statistics.

PRODUCTION VALUES DURING THE SHOOT

One of my definitions for production value is that it is the quality which makes the finished show look different from the script. Shooting is where this transformation begins. The director and camera person work with composition, light, camera movement. The producer concentrates on the talent. But the interaction should add up to a jolt of energy that brings a scene rolling off the page and onto tape infused with a life of its own. How does it happen?

The first step is to begin with a consciousness that a shoot is anything but glamorous. It is among the most grueling and difficult tasks in the business. Understanding that fact and approaching the task with a professional attitude tempered by kindness go a long way in alleviating the pain. Kindness means that all participants come to a shoot prepared to give their best effort to execute the plan as conceived and ready to improvise if necessary.

Preparation is the necessary ingredient. Make sure that the talent is well rehearsed and comfortable with the camera at least a day before the shoot. Preset the lighting as much as is possible. Decide on the blocking before crew and talent are gathered to watch you make your decision. If a show is so complex that it requires testing of camera moves, lighting adjustment, and rehearsal of lines, then schedule a practice day. There is nothing more enervating or debilitating than hanging around a set *waiting*. And somehow, a camera captures nothing quite as well as the look of fatigue or exasperation in the talent's eye.

One of the joys of shooting is that the obverse is also true. If everyone comes prepared and the shoot goes well, the energy level grows in quantum leaps.

The lighting, designed in accordance with the concept, puts everyone in the mood that the finished tape is supposed to convey. The director has a precise mental image of how the cameras and talent are to move, and clearly conveys that information. The crew is inspired by excellence and confidence and gives more than 100 percent effort in response to being treated like professionals.

EDITING

Editing is where it all comes together—the scenes executed exactly according to script, the mistakes that need correcting, the fortuitous accidents that beg to be acknowledged and worked into the program. High production value comes from editing that utilizes clever design, interesting movement, beautiful color, or animation.

First put the show together the way it was originally planned. The places where you find your mind wandering are the places that need work. Try cutting down if not cutting out completely. Most programs suffer from being too long. Even if a scene contains critical information, consider chopping it. If it is tedious, no one will pay attention anyway.

If a boring scene can't be eliminated, can it be reworked in a more interesting way? Some dynamic graphics? Some stock-shot material? Don't completely discount a reshoot. If replacement is not viable, can adding sound effects or music help?

Editing is about logic, rhythm, and patterns. There are infinite possibilities for arrangement. Experiment. Keep working at the dead spots until a smoothly unfolding whole is produced.

SUMMARY

Think about the medium.

Television addresses its audience via a relatively small screen which is usually viewed in a well-lit room—an environment that induces only partial concentration. The variety and vitality that emerge from that small screen are all that keep anyone's attention, so there had better be plenty of each. Private television may have an advantage because it usually addresses a captive audience. The shows are designed for specific groups of people who have a vested interest in acquiring the information presented. But their need to know will buy goodwill only until the first clunker of a sequence. There is too much else competing for their time and attention.

Private television is the promise of custom design. It is the best of media or television production techniques applied to the presentation of specialized information. If the information is important enough to turn into a television program, it is important enough to do well. High production value is the standard by which it will be measured.

12 Program Development

ROBERT T. HIDER

Many people think that program development is broken into three steps: preproduction, production, and postproduction. However, the most important decisions to be made on any program are those completed long before the final draft of the script is written. These decisions are the crucial ones: Should this program be produced or not? Can we be sure that videotape is the best way to present this information? Examining the motives behind every proposed presentation is the way to streamlined production and quality programming. You would be wise to go through the following six-step process before launching into any video production:

1. Needs analysis
2. Determination of objectives
3. Determination of content
4. Analysis of presentation methods (media split)
5. Treatment
6. Script and visualization

The first four steps will be examined in this chapter. Items 5 and 6 will be covered in detail in Chapter 13.

NEEDS ANALYSIS

The major question to be answered in this stage of the program development is: What do the viewers need to know about this subject? This question may be asked in a number of ways depending on the particular organization. For example, an educational institution might rephrase it: What do the viewers need to know about this subject in order to pass the course? If the organization is a hospital and the audience consists primarily of patients, the question might be: What do the viewers need to know about this subject in order to get well or to feel more comfortable about being in the hospital? A religious organization might ask: What does the congregation need to know about this subject to improve their spiritual awareness? Whichever way the question is worded, determining the answer to it will enable you to focus on what specific information, from the broad universe of knowledge, will be most relevant to the viewers.

Also within the needs analysis section, you should identify your audience. Does it comprise management? laborers? new-hires? senior employees? first-year college students? seniors? kids in kindergarten? Or is the program to be directed toward a general audience—all levels of people in the organization?

The next step is to find the expected viewers' current knowledge about the subject. The difference between the current level of knowledge and the desired knowledge is the information that you want to cover in the program.

These is no substitution for firsthand investigation of what the audience knows about the subject. Interviewing segments of the audience does the following:

- It demonstrates that you truly care about how the information is presented.
- It will expose you to the actual terms and jargon that are being used in the field.
- It will help you pretest a few production ideas.
- It develops a network of individuals on whom you can call in the future for program ideas for new programs.

Analysis of the needs of an audience may reveal, on the one hand, that a proposed program is not necessary because the potential viewers already have the information. When employees are asked to watch a program which management perceives as necessary but which in fact is repetitious and uninteresting, the viewers are understandably bored and sometimes even angry. Remember, the success or failure of private television programs also depends on what the audience wants to know about—wants to hear about. This is as important a factor as what management wants to transmit.

There are some other things to consider in needs analysis, especially when the program is for training. You must consider who is going to use the tape. Will it be used by an instructor in a classroom, or will it be used for self-study? Will there be just one viewer at a time, or will there be a large group? If an instructor will be using the tape, it is important to determine how knowledgeable he or she is in the subject matter and also how proficient he or she is as an instructor. If, for example, the instructor is an expert on the subject, the projected video program may not have to be so technical. But, if the instructor is not well-versed, the videotape should be designed to impart most of the technical knowledge, leaving only questions and discussions to the instructor.

DETERMINATION OF OBJECTIVES

The successful completion of any task requires that clear objectives be established before the work begins. This step is especially vital in the development of video programs that are to be used for anything other than entertainment. If your objectives are clear and understandable, they will determine the specific content that should be included in the program and the presentation method to be used. Another reason for determining clear objectives does not really surface until well after the program has been completed and evaluated. Those evaluations should be compared with the original objectives to see if indeed they were met.

Almost all private television programs are a learning vehicle in one way or another. In fact, a learning objective should be written with each program. Put simply, this statement is: "What will the viewers be able to do at the completion of the program that they could not do before?" Determine what you want the target audience to be able to do, or do better, or do differently. *Doing* means *performing*, and it is expressed in such verbs as identify, plot, write, complete, assemble, and calculate. Too often, objectives are expressed in more nebulous terms, such as "will know" and "will understand." Such vague objectives are of little assistance in the program development process. Educational institutions define teaching objectives in terms of what the student should learn, and programs can always be evaluated in those terms. Private television can do the same.

In identifying your program objectives, first determine what overall objectives should be accomplished by the program. Then, for each overall objective, list the subobjectives, such as the knowledge and skills necessary to accomplish the overall objective.

Case Study 1. If one objective is "that the viewer will be able to fill out and complete an income tax return (Form 1040) correctly," two subobjectives will be:

1. The viewer will be able to gather all data and facts related to the taxpayer.

2. The viewer will be able to analyze and interpret the facts as prescribed by current law.

If we take these subobjectives a step further, they can be applied to each of the various aspects of the tax return—"Gross Income," "Itemized Deductions," "Sales and Exchange of Property," etc.

The better defined the objectives, the easier it is to create the program in a logical, unconfusing manner.

Once you have determined your objectives and subobjectives, you will have established the framework for your program. You will know what the program should accomplish and you will have the basis to continue to the next steps: determining the program content and presentation methods.

DETERMINATION OF CONTENT

What must be covered in order to achieve the program's objectives? To answer this question, you should first differentiate between the *need-to-know* and the *nice-to-know*. For example, if the program's objectives include "to be able to operate a jackhammer," the operating steps will be need-to-know information. The history behind the development of the jackhammer will be nice-to-know information.

If the information to be presented contains a bewildering amount of figures, try to decide which are the most important and which are least important. Then you can allocate program time and emphasis accordingly.

At this point it will be useful to consolidate the necessary information into a written outline arranged in order of presentation. This sequencing depends on two factors: *continuity* and *complexity*. The program should flow logically from topic to topic and from the basic to the more complex.

Look at your program in its entirety. A training program, for example, has an introduction, the acquisition of new knowledge or skill, the application and demonstration of this knowledge, and a conclusion. This is true not only for the overall program, but for each section or topical area. Each module should bridge smoothly from the preceding one to the succeeding one. The individual topics should be arranged so that the overall program is an integrated unit, not merely a group of ideas presented in a specific time frame.

Program designers often make the mistake of trying to cover a complex procedure before the viewers have mastered the basic skills. Sometimes a simple digression away from the basics can be innocent enough and sound interesting, but it may impede and perhaps stifle the viewers' understanding. In your programs, sequence the material from the basic to the complex with smooth transitions for continuity.

PRESENTATION METHODS AND MEDIA SPLIT

Too often, organizations treat their audiovisual departments as merely an equipment facility, and make the mistake of thinking a program must be done on video long before they have established the need and objective of the program.

Even if you have done your needs analysis, defined your objectives, and decided the content of a presentation, you still do not know *how* you should properly present the information required. This point cannot be emphasized enough. If you are a video person, one of the most difficult tasks you will encounter is to say, "No, I don't think that is the proper use of this medium." In other words, you are turning away business—and that hurts! But remember: Bad video hurts more than no video.

To determine which medium is the best for your presentation, several questions must answered.

1. Can your selection of a medium assist the presentation of your objective?
2. What is the time available to produce your presentation?
3. What is the operating or allocated budget for the project?
4. What is the size of the audience?

5. What is the sophistication level or demographics of the audience?
6. What is the shelf-life of the program?
7. Where will the presentation-program be shown or distributed?
8. What equipment is available to display the presentation?
9. How many copies of the presentation will be needed for the distribution you seek?
10. Is instantaneous or simultaneous reception by audience members a factor?
11. What people resources are available to aid in the production of the presentation?
12. To what degree is the material of a confidential nature?
13. Is there anyone who might be offended, upset, or shocked by the presentation?
14. To what extent is it necessary for the audience to have repeated exposures to the presentation?
15. To what extent do you want the presentation to appeal on an emotional level to the audience?

The key elements that precede any actual production have been presented here. The preparation of treatments and scripts, like any creative endeavor, is a very personal process. In Chapter 13, Diane Sharon presents a variety of styles and approaches to writing treatments and scripts.

One last question may come to mind as to how much time should be devoted to the preproduction stage as opposed to the actual shooting and editing of a presentation. While it is difficult to come up with rules of thumb, certainly the preproduction stage (including the preparation of the treatment and script) at least should match moment for moment with the production and editing. Depending on the complexity of the program, it may even exceed it by a factor of 2 or 3. It all comes down to the axiom: The better prepared you are, the higher the probability of success.

13 Scripting

DIANE M. SHARON

Writing a script is harder than it looks and not as hard as you think. Like hunting for big game with a 35mm camera, scriptwriting works better if you do your planning ahead of time, and if you approach the project as a professional.

Writing a video script is different from writing a report, a speech, or a story. A script requires the writer to write for the eye as well as for the ear. Video is *visual,* and the key to writing for television, even private television, is, "Don't just tell it if you can also show it!"

With this thought in mind, creating a successful mixture of the verbal and visual elements of a videotape requires not only imagination, but also organization. This chapter describes the three steps involved in scriptwriting from initial idea to finished screenplay. They are the approach, the treatment, and the writing.

THE APPROACH

The subject of your program is a major factor in determining the approach for your video script. The approach addresses the format of the program. Will it be a scripted talk in a studio setting? A round-table discussion? An interview? A documentary? Or a magazine format? This is where you decide. There are four basic approaches to a video production. They are discussed below. Your program might be primarily one or another, or it might be a combination of all four.

"Talking Head." Almost any topic lends itself to the "talking head," the expert or executive behind a big oak desk or in a mock-up of a book-lined room, discussing the outlines of, and implications inherent in, a particular set of circumstances. Occasionally this approach is appropriate, as with a high-level problem being addressed by the company president or the chairperson of the board.

In rare cases, the person at the top is a gifted public speaker as well as a sharp business leader. But, for the most part, the big attraction is that it is the chief speaking—a novelty that can wear off quickly when the delivery is less than inspiring. Your best bet when you use a "talking head" approach, is to *keep it short.* Whether you prepare the remarks or the speaker composes his or her own talk, some suggestions may be helpful.

Keep the concept simple. Keep sentences short. Vary the placement of subject and verb from time to time. If at all possible, get your speaker to throw away the prepared script and talk off the cuff. With the ideas still freshly organized in his or her mind, the result is often a more spontaneous delivery. This is especially desirable in a round-table format. On the other hand, if the individual is paralyzed by the lights and setting of a television production, putting his or her remarks on TelePrompter or on cue cards may be the best approach. Other formats in the "talking head" genre include the fireside chat, the studio interview, and the roundtable discussion. Think about placing the subject in unusual surroundings: the company president at the factory, or the marketing executive in the aisles of a grocery store, for example.

Documentary. A documentary depicts a real person within a realistic environment. As an example, a "talking head" may be an economist discussing inflation. A documentary may be a supermarket shopper talking about personally experiencing the rising cost of feeding a family.

Think about testimonials as a communication tool, especially within a documentary format. Real people in real situations talking about their real feelings make the best stories. And what you're doing when you make any audiovisual program is telling a story. Even the best drama borrows heavily from life for credibility and impact. If you can find a way to harness this impact for your video presentations, you will have your audience in the palm of your hand.

Optimally, you interview auto workers on an assembly line, not in the studio. You record their words, however rough-hewn, for their sense, not necessarily for the speakers' grammar. Because the auto workers will be expressing the views of others like themselves, the words carry credibility. You might intercut the comments with "talking head" experts, and you might reinforce the points you are making with vignettes or role plays. But you will capture the lion's share of credibility using real people talking about real feelings and experiences.

Vignette. The vignette can either be scripted or improvised, but it carries the same criterion for validation as the documentary: It is true to life?

Vignettes are used primarily to illustrate situations for an audience—literally, to dramatize events. They allow more control than do documentaries, because the exact dialogue and story development are determined before a foot of tape is shot. Vignettes can be useful in demonstrating correct procedures, as in role plays, or in human relations situations, such as personnel or sales training. Sometimes humor can help to make a subtle point; parodies or takeoffs are effective in establishing rapport with an audience; and, of course, humor contributes entertainment value to any program.

There are special considerations involved in vignette-based programs. Whether dramatic, humorous, or role play, vignettes require professional actors, theatrical direction, and realistic dialogue. Just as a theater play differs in flow, pacing, and perspective from a novel or news story, the vignette format requires a different approach to scriptwriting from that used in documentary, "talking head," or narrative.

Even parodies or takeoffs derive their humor from the kernel of truth which they understate or exaggerate. Nothing is a bigger turnoff than stilted dialogue or melodramatic speeches masquerading as realism. When you do decide to use the vignette to make your point, make sure you use a writer whose specialty is drama. Dialogue, dramatic tension, resolution—all the elements of good theater must be present in your video vignette about widgets. This is no mean feat.

Narrative. Each of these approaches might benefit from additional material as connective tissue, and, of course, many programs consist of narration only, given as a voice-over or delivered by an on-camera narrator.

Writing narrative scripts is fairly straightforward and requires primarily that the writer take the perspective of the intended audience members and "walk them through" the information to be conveyed in the program. Keeping a conversational tone is the best advice for any writer of the narrative format. If you write the narration in the rhythm of speech, a narrator can read it aloud easily and understandably.

Start by establishing a common basis for your audience. Begin at the beginning, and proceed in logical steps through your outline. If you're writing narrative transitions for other material, make sure you know exactly what you're leading in and out of, so your narration contributes to the flow instead of fighting it.

If narration is to be the basis of the entire piece, make sure you have a good sense of the visuals you will be using. Don't write "wall-to-wall" copy—allow the natural sound of the locations or events in the video to break the narration, and, whenever possible, to move the story forward.

Whatever approach you take in your proposal, you will have to describe it to your clients in the form of a treatment (see the next section) so that they can see what you mean. Some examples appear in Figures 13-1, 13-2, and 13-3.

As you develop your approach, it helps to think about a strong underlying visual concept or "hook" for the program. Try to visualize the program message, and then use that visualization in your script. The client may use a phrase or term in explaining the project to you which may strike your imagination. The product may lend itself as a design element. Perhaps the physical layout of the product may give you a visual cue.

Case Study 1. In this case, the client wished to explain a sophisticated routing system for telecommunications. To introduce this complex issue, the program began with shots of railroad tracks in a railroad yard and demonstrated in time-lapse photography how various trains could be switched from one track to another. This was a graphic representation of physical movement which the audience could relate to far more easily than to the movement of electrons on a telephone circuit.

Sometimes, an analogy may also be your visualization. To explain an import-export relationship, you might be looking at the same coin from two different sides. This coin motif is appropriate here on many levels. Besides explaining the program concept graphically (i.e., a coin is an obvious element

---

TREATMENT

Approach

The videotape project will consist of two 20-minute segments. The first will focus on the *International Department,* and the second on the *Corporate Banking Department.*

Our orientation in both segments will be *documentary,* using actual people, places, and events wherever possible. Where we cannot, we will use descriptions by the principals, in their own words, of the transactions or relationships under consideration. Voice-over narration will be used to integrate elements within each segment, and to clarify functional interactions where required.

Each tape will examine the department's mission, how the department functions, and how it interacts with other bank departments to provide the best possible services for its customers. Both segments will be similar in outline, and the message in both will be clear that the challenge at the bank is to work together, beyond departmental boundaries, to serve bank customers. The need for *active marketing efforts,* and for an up-to-date knowledge of all banking products and services, will also be emphasized.

Each segment will have the following three elements:

1. Overview:
   (approximately 5 minutes)
2. Relationship management:
   (approximately 10 minutes)
3. Summation and conclusions:
   (approximately 5 minutes)

Figure 13-1 Documentary treatment.

of commercial trade), it also allows you to play with the concept in your script: "Let's look at this from the other side of the coin," or, "Your decision requires more thought than just a flip of a coin."

TREATMENT

The treatment is a short proposal in prose form that describes your script idea to the client. It should include all the information gleaned in the prescripting interview: the objective of the videotape, the audience, the running time, etc. (See Figures 13-1 and 13-2.)

The treatment also gives as much information about the visual aspects of the program—its locations, the characters or company officers involved, the types of studio sets that will be used. (See Figure 13-3.) Anyone who

---

TREATMENT

Objective

This 15- to 20-minute videotape will introduce bank branch personnel to the Computerized Teller System.

The tape will demonstrate this unique, new, computerized operation and explore what it will mean to tellers, managers, and bank customers. The emphasis of the tape will be on how the system will benefit all who work or bank at a branch bank.

The tape will portray the system as more than just a way to automate current procedures. The tape's message is that the system is a new and better way of life for bank tellers and their customers.

Approach

The tape will be in *documentary* style, taped in actual locations where the system is in operation, and featuring interviews with the tellers, managers, and customers who have been using it since May 1979.

A *narrator* will introduce the system and provide the unifying thread for the tape. The narrator's voice-over will intercut with interviews with tellers, bank managers, and customers, who will share with the audience what the Computerized Teller System has meant to their jobs and their lives. In every possible instance, we will use real people and actual equipment in operation to create an ambiance of credibility and immediacy.

Figure 13-2 Narrative treatment.

reads the treatment should be able to get a good idea of exactly what the program is about and what it will look like.

Because a treatment defines all the program elements, it should be submitted for approval to the producer, the director, the client, and anyone else responsible for that particular program. All treatment signoffs should be obtained before the script is ever begun. It is easier to change a treatment than it is to change a script, and changes will be even more difficult and expensive once the program is in production.

The treatment phase of a script is the time to clear up any ideas that may seem confusing to the writer or the client, and to make sure everyone is in agreement about the style, tone, and approach of the project. On the producer's side, it is also the time to obtain any initial clearances for location shooting, if required, and to begin to budget out the production.

In many proposals, you will also have to provide an outline of the material you will treat with the approach you have selected. This outline will serve the purpose of guiding you through the "agony and ecstasy" of writing the script, so make sure it is detailed enough, and realistic enough, to enable you to do your job.

> **TREATMENT**
>
> The *vignettes* which will be portrayed on the tape, as well as the major protagonists, will be based on actual incidents and people as we cull them from our own experiences and from focus-group interviews with our target audiences.
>
> As the tape is presently visualized, we will focus on women, each representing one of the seminar's target audiences. The younger woman is a rising executive in a medium-sized company. The older woman is married, her children are older teenagers, and she is starting to realize that her active mothering years are ending. She is beginning to think about what her next stage will be.
>
> They meet for lunch to discuss the fund-raising event they are planning for a local hospital. The women have never worked together before, although they know each other slightly from their work on behalf of the hospital. The luncheon offers an informal occasion in both their busy schedules to get together to plan their fund raising, and it incidentally gives us an opportunity to eavesdrop on their conversation.
>
> They introduce themselves, and the life situations of the women become apparent. Their questions about their ability to handle different financial and logistical responsibilities also come within the fund-raising context, and each supports the other's abilities on the basis of what they've learned about each other in their luncheon conversation. It becomes clear to the audience that both women know more about money, finance, and organization than they thought they did.
>
> The younger woman has recently gotten a promotion and raise, events that have aroused her feelings that she needs to begin to take her financial life seriously (maybe she has just reached 30 or 35—a major birthday, as well). She expresses her feelings of intimidation: "My father always took care of insurance—we'd get a couple of shares of stock from my grandmother on our birthdays, and savings bonds on graduation. I never had to know anything about handling money."

Figure 13-3 Vignette treatment.

If you've been talking to the client all along, you should have no problem eliciting the information you will need to do the script outline. In fact, if you're skillful, the client can verbalize the outline for you, and you will simply have to write it down.

Ask the client what topics will need to be covered by the program, which are the most important and which are less. You will find, most often, that the material organizes itself logically, and often the client's way of thinking about it is a valid basis for an outline. After all, your client has been thinking about the subject of this videotape for longer than you have, and he or she has had to come up with a variety of handles for the topic. Chances are that your outline is lurking in the way your client describes the situation.

A word about preparing treatments and proposals. Don't get carried away. As a rule, I keep them to under ten pages, and, if possible, to five pages or less.

As an outside writer, I will ask for a token payment for preparing the proposal. This fee can be built into the price of the script. The work that you do preparing the proposal is obviously germane to the script itself, and is an integral part of the scriptwriting process. Allocating some payment to this phase of the work acknowledges its importance and ensures the client's seriousness of purpose.

Sometimes a client is not prepared to put dollars down at this early stage. In such a case, a proposal may be a two- or three-page letter, summarizing the project but not including an outline or sample script, and proposing an approach in the most basic terms. As a writer, whether you're on staff or free-lance, your time and creative energy have a value. It's amazing how the demand for a full-scale proposal suddenly seems less pressing when a specific fee is attached to it.

WRITING

Writer's block often is said to be all in one's head. One is also told that the surest cure is just to write the first paragraph. Both statements are true, but what they leave out is the element of discipline required in overcoming that initial resistance.

The best way to start writing a script is to wrestle yourself into a chair with a typewriter or a yellow pad in front of you and simply to announce to yourself that you are not to get out of that chair for a cup of coffee or a chocolate bar, or to make one last phone call, until the first two pages of script are completed.

The first step is to divide a sheet of typewriter or yellow lined paper into two vertical halves. The left-hand side is for video instructions, and the right-hand side for audio. In working with many private and broadcast clients, I've seen many formats for the script, but the standard one looks like the example in Figure 13-4.

Double-space the audio, single-space the video. Left to your discretion are when to capitalize, how often to paragraph, and whether or not to number each change in scene.

Many writers abbreviate key terms in their scripts. Here are some of the more common terms and their abbreviations:

Indicate whether the speaker is visible *on* *c*amera (OC) or whether you will hear only the narrator's *v*oice *o*ver another picture (VO). When you want to indicate that you will keep the natural sound which accompanies the picture on film or tape, mark the scene "SOF" or "SOT" for *s*ound *o*n *f*ilm or *t*ape. If sound effects are to be used, mark the scene "SFX."

There are various ways of describing how one or more persons are to

13-8   PRODUCTION WITHIN PRIVATE TV

24. FADE TO BLACK	(MUSIC: UP WITH FEELING OF ENDING/BEGINNING AS WE MOVE INTO NEXT SECTION)
	(MUSIC: QUIET, SOMBER)
25. FADE UP ON WINDOW AS HAND PULLS ASIDE CURTAINS, OR HAND RAISES SHADE ON WINDOW. FACE OF WOMAN APPEARS, PEERS OUT.	(NARRATOR: VO) We all expect to be working for a long time . . . and we have plans and dreams we're building for ourselves and those we love.
26. INSIDE WOMAN'S APARTMENT, WOMAN IN BATHROBE, ON CRUTCHES WITH BRACE ON LEG; LAP DISSOLVES OF HER FROM THE REAR OR ¾ VIEW; FEELING OF SLOWNESS, LONELINESS, PAIN, IN MONTAGE OF PHOTOS AS SHE GOES TO SHELF, PULLS OUT A BOOK, MAKES HER WAY TO A CHAIR, SLOWLY LOWERS HERSELF DOWN, LEG STRAIGHT IN FRONT; SHE PUTS CRUTCHES DOWN, OPENS BOOK, READS WITH WINDOW LIGHT FALLING ON HER HAIR, GLINTING OFF BRACE.	(MUSIC: CUT OUT FOR A BEAT. SLOW, HEAVY CELLO) (SFX: AS NEEDED.) But there is always the chance that an unexpected illness or accident might interrupt our earning power for a lengthy period.
27. DOORBELL RINGS, SHE CLOSES BOOK (WE SEE HER HANDS CLOSE COVER OF BOOK, REACH FOR CRUTCHES); SHE SETS CRUTCHES ON FLOOR, PAINFUL SERIES OF SHOTS AS SHE LIFTS HERSELF OUT OF CHAIR, MAKES HER WAY TO DOOR, OPENS DOOR.	If you should suffer such an accident or illness, you can be reassured, knowing that your income is protected in large part by the Long Term Disability Plan.

Figure 13-4 Sample script layout.

be photographed in a scene. An individual may be in close-up (CU), medium shot (MS), or long shot (LS). When more than one person is in a scene, you should specify a two-shot (2-shot), three-shot (3-shot), or group shot (WS or wide shot).

How detailed you are in your stage directions and how specific you are about your picture depend on the assignment and nature of the program. If you're interviewing someone in a factory, you need not be more specific than to say "Joe Worker—OC, factory background." If, however, you are specifying the picture for a video montage, you might want to be more detailed, as shown in Figure 13-5.

My approach is to write the audio portion first, and then to go back and fill in the video. Chances are, you've been visualizing the scene in your head as you write, and it may be easier to write down first the audio and then the video.

While you're banging out your first draft, you should be running your fantasy tape of the program through your head. With experience, you will be able to sense edit-points and places where emphasis or a change of scene is required. The fine art of building these into your script is called *pacing*.

Unless you're building a montage, a scene should be on screen for a minimum of 3 seconds. On the other hand, depending on the flow of the program, don't hold a visual on the screen without some change of perspective (from medium shot to close-up, for example) for more than a minute at most. Even with a "talking head," indicate a move-in or pull-back on the camera at regular intervals to lend as much visual interest as possible. Close-ups emphasize points. Wider shots lend perspective.

Be alert to graphics such as photographs, charts, or even interesting set elements, in planning the pacing. Always think of yourself watching the program. At about the time you find yourself mentally turning away from an image, you should indicate a change of perspective or a change of scene. Pacing is where a writer's imagination and inner ear really count.

Most of your programs will be scripted *before* you shoot a foot of film or tape. You can decide in a vacuum who will say what in which order, what your transitions are, and what the settings will be, especially if all your material will be shot in a studio or some other controlled setting. One caveat: When you're writing the script in advance of the shooting, make sure that what you're visualizing is realistic and actually exists.

I learned this the hard way. I was planning an energy conservation tape, and my client wanted to highlight the steam traps that the factory had installed. To hear the client talk about them, steam traps were dramatic visual elements, and in my draft script I indicated about 2 minutes of various steam trap visuals. Luckily, during my survey of the factory I asked the plant engineer to show me a steam trap, only to discover that it's a static little valve *inside* the steam pipe! I immediately built in some graphics to cover that 2 minutes and to demonstrate visually what that steam trap did inside the pipe.

The one kind of program you script only *after* you shoot most of what you need is the one you've decided to do in documentary style. The preparation you do beforehand consists of preinterviews and surveys, followed by a very detailed outline consisting of the elements you expect to have, with the substance of the information they'll convey. After you shoot your

1. MONTAGE OF PHOTOS FROM TYPICAL FAMILY ALBUM—OLD-FASHIONED BLACK CORNERS, BLACK PAGES (PAGE-TURN WIPE MIGHT BE NICE—OTHERWISE PICTURES COMING TOWARD CAMERA FROM FOUR QUADRANTS OF SCREEN)—MELANGE OF LIFE EVENTS AND DAILINESS—INFANTS BEING BATHED, FIRST STEPS, FIRST TOOTH, CHILD IN A CAST, PEOPLE CAUGHT OFF-GUARD, SCRUNCHED-UP FACES, PHOTOS POORLY FRAMED, CHILD IN MID-SOMERSAULT, AUNTS, UNCLES, GRANDPARENTS, PARENT CAUGHT IN HAIR ROLLERS, CHILDREN CLOWNING AROUND, SWIMMING PARTY, BAR MITZVAH OR FIRST COMMUNION—ETHNIC DIVERSITY AMONG FRIENDS—SCHOOL PLAY, GRADUATION, WEDDING. MONTAGE OF HUGS (AT WEDDING, OR IN GENERAL) & BIG SMILES.

(MUSIC: CALLIOPE FEELING, START MODERATE, BECOMING FASTER AND FASTER)

(SFX: VOICES OF FAMILY VIEWING ALBUM. PHOTOS, EXCLAMATIONS, CONVERSATION, MEMORIES, ETC.)

2. FINAL SEQUENCE IS BRIDE HUGGING FATHER, BIG BROAD SMILE OVER HIS SHOULDER. PHOTO GRADUALLY BECOMES MORE & MORE CLOSE-UP, MORE & MORE GRAINY, AS BIG SMILE FILLS MORE & MORE OF FRAME UNTIL IT LOSES FOCUS—

3. GRADUAL FADE TO BLACK.

(MUSIC: CALLIOPE SOUND NOW VERY FAST, ENDS WITH A WHIMPER MID-PHRASE)

(NARRATOR: VO)
Sometimes, unexpectedly, the rhythm of your life can skip a beat . . .

(MUSIC: CALLIOPE RESUMES, SOFTER, SLOWER)

(MUSIC CONTINUES UNDER NARRATION)

Figure 13-5 Montage description within script.

THE PEOPLE: THE TELLERS

Several tellers can reinforce how easy the new banking system is to learn, and how much better it is to work with. We will see the system working as they describe how it has simplified and improved their work lives. As we see these operations in action, they will mention such benefits as the following:

- *Less overtime,* thanks to computerized proofing the day's transactions in less than one minute—an operation that could take as long as an hour and a half to do manually
- *Fewer errors,* which means less pressure, ease of proofing
- *No more running around behind the counter,* because the system handles everything right at the *work station*
- *More pleasant work environment*
- *Less harried dealings with customers*—time to sell other banking services
- *Freedom from the repetitive chore of counting out cash all day,* thanks to the automatic cash dispenser
- *Enhanced self-image*
- *Reduced exposure to hold-ups*
- *Task modification:* e.g., in head teller's job, increased orientation to customer services, interaction with people, problem solving

---

Figure 13-6 Outline for documentary presentation.

footage and review your interviews, you can put together an actual script from verbatim transcripts of those videotapes. Figure 13-6 shows an example of a preliminary shooting outline for a documentary style of program.

Figure 13-7 shows the final script, which was written by cutting and pasting "on paper" first, from verbatim transcripts after shooting. Then the videotape was edited to conform to the script.

Notice that people never speak as clearly as they write. Remember that when you see the finished piece, the verbal style of the interviews will sound just fine—your ear is used to making allowances that your eye would never tolerate.

The key in any scriptwriting exercise is *flexibility*. If you're writing a tight script before shooting, you may find your words being modified when they are spoken by actors. That's fine—often they will sound more like actual speech. Certainly, when you are editing documentary material on paper, you will find material which doesn't "cut together" once you get into the editing room, and you'll have to modify your original script. These gray areas of flexibility are often where the most creative video work takes place. As a scriptwriter, you should certainly expect them, and even plan for the "polishing" opportunity they provide.

VIDEO	AUDIO
28. 2-SHOT	REPORTER (ON-CAMERA): When the system was installed, how did the tellers take to it? SMITH (ON-CAMERA): Well, they went through a whole process. Approximately six months before it came in, we sat down, we had meetings, we told them what it was about, and we sent them out to training for two weeks, and we had a flying squad they came in for the transition period. Then half my staff came back and worked with the flying squad, and I had a very smooth transitional period.
29. CLOSE-UP OF TELLER TALKING (SAME AS IN #27).	TELLER (ON-CAMERA): It's really great. It helps everything. Everything is faster, efficient, it's really good. REPORTER: How hard did you find it to learn? TELLER: It was easy to learn. I think the problem was coming back into the branch, taking the first customer. It was like being a teller all over again. It was really nervous. But after that, it was so much easier.

Figure 13-7 Transcript of documentary presentation.

CONCLUSION

No one can tell you how actually to write a script. After the initial spadework, the proposal, the outline, and the research, your own writing ability comes into play. There are, however, rules of thumb, proven by my experience, that may help you in approaching your own scriptwriting task.

1. When in doubt, cut it out. Briefer is better—assuming you aren't mangling meaning in the cutting process.
2. Tell 'em what you're gonna tell 'em, tell 'em, and tell 'em what you told 'em. This may sound like a contradiction of rule 1, but in fact, it is the

first rule of storytelling. Begin with a quick summary statement of what the program is about. Then go on to the substance of the program, and conclude with a brief statement summarizing the central point. Television news programs do it; even commercials do it. Borrow this rule from the mass media pros.

3. Keep the language simple. Avoid jargon. Avoid convoluted sentences. What you're writing is meant to be spoken out loud, and you'd be surprised at what can be conveyed with simple declaratory sentences. Try reading your words aloud for cadence, rhythm, pronounceability, and flow. Aim your presentation language at an intelligent audience. You can be clear and concise without being patronizing.

4. Keep the concept in focus. After you've been through the exercise of determining both audience and objective, you should have one clear idea that you're trying to convey in your program. Stick to it. Don't muddy the waters with other ideas, "by the way," or try to piggyback other objectives that don't really belong with your primary one.

5. A picture is worth a thousand words. When you fall all over yourself trying to explain a complicated process, ask yourself whether a graphic, or footage of the item, can make your point better than words. That flexibility is one of the blessings of video. Take advantage of it.

To repeat, scriptwriting *is* harder than it looks. That's because there are a great many elements to keep in mind and no one can tell you all the rules. But if you're organized, writing a script is not as difficult as you think.

State your objective. Know your audience. Get the specifications of the program's length, shelf-life, and deadlines. And make sure you know your budget limitations. Then write a script outline. From this point on, each step flows from the one before, with discipline as the taskmaster. Fill in your information gaps. Decide your approach. With all the elements in place, wrestle yourself into a chair and write your script.

Ultimately, the success of your script depends on how well you understand your client's objectives and how competent you are in translating them into a video script. In the final analysis, your own creativity and experience are your best allies.

14 Set Design

MARILYN REED

Scenic design provides much more than a decorative background. The scenic designer gives shape and continuity to the action, provides physical support for the actors, and establishes the environment for the action. To meet these demands, a scenic designer needs the imagination of a creative artist, the skills of the stage artisan, and the sense of production of the actor and director.

The designer as creative artist must be articulate in line, color, and form, a skill achieved through the study of design, drawing and painting, and period furnishings and decor. To create a set that can be fully realized, the designer, as stage artisan, must be knowledgeable in scenery construction, limitations of material, and lighting techniques. A good understanding of directing and staging enables the designer to choose the appropriate visual expression to support the action of the script.

This chapter is in two sections. The first addresses itself to *creating a design strategy*. It discusses the formulation of a concept, research, studio and equipment parameters, developing sketches and plans, sources of materials, and budget.

The second section of the chapter is a nuts-and-bolts discussion of common *structural elements*, that is, flats, drapes, cycloramas, and set pieces.

Furthermore, there is a survey of typical costs related to set production.

CONCEPT

The set designer usually becomes involved after a script has been written and approved. At this point, the producer, director, and writer should have a clear idea of the objective of the program and an appropriate format for presentation to the specific audience. Mutual understanding of the production concept by everyone involved is the key to a successful program.

At this point, the designer must be concerned with the practical as well as the conceptual. The amount of time, money, and personnel available will affect design decisions. Facilities scheduling is also a factor. The set that can be assembled at the shooting location over several days will be very different from the one that has to be assembled in only a few hours. These guidelines must be established before a designer can really begin to solve the problem of what the design will be.

There are a few questions that should be asked before designing a set.

14-2 PRODUCTION WITHIN PRIVATE TV

What Is the Style of the Production? Will the show be done as a documentary, a farce, or a serious drama? The style of the setting must support the style in which the information is being presented to the audience. Whether a set is a realistic office interior or an abstract fantasy is a matter of style. These choices can tell your audience a lot about the subject in an instant. A set depicting a construction site painted in bright, cartoonlike colors prepares the viewer for a comedy. The same set painted with the reality of a photograph sets a more serious tone and would be appropriate for a program on building safety.

What Is the Locale? The action of the script must take place somewhere. Usually the writer establishes the time and place appropriate for the action. The location is made visual through architectural detail, choice of props and furniture, interior décor, backdrops and vegetation. At first glance, the viewer should know where the action is taking place and have an indication of the station of life of the characters.

What Is the Mood? This is probably the hardest concept to discuss because it is not created by any single element. It's that quality which is the result of all the elements of a production combined to produce an emotional reaction in the viewer. It is the atmosphere inherent in the script, and it is transmitted by not only the set, but the lighting as well.

What Space Relationships Are Needed? The designer needs to know how many actors will be on the set during each scene and what kind of movement the director is planning. Is the actor to move across the set or walk toward the camera? The position of doors, stairs, ramps, furniture, or set pieces will dictate the positions of the actors and their paths of movement.

How Many Camera Angles? The size and center of interest of the set are influenced by the number and location of cameras required by the director. Knowing how much of the set will be seen is a time-saver when determining where and how much finishing detail is needed. The extreme example is the exquisitely designed multilevel interior with fireplaces, exotic window treatments, fancy moldings, and lots of furniture for a scene with one actor shot from one angle. Of course the set looks superb in the studio, but no one outside the production staff will see it. The viewer will see only one small portion of the total set. (Figures 14-1, 14-2, and 14-3).

Case Study 1. This program used one actor as spokesperson to outline the changes in the new tax laws regarding capital gains taxes. The new regulations were of vital interest to our public audience of investors. The subject matter was highly technical and detailed. We worked to balance the complex information with a warm, informal approach in the visual presentation.

We chose to set the action in a library or den. The set was designed with two acting areas and allowed for on-camera movement between them. This layout gave the director many possibilities for visual variety.

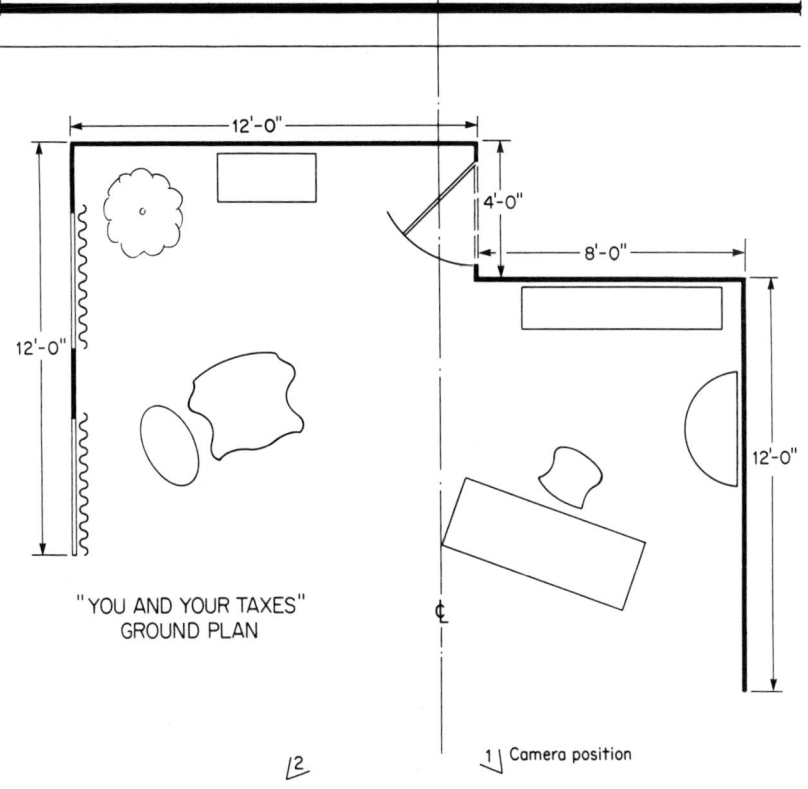

Figure 14-1 Ground plan of den set, reduced from drawing using ½ inch to the foot.

The wing chair provided for both seated and standing positions. An upholstered wing chair was chosen for its fabric texture and its lack of sharp highlights that reflect off the usual leather coverings. The cherrywood cabinet complemented the chair in line and color.

The windows were placed so that the motivating light would fall on the set and in the shot. The lighting director used a pattern of tree branches with amber gel at a low angle through the window to establish the season and time of day.

The desk area gave the flexibility of playing scenes seated or standing behind the desk or seated on the front edge of it. The leather-topped, walnut Chippendale desk had the elegance that would appeal to our audience. The line and detail of its cabriole legs made an interesting foreground in several shots.

Every piece of furniture or accessory was selected for what it contributed to the total picture—from the desk chair with its crest rail to the landscape painting with its subdued tones and rich gold frame.

14-4 PRODUCTION WITHIN PRIVATE TV

Figure 14-2 Portion of den set near window. (*Merrill Lynch Video Network.*)

With the answers to the preceding questions, the designer should have a clear idea of the production concept and the physical requirements of the show. With script in hand, she or he now heads for the design studio, a quiet office, or the kitchen table to start transforming concepts into solutions.

RESEARCH

Researching a design idea is itself a very creative endeavor; if there is time, it can be a lot of fun. A good place to start is the *swipe file*. This is the place where you've stashed all those great examples of period and contemporary interiors, color ideas, graphics, and lighting clipped from magazines and newspapers. For a quick update of your file, thumb through recent issues of *Architectural Digest, Interiors, Interior Design,* or *Arbitare.* Don't overlook the inspiration to be gleaned from the advertisements as well as the feature articles. Be open-minded. Great ideas have emerged while rumaging in a scrap-metal yard as well as in the local art museum.

There are sources of design everywhere. Looking at paintings is good for developing ideas on style or mood. Interesting ways to handle furniture can be found in department store arrangements. Even window displays can offer inventive ways to use simple materials to communicate an idea

SET DESIGN 14-5

Figure 14-3 During most of a program, only a small portion of the set is seen. (*Merrill Lynch Video Network.*)

or a mood. And that scrap-metal yard is a place to discover construction materials that can be reused in a decorative way.

STUDIO AND EQUIPMENT PARAMETERS

Realize that the placement of scenery must allow space for placement of cameras, personnel, and equipment, such as the TelePrompter system and its operator, slide projectors and graphic stands, and the producer with access to a monitor. Approximately one-third of any studio space is usually devoted to the behind-the-scenes equipment and people.

Think about the lighting. Where are the places that lighting units can be hung in relation to the set? Positioning the set so that a "window" is very close to a studio wall eliminates the option of creating a lighting effect through that window. Referral to a studio floor plan (commercial studios are happy to supply them) that includes lighting positions is the one way to be sure that the lighting director will be able to light the set to its best advantage. Better still, create a master floor plan indicating lighting grid, microphone jacks, and electrical outlets. After a program has been designed, keep the records on file; you may have to recreate a set someday in the future.

The picture quality of your set is governed by the ability of the video system to handle contrasts in reflected light and color saturation. Ignoring this limitation is the cause of many picture difficulties. Use of colors, such as canary yellow, that are too bright will usually cause bloom or flare and distortion of the other colors in the scene. Very dark colors, such as chocolate brown, and dark fabrics will show no detail or color.

In working with video, the designer must choose colors and furnishings carefully to be within the reproductive range of the camera. Avoid very reflective materials, such as mylar, unless a special effect is desired, as they can negatively affect the color balance of the video camera. Reflective accessories, such as brass lamps and chrome chairs, can be sprayed with a Krylon removable spray to reduce the "hot spots" created by studio lights.

SKETCHES AND PLANS

A ground plan and sketch or model define the scope of a production set. They must be approved by all members of the production staff whose work they influence, such as the producer, the director, and the lighting and costume designers. The scene designer must satisfy these other team members without sacrificing the original design concept.

Rendered in scale, color, and perspective, the sketch shows the atmosphere of the scene, which is difficult to do in a scale model. Since the focus of any production is the action, the sketch must show the relationship of the human figure to the set. All architecture, furniture, draperies, pictures, and other decor must be shown. Additional sketches can be made to show the lighting changes and picture composition of key scenes.

The sketch is accompanied by a ground plan. This drawing shows graphically the arrangement of the set on the studio floor. Outside of the script, it is the document probably referred to most often. The designer uses it to explain the final design. The director studies it to refine the staging. The carpenter refers to it to position the scenery in the studio. And the lighting director needs a ground plan to begin working out the lighting design.

Ground plans are drawn to scale. Usually, ½ inch on the ground plan equals 1 foot in actual size. If the space is very large, a ¼-inch scale may be easier to work with. There are graphic conventions used in drafting a ground plan that make the information more easily understood. (See Figures 14-4 and 14-5.)

In addition to showing the exact size and location of all architecture, walls, set pieces, and furniture, show also any steps, platforms, and raised areas. All details of the permanent studio structure must be included to avoid such mishaps as blocking access to all microphone inputs or electrical outlets. The plan is dimensioned from two reference lines. One is the center line and the other is known as the set line, which is located in front of the set and perpendicular to the center line. It is not necessary to dimension everything on the ground plan.

SET DESIGN **14-7**

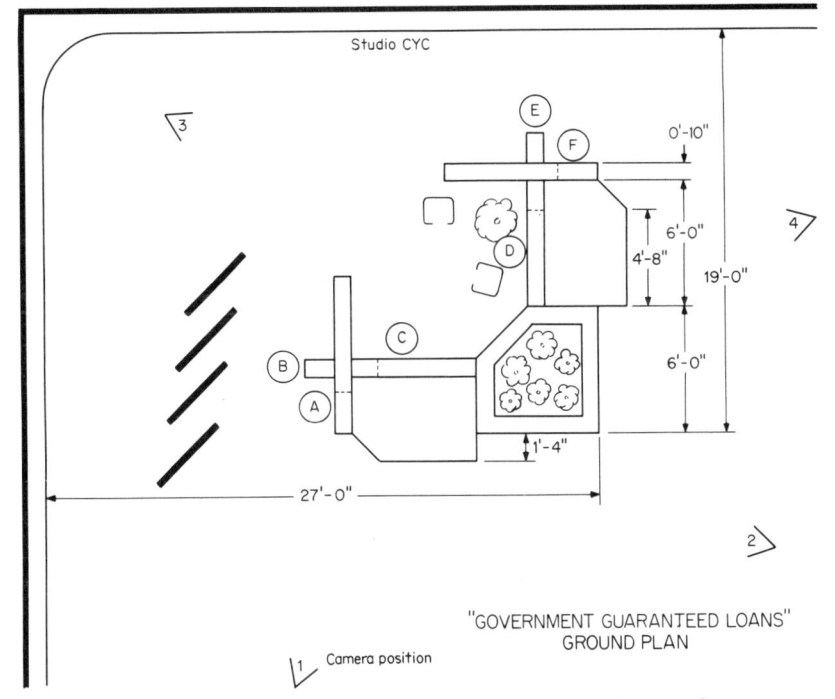

Figure 14-4 Ground plan, reduced from drawing using ¼ inch to the foot.

Most of the detailed measurements will come from additional drawings. What should be indicated are the points necessary for the carpenter to locate the set in the studio; straight runs of scenery; and the horizontal and vertical measurements of platforms and stairs. Also, each piece of scenery is labeled for identification in other drawings.

A designer may elect to build a scale model of the set in addition to making the sketch, especially if the set involves complex three-dimensional scenery. It gives a true indication of the space relationships of scenery and furniture. If the designer is also to supervise construction, building a model helps to identify possible construction problems. A model can be used by the director to plan actor and camera movement. It definitely aids the producer and the director in visualizing the total effect.

After presentation and approval of the ground plan and sketch or model, it's a good idea to test the design elements on camera. Before you buy 15 gallons of a particular blue paint, put a sample on a piece of cardboard, put some light on it, and see what color the camera thinks it is. That blue may look very blue-green to the camera and not at all what you had in mind. You may find you need to add red to achieve the desired color on the screen. Even with the improved quality of today's color cameras, the color picture on the monitor may not entirely agree with the color as the eye sees it.

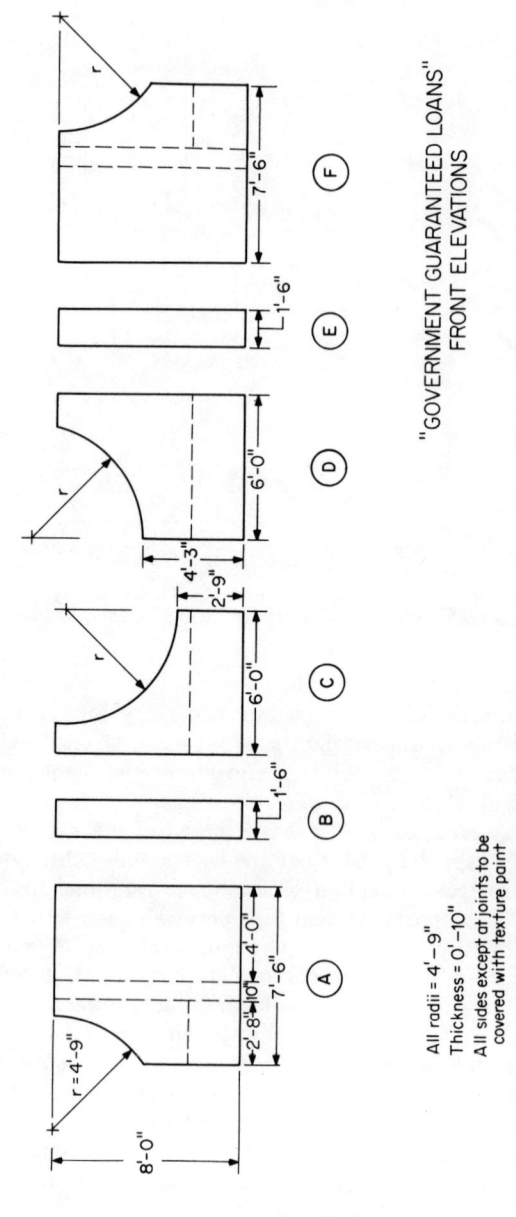

Figure 14-5 Front elevation of ground plan in Figure 14-4, reduced from drawing using ¼ inch to the foot.

"GOVERNMENT GUARANTEED LOANS"
FRONT ELEVATIONS

All radii = 4'-9"
Thickness = 0'-10"
All sides except at joints to be covered with texture paint

14-8

In testing fabrics, wallpapers, moldings, or even a style of scene painting, the test conditions must be close to what they will be during the shoot to achieve accurate results. Things to be concerned about are (1) the distance of the camera from the object being tested, (2) where the camera will be focused (since it probably won't be on the molding), and (3) the amount of light. These all affect the amount of detail that will read on camera. Of course, if you have the luxury of building the complete set, lighting it, and then making changes, that's great—but very unlikely. Whatever on-camera testing can be done prior to building the set will save time, money, and gnashing of teeth when revisions have to be made at the last minute.

Before the building can begin, every detail that was not included on the ground plan must be put down on paper. (See Figure 14-6.) This means drafting a set of front elevations, doing individual drawings of any complex detail, and providing paint samples. With these, the designer can explain the set to the carpenter and crew who will build and paint the set. If this process is followed, there should be no questions for which the designer doesn't already have answers on paper.

As the ground plan gives a composite horizontal view, the front elevations give a disassembled vertical view of each component from the front (as shown in Figure 14-7). The elevations are drawn as though the set were flattened out and each piece viewed head on. These drawings are done to scale, showing the size and shape of each piece of scenery. The dimensions and labels should correspond to the ground plan. Although the amount of detail that can be shown clearly in a ½-inch scale drawing is limited, it is necessary to indicate the position of all architectural trim, pictures, and light fixtures.

Anything that may require special construction for support should also be indicated on the elevation. Notation of radii and centers of all curves is also necessary for the carpenter. Special surface coverings or finishings should be specified.

SOURCES FOR MATERIALS

Locating furniture sources outside the firm means searching furniture showrooms, wholesalers, and antique or art galleries that are willing to rent on a weekly basis. Calling the companies listed in the Yellow Pages will yield a list of possibilities. Here are some possible headings you might try:

Antiques—Dealers

Antiques—Reproduction

Art Galleries

Craft Galleries and Dealers

Display Materials

Furniture Dealers

14-10 PRODUCTION WITHIN PRIVATE TV

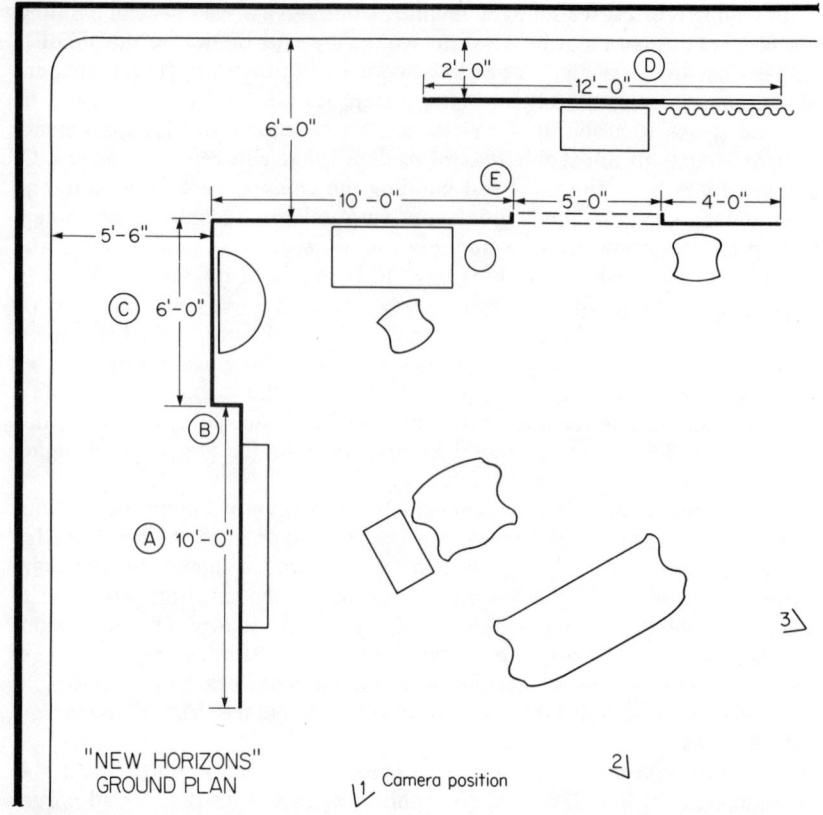

Figure 14-6 Ground plan for multidimension set, reduced from drawing using ½ inch to the foot.

Showrooms

Furniture Renting and Leasing

Theatrical Equipment and Supplies

Television Studio Equipment

If you don't have time to explore, try contacting still photographers, commercial production companies, local television stations, or college theater departments for leads. These are headed by people who likely have had to dress sets and can steer you in the right direction. When you go out to look at the furniture, note dimensions and make sketches or take instant snapshots. This tip is particularly important when matching furniture coming from various sources. It also makes it easy to show the director what the possibilities are before the furniture arrives in the studio.

Other sources of materials can be found in interior design, theatrical, and display houses. These businesses have a lot in common with scene

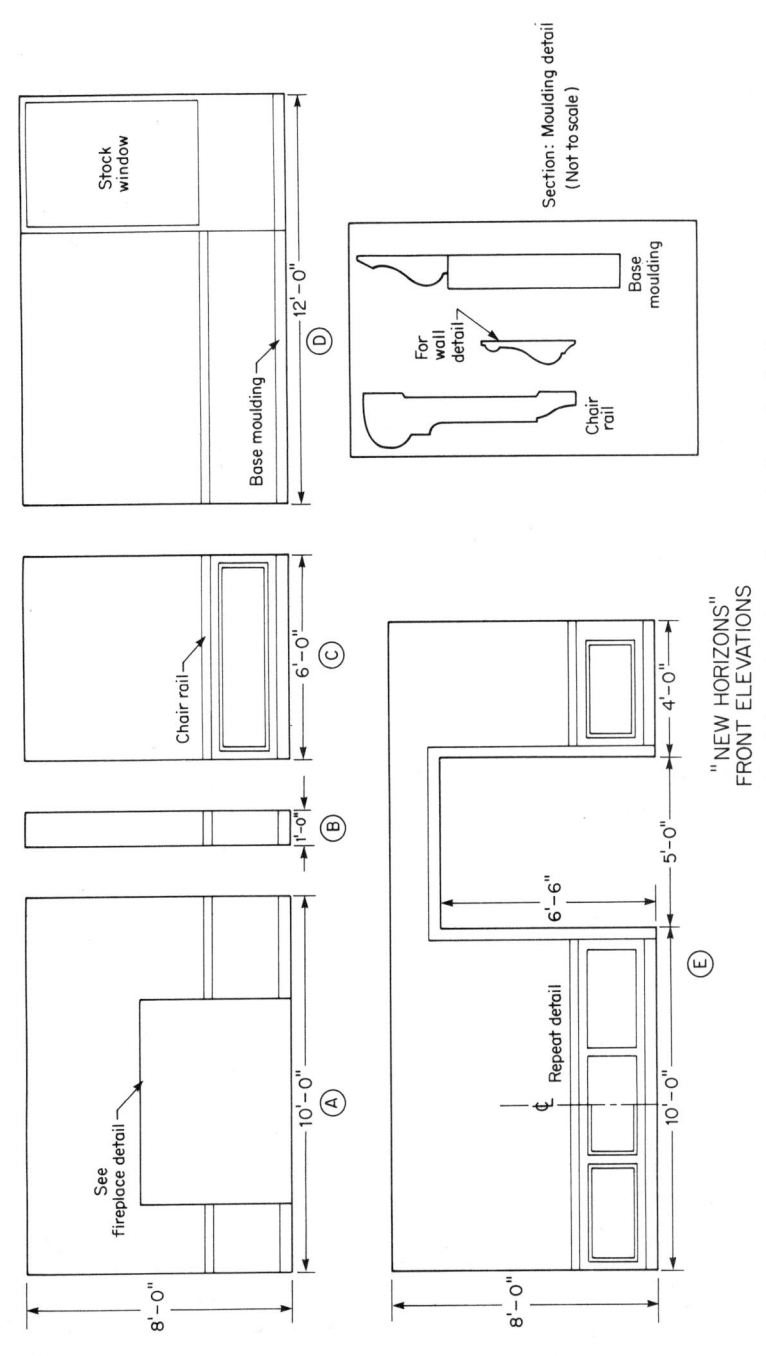

Figure 14-7 Front elevation for multidimension set, reduced from drawing using ½ inch to the foot.

14-11

design. The interior design field has sources for decorator fabrics, textured wall coverings, screens, furniture, and lighting fixtures. Theatrical supply houses can provide hardware, tools, paint, and scenic fabrics. Decorative panels, cardboard tubes of varying diameter, or Christmas decorations for a shoot in July can be found in display houses.

BUDGET

The budget is always a key issue with the producer. Moreover, looking at a ground plan doesn't always give you the answer to "How much will this cost?" Consideration must be given to each element of the set. Can the needed furniture be found or borrowed, or do original set pieces have to be built? (Figure 14-8.) If furniture has to be rented, can it be good-looking residential pieces, or are quality antiques necessary to achieve the desired effect? (See Figures 14-9 and 14-10.) The following cost breakdowns show how the answers to these questions affect the bottom line.

For the program *Assessor Training*, the costs were as follows:

Labor	$720
Lumber	100
Paint	150
Total	$970

For the program *Government Guaranteed Loans*, the following was the breakdown of costs:

Labor		
Carpenters	$ 420	
Scenic artist	400	
Production assistant	800	(build)
	335	(strike)
	$1,955	
Materials		
Lumber	$ 696	
Paint	530	
Flooring	550	
	$1,776	
Furniture rental	$ 365	
Total	$4,096	

SET DESIGN   **14-13**

Figure 14-8 Here, stock flats were used and all the furniture was obtained from sources within the company. (*Merrill Lynch Video Network.*)

Finally, the program *New Horizons* cost:

Labor	
Carpenter	$ 120
Production assistants	445
	$ 565
Materials	
Lumber	151
Paint	55
Carpet	496
	702
Furniture rental	1,944
Total	$3,211

STRUCTURAL ELEMENTS

Versatility as a designer comes from a knowledge of scenery construction and of the use of various materials and their limitations. The more you

14-14 PRODUCTION WITHIN PRIVATE TV

Figure 14-9 Construction of original pieces is expensive in labor and materials. (*Merrill Lynch Video Network.*)

know about current materials and techniques, the better you'll be able to apply them in original ways to express your ideas. A designer must keep in mind the demands placed on scenery for studio use. It must be portable, since it may have to be built outside the studio. It must have enough structure for support while retaining its portability, and its construction must be economical. This means balancing cost against the weight and structural demands and the versatility of the scenery. Higher costs are affordable if the scenery can be reused.

Decisions about scenery reuse should be made before construction begins. The amount of storage available may affect the way a unit is built. For example, a stair or wall unit can be stored more easily when it is built in sections that can be bolted together when in use. If you're building a piece of scenery that you know you'll never use again, you may choose to use a cheaper grade of lumber or spend less time finishing the unit. A good inventory of stock scenery means that sets can be put up quickly and economically and that the storage space is being used wisely.

FLATS

Most scenery built is framed flat in wood and covered with whatever material suits the demands. The frame is usually of 1-inch x 3-inch white pine.

Figure 14-10 Use of stock flats kept the labor and materials costs down. The expense of the carpet was justifiable because of its reusability. The major cost was the rental of good-quality antiques to achieve the desired look. (*Merrill Lynch Video Network*.)

The method of frame construction will vary slightly depending on the covering material. Common materials used for covering flats are cotton duck, muslin, plywood, and composition board. The choice of materials is based on the cost versus the longevity. A flat covered with duck or muslin is the least expensive, very lightweight and portable. The disadvantages are that these materials can be torn easily and show wear if not handled properly. Using plywood for the face of the flat and covering that with muslin makes a very durable piece of scenery with a good surface for painting. It is also more costly. Facing flats with composition board is a compromise between the two methods in terms of cost, but the main problem is the tendency of composition board to warp in humid weather. Any warping of the flat will make lighting very difficult and uneven on the wall surface.

Flats can be built to any irregular shape or dimension. But, if you wish to establish a policy of storing and reusing scenery, build and make use of standard-size pieces (Figure 14-11). The most economical way to build a 10-foot-wide wall is to make two flats 4 feet wide and one flat 2 feet wide, and to hinge or clamp them together on the back. The front seam is masked by covering material or wide masking tape. The standard for the height of the flat is determined by the height of the studio ceiling, but it is usually 8 feet.

Flats can also be framed, with openings for doors or windows. The archi-

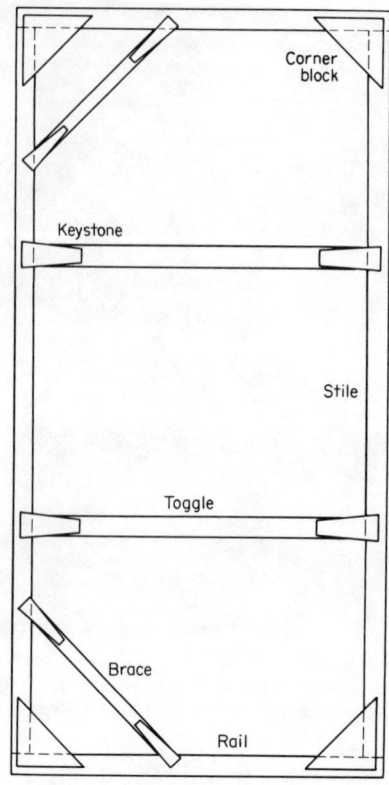

A STANDARD FLAT
Rails, stiles and toggles cut from 1" X 3" pine
Braces cut from 1" X 2" pine
Corner blocks and keystones cut from 1/4" plywood
Front of flat to be covered with muslin

Figure 14-11 Components of a standard flat.

tectural parts are slid into the opening and attached to the flat frame. The other method is to build the window sash or door with the casings into the flat. These are referred to as window or door units. This method is preferable for most studios unless most sets call for period interiors with no standard window or door treatment.

Scenery will not stand by itself without additional support. The simplest brace is L-shaped, made of 1 x 3 inch pine and attached to the flat frame with hinges. If the braces need to be semipermanent, clamps or loose pin hinges can be used. A sandbag is placed over the end of the brace to provide stability.

The next group of unframed scenery includes curtains, drops, and the cyclorama (or "cyc"). They all provide a large area of scenery with minimum structure and maximum portability. Because they are unframed, they need to be hung from a pipe or batten for support.

DRAPES

Curtains are the simplest of the group. They are often used in combination with other scenic elements as masking or a background, but can be used effectively alone. They can be hung in straight folds or draped. Lighting can be used to accentuate the folds or to project a pattern on the surface. Large drapery panels are made by sewing widths of fabric together with vertical seams. Such seams are less conspicuous because of the vertical folds of the curtains. Since the direction of the weave is lengthwise, they will hang better and there will be less strain on the seam.

There are a variety of fabrics available for curtains depending on the specific use. Are they to be opaque, translucent, or transparent? Are they to be stock, or is this a one-time effect? Answers to these kinds of questions will determine the best fabric to use. Velour is often used because it is opaque, has a rich texture under light, and drapes well. It is also expensive. Some economical alternatives are duvetyn or flannel. Translucent shapes (with a soft light behind) add a realistic touch to an office setting.

CYCLORAMA

Most studios are equipped with a cyclorama. It is usually shark's-tooth scrim or bobbinet hung ceiling to floor covering two or three walls. It is hung on a curved track to conceal the corners of the studio. This is the easiest way to create a neutral background, or it can be used as a backing for interior sets.

A white cyclorama is the most versatile for lighting. Pools of light, a wash of color, or patterns projected onto its smooth surface can change its appearance radically and quickly. The cyclorama lights must be controlled separately from the acting-area lights and the set must be far enough in the foreground to eliminate visible shadows on the cyclorama.

SET PIECES

Of the three-dimensional scenery, the most frequently used are platforms and stair units. Platforms are framed out of 2 x 4 inch lumber and usually covered with plywood. The 2 x 4 inch legs can be nailed or bolted to the frame. The height of the platform can be varied easily by changing the legs. Platforms are often used to raise the acting area so that the camera can be at eye level with a seated performer—allowing the communication between performer and viewer to be more direct. Platforms can also be

14-18 PRODUCTION WITHIN PRIVATE TV

Figure 14-12 The logo was cut out of Fome-Cor and hung in front of the studio cyclorama. (*Merrill Lynch Video Network.*)

used to create a more interesting background by establishing more than one level. Stairs can be used to create the illusion of another level or of rooms beyond what the camera sees. They also add depth and dimension to the set.

Action on stairs is difficult in most private television studios because of the ceiling height. An actor 5 feet 10 inches tall can use only about four stairs before the lights are a mere arm's length away. Noise as actors move can be a problem on both platforms and stairs. If the final covering is to be carpet, it may eliminate the problem. If not, homosote, a type of composition board, can be nailed to the plywood and then carpeted.

Specially designed three-dimensional set pieces are generally framed in 1 x 3 inch or 2 x 4 inch lumber depending on the weight they must support. (Figure 14-12.) The covering material can be plywood, Fome-Cor, or fabric. If the piece is to be painted, it can be covered with plywood and then muslin for a good painting surface. If the final covering will be fabric or contact paper, the frame can be covered with Fome-Cor. This material is polystyrene foam laminated between paperboard facings. It is lightweight, yet rigid and strong. It can be cut easily with a razor blade, and the surface is suitable for markers, inks, or silkscreen. (Figure 14-13 shows one of its uses.)

SET DESIGN 14-19

Figure 14-13 A logo was used as the basis for a set design. (*Merrill Lynch Video Network.*)

Accessories and Set Dressing

With construction underway, the designer can concentrate on the set props—those little items that add texture and interest to a set. They may include lamps, clocks, paintings, flower arrangements, bookends, vases, nonfunctional furniture, carpets, and desk accessories.

The first place to look for such items is around you. Find out which department handles the office furniture for your firm. Often there are side pieces that are not being used that can be borrowed for a production. Make friends with people in the office who have attractive office accessories, plants, or pictures that may be of future use in a set.

If the set will be used only once or twice, it may pay to rent certain props instead of buying them and then worrying about storage space. Keep details simple. Backgrounds that are too busy can be distracting. Be careful of using pictures of heads the same size as the head of your performer, and avoid printing that is *almost* readable, because the viewer may try to read it and be distracted from the program topic. Make it easy for the camera operator by making sure no plants, lamps, or architectural details will appear to "grow out of a person's head" on the screen.

CONCLUSION

What makes a good television set? A quality set has good composition, a feeling of balance on the screen. A good set is within the contrast range acceptable to the camera, providing it with sufficient separation of color.

The final step in the design process is the evaluation of the design conceptually and technically. Recall the original intention of the program. Does the design solution help or hinder the communication of the idea? Does the set work well in relation to the staging and lighting? Were the actors and the director able to work well within the set?

The only way to find the answers to these questions is to look at a television monitor. The appearance of the set to the eye is not important. Remember, it's how it looks on camera that counts.

15 Graphics

JAN THIEDE

The design of graphics has developed from felt boards with stick-on plastic letters to computer-generated, multicolor, multidimensional extravaganzas. Art materials, techniques, and video equipment are developing and improving so quickly that there is almost no limit to what graphics can be designed. But, regardless of the video facility where we work, we need to fully understand the role that graphics play and how best to prepare them. This chapter will include a discussion of:

Graphics and their purpose

Developing graphics

Transferring artwork to videotape

Thoughts for potential artists

GRAPHICS AND THEIR PURPOSE

Graphics can be used in a variety of applications. Their use can range from opening titles to product shots to inserts consisting of charts or graphs to set design pieces used as background shots. Graphics can incorporate words or type, symbols, logos, numbers, human figures, or any combination thereof. Because of the wide possibilities available, it is necessary for the graphic artist first to understand the purpose of graphics as they relate to a specific project. There are four uses of graphics:

- Identification
- Highlighting key information
- Simplifying complicated material
- Entertainment

Identification

This is the most common use of graphics within private television, as seen in Figure 15-1. It includes the title and credits of programs. Graphics inform the viewer who the speakers are or what objects are being viewed. They place time and location. In short, they give the viewer a frame of reference.

Figure 15-1 Identification graphic for miniseries. (*Merrill Lynch Video Network.*)

Highlighting Key Points

Private television has often been likened to the telegraph, which sends crucial information rapidly—details to follow. Because most private television programs are short (generally under 20 minutes) and yet cover a great deal of information, it is important to reinforce or highlight information with graphics. (Figure 15–2 shows an example.) Generally, this is accomplished with words or numbers on the screen repeating (in a simplified fashion) the data that are being spoken. For example, if a talent on a program is discussing the growth of an industry and citing yearly figures to support the statement, a graphic insert that states "Industry average up 400 percent in 10 years" highlights and reinforces the spoken word.

Of course, graphics need not be static; whether they are used for identification, highlighting, simplification, or entertainment, they can be dimensional, contain movement or both. In the above example, if the industry is the petroleum group, a stack of oil barrels might rise in the background as the words or type are fixed in the foreground.

Simplifying Complicated Material

The nature of private television is such that complicated materials must be discussed. Generally, broadcast television does not have to deal with

GRAPHICS   15-3

Figure 15-2   Highlight graphic. (*Merrill Lynch Video Network.*)

the problems of showing an equation (and how numbers are substituted within the equation) or a manufacturing process flowchart or the configuration of a hydrocarbon molecule. Private television cannot avoid the issue, so, rather than ignore it, the graphic artist must know how to simplify complex material. (Figure 15-3.)

Fortunately, there are some tricks to the trade. While some of them will be highlighted later in the chapter, let me point out a few of them here.

1. Have the client identify *the* most important elements that must be included and try to eliminate the rest.
2. Have the script writer introduce the material in stages and the graphics show the material from its simplest to its most complex form.
3. Make full use of shading, color, pop-ons (information appears suddenly), and reveals (information appears gradually). The viewer's attention can be controlled by presenting in light colors or white the information under discussion and graying down or darkening the rest of the image.
4. Don't hesitate to use arrows or other pointers to guide the viewer.
5. Convey to the director that the graphics should stay on the screen longer than might be customary.
6. Always have an impartial viewer check the graphic for understanding before it is recorded. An artist and the client can get too close to the topic.

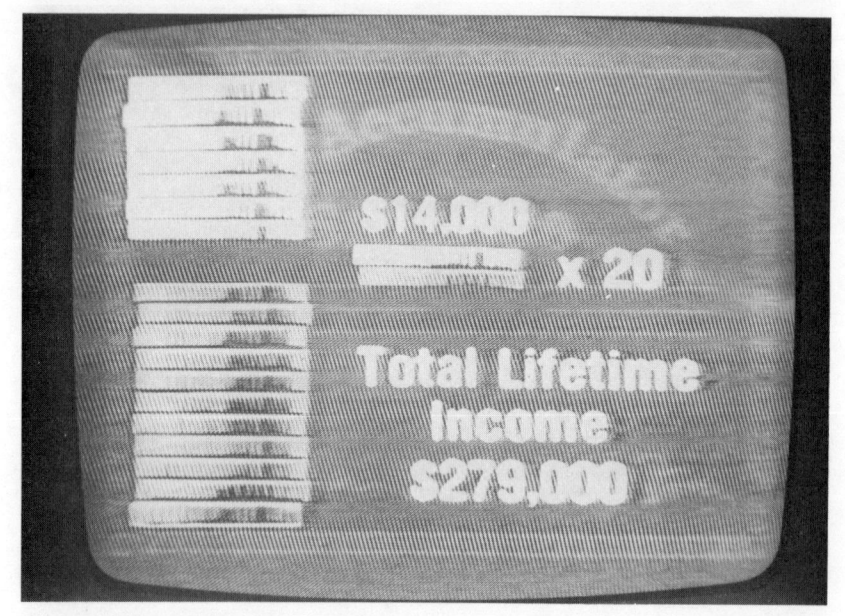

Figure 15-3 Complex graphic; during the actual program, the coins moved and the words faded. (*Merrill Lynch Video Network*.)

Entertainment

This is a broad category. It covers graphics ranging from set design pieces (adding color and interest to a set) to animated figures used for comic relief. It is here that the graphics artist can add a sparkle or style to the program. Entertainment graphics, like the one in Figure 15-4, do not necessarily convey information directly related to what is being said, but, rather, can contribute to the visual enhancement of the program.

DEVELOPING GRAPHICS

After you have met with the writer, director, and client, it is time to start developing and executing your ideas. There are two steps to employ. The first is the storyboard, and the second is the mechanical. From this point, the chapter will focus on graphics that are prepared by an artist. Other forms of visual support, such as photographs or chalkboard illustrations, are outside the realm of this discussion.

Storyboards

One of the best ways to develop your ideas is to sketch them. This process is known as *storyboarding*. An example is shown in Figure 15-5. The more

Figure 15-4 Entertainment graphic. A slide on a rear projection screen is used as part of the set. (*Merrill Lynch Video Network.*)

complex the content of the program, the more you need to commit to paper the design, words, and format of the graphic.

There are special pads designed for this purpose. They have outlines of television screens in the proper proportions and areas for notes or narration marked right on them. If you don't have these pads, rule several 3 x 4 inch boxes on an 8½ x 11 inch piece of paper, leaving space underneath for comments, and have a stack duplicated.

Make a sketch to illustrate each graphic in a separate frame or box, and number each frame consecutively. Match the number of each frame to the corresponding copy of narrative in the script for easy identification. Beneath each sketch, make notes, which might include typefaces, colors, camera moves, fades, pop-ons, or special effects that may occur. A completed storyboard is shown in Figure 15-5.

Meet with the client and director, get their opinions, and make any suggested changes. If a meeting is not possible, make photocopies of your storyboard and be sure the appropriate people see them, note any changes, and get the storyboard back to you as soon as they possibly can.

It is to your advantage to have an approved storyboard as early as possible. Now is the time to make any corrections on the storyboard and either resubmit it in rough form or redraw the entire thing, tightening up the drawings and typography and perhaps even adding some color with felt-tipped pens.

Require at least two approving signatures on the final storyboard before moving on to the next stage. These should be from the director, and either

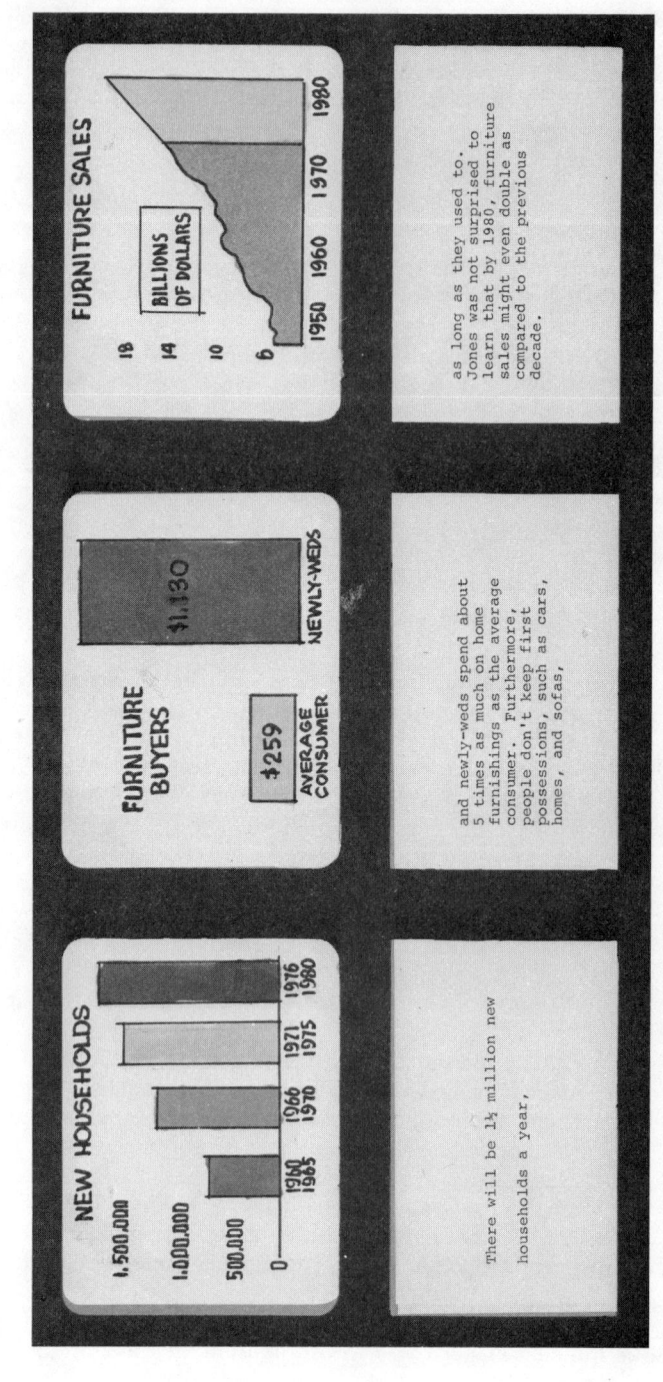

Figure 15-5 Completed storyboard. Detailed sketches enable all parties to visualize and approve graphics that will be in the program.

the producer or the client. It might be a good idea to have a rubber stamp made up with the words "Approved by" and "Date," with spaces for two signatures. Be sure to get these signatures on the final artwork or mechanicals as well. No matter how careful we are, under deadline pressures, a misspelled word or reversed arrow is going to sneak by occasionally.

Mechanicals

After the storyboard has been approved, a precise layout on tracing paper should be prepared in a 3 x 4 proportion. You see, all television sets are in the proportion of 3 units high by 4 units wide. A graphic must be planned within this 3 x 4 space. On the tracing paper, sketch in the type size and the design of any symbol or figure. The tracing paper sketch can now act as a reference for your final mechanical.

A mechanical is finished artwork prepared as the original for reproduction. The mechanical may then be shot on 35mm film and projected as a slide (directly into a television camera via a device called a *TeleCine* or on a rear projection screen); it may be shot onto a sheet of Kodalith film (a black-and-white negative film) which can then be gelled with colored gels, placed on a light box and shot directly by a television camera; or it can be used as an original artboard (Figure 15-6), which can be shot directly by a TV camera.

The actual mechanical should be done on a piece of bristol board or illustration board. Type (for words or numbers) can be set by a typographer or placed on the board with rub-on letters (see Figure 15-7). You can work either with black type on white board or white type on black board. If you have prepared a mechanical in one format and need the other, a negative photostat can be prepared to give you the desired result.

If you are new to the world of graphics, it may be easier to have all type, figures, and logos on separate sheets of paper, which are then pasted in place (known as a *paste-up*). Cut lines and light pencil marks are permitted at this stage. Neatness isn't the priority here, but accurate placement and straight lines are. If a line is slightly off horizontally, it will be highly exaggerated through the video lens. Very fine lines and lots of details in your illustration should be avoided as they will either break up or not register at all. For this reason, bar charts and line charts should be done with wide-rule charting tape, and the type should always be large.

Don't forget that completed black-and-white mechanicals most also be approved by the director and the client or producer. If possible, have them sign their approvals on photocopies and not the originals.

TRANSFERRING ARTWORK TO VIDEOTAPE

Preparing artwork for videotape requires skills and knowledge that are somewhat different from those needed for the print medium. Both do require

15-8   PRODUCTION WITHIN PRIVATE TV

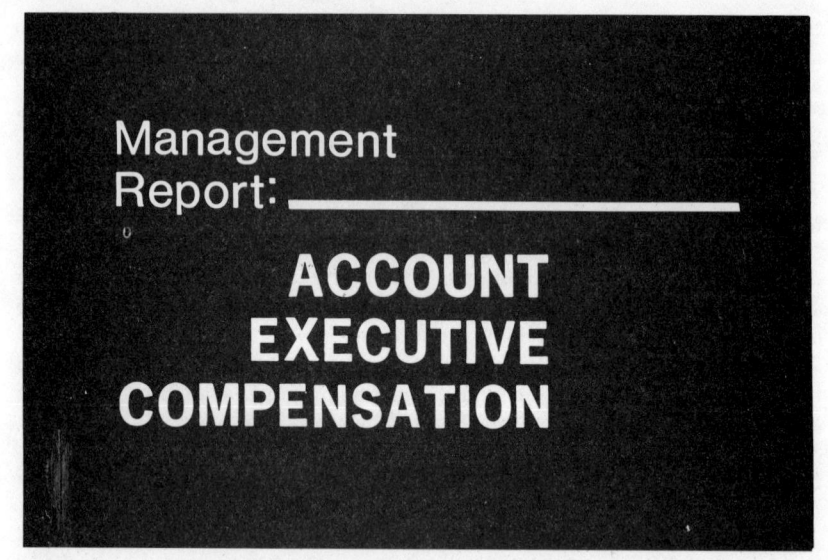

> **Management Report:** _____
>
> # ACCOUNT EXECUTIVE COMPENSATION

Figure 15-6   Art card mounted on black illustration board, to be shot by the television camera as a graphic insert. (*Merrill Lynch Video Network*.)

creativity, attention to detail, and the desire to prepare clear and understandable graphic material; however, television imposes its own limitations, and the graphic artist must be aware of them. This section will attempt to provide helpful hints that are applicable to any size video facility.

The Safe Title Area

The first thing you notice when you stand in a control room is that no two television monitors show you the exact same image. One will show you more detail or information on the left; another will show you more information on the top or bottom.

Different manufacturers, different engineers, different care and handling will affect how much the viewer will ultimately see. While the loss will be different for each monitor, an average loss has been computed, resulting in creation of a *safe action area* and a *safe title area* (also known as *TV safety*). As the names imply, a safe action area is the field in which the viewer will see most of the action; more critical elements, such as titles and product shots, should be placed in the safe title area.

If a television screen has an area of one (1) unit, then the safe action area is equal to approximately 0.81 unit (centered), and the safe title area is approximately 0.64 unit (centered). (See Figure 15-8). As a rule of thumb, leave 10 percent free space around the graphic. This way, you know it will be in the safe title area and thus be seen by all viewers.

Another approach is to borrow your television engineer's registration grid and photograph it at various distances (some longer, some shorter)

Figure 15-7 Rub-on letters. They are applied to illustration board with burnishing tool. (*Courtesy of Jan Thiede.*)

onto 35mm positive slide film. When these slides are processed, put them into your TeleCine unit to determine which slide (or, actually, grid) accurately matches TV safety for your system. When you have found the best one, take the slide out of the mount; tape it onto the focusing ground glass of your 35mm camera. Thus, when you look through the viewer to shoot your graphics, you can frame them within the grid, and you can be sure they will fall within your TV safety area.

Type Fonts

With the thousands of typefaces available from photo typesetters and rub-on letter manufacturers, there is no need to use difficult-to-read letter styles. On television, extremely thick letters fill in and the ones with thin lines disappear. Outlined or striped letters can buzz annoyingly; very fine serifs can get lost; letters with bold serifs, if spaced too closely, will run together; and type that is extremely unique in its design can be very hard to read.

When choosing a typeface, look for one that has an equal amount of white and black space in and around the letters. This balance will help ensure easy readability. The spacing between letters should be greater than what would be considered normal for print or slides. Samples of good and poor typefaces are shown in Figures 15-9 and 15-10.

This is not to say that, for emphasis or a special effect, an unusual type-

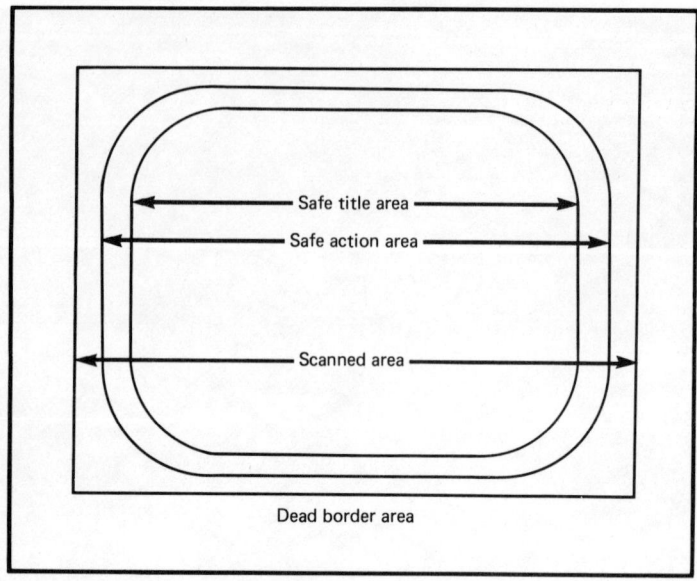

Figure 15-8 Action and title area.

face can't be used successfully. But be sure to put a sample of the type in front of the graphics camera before committing yourself, in case it doesn't work.

When ordering type, try to use letters that are ¼ inch high or larger. Type size is referred to in points. A point is .0138 of an inch (approximately 1/72 of an inch). As a result, 24-, 36-, or 48-point type is fairly convenient when ordering rub-on letters. Usually, the maximum size for hot type or machine-set type (one of the least expensive types) from a typesetter is 14-point; therefore, that is the size to order. It can then be enlarged photostatically. It is better to make things too large and thus too readable than the other way around.

Eight to ten lines of type should be the maximum amount of information on a screen at one time. You don't want to put so much on the screen that the type must be very small in size, nor do you want to give the viewer an opportunity to read ahead of the narration.

Color

The use of colors can be tricky. A good system is to set up a lot of color papers in front of the graphics camera and check the color monitor. A series of blues can range from blue-green through a blue hue to blue-violet and still appear identical on video. Red hue is not generally a good color inasmuch as it usually is quite noisy (the grain appears active) on the screen.

Figure 15-9 Samples of legible type.

If it must be used, red generated through the switcher or a character generator may work better than a red gel or colored paper.

For copy and artwork, blue, green, or any dark color recedes, so they all are good choices for backgrounds, headlines that are set in a larger typeface, watermarks, or items that are not being emphasized. Yellow, white, light green, and orange will catch the eye and should be used to highlight areas. This is also true of boxes around an element, arrows, flashing letters, or movement.

Camera Angle

To ensure that the information will appear straight and properly centered on the screen, you can use a grid pattern or select a horizontal wipe on the switcher, making sure that all the letters from your graphic sit evenly on the wipe. Camera angles play an important role here. If your camera is pedestaled up too high or is sitting too low, you will get what's known as a keystone effect. If the camera is positioned too far to the left or right, the card will appear to be bowed.

Art Cards

As mentioned, the simplest and most common graphic is the art card. A workable size for this format should be established to comply with recording conditions. A 15 x 20 inch horizontal board with the artwork 9 x 12 inches centered (our 3 x 4 proportions) is a good size and will allow plenty of shoot-off (the area surrounding the graphic). The most basic art card is

Figure 15-10 Samples of poor type.

white lettering on a black background. White rub-on lettering on a black Bainbridge board; white tape underlining a headline; or a logo pasted on will be sufficient for many graphic needs. If type is set by a typographer, a paper negative stat can be made and in turn pasted in the center of the black board. Shapes can be cut from colored paper or adhesive-backed color film and mounted on a black or colored background.

Not all art cards will consist of lettering or symbols on a flat background. If there are brochures, photographs, booklets, forms, or other ancillary material, they can be mounted on boards and these boards mounted on larger color boards with small supports behind them. When attached to a cyclorama or background surface, they will appear to float away from that background, adding perspective and dimension to the shot.

When it is necessary to include a white printed form (such as a sales order form) or newspaper clipping as a graphic and the print from the back shows through, you can cure this problem by flush-mounting the original on a piece of black paper with spray cement, and then mounting it on the background you intend to use.

Reveals and Pop-Ons

One of the standard methods of presenting information by graphics is through the use of *reveals*. Here, new information either dissolves or pops on. Let's look at how this can be accomplished on a Kodalith. Understand that this approach requires that the graphic be prerecorded and edited into the main body of the program.

Figure 15-11 Masks on Kodaliths. (*Merrill Lynch Video Network.*)

Proceed by securing the Kodalith to a light box. Black masking tape is best for this purpose. Continue using black masking tape and black paper to block out any peripheral light. You may get white edging if you don't. The record camera must be carefully aimed and focused directly on the Kodalith. Areas that are meant to pop on the screen should also be covered or masked so there is no light seepage.

First, record the graphic with all the masks in place. Be certain that you have plenty of usable footage. Second, don't stop recording. If you do, there will be visible shifts at the edit-points. Third, peel the masks off the Kodalith, making certain that you do not move the light box or the camera. Record more footage than you need. Finally, when you edit all the stages together, the uncovered sections will appear to pop on the screen. (See Figure 15-11.)

A simple reveal method can be accomplished by making tracks around the mounted Kodalith with strips of black mat board (see Figure 15-12). By fitting black mat boards into these tracks, you can then pull or slide the boards along the face of the Kodalith, exposing or revealing different portions as you do so. This effect is similar to the effect you can get from a switcher known as a *wipe*.

Character Generator

The format that probably best reflects the future of the graphics field is the computer character generator (Figure 15-13). These marvelous machines

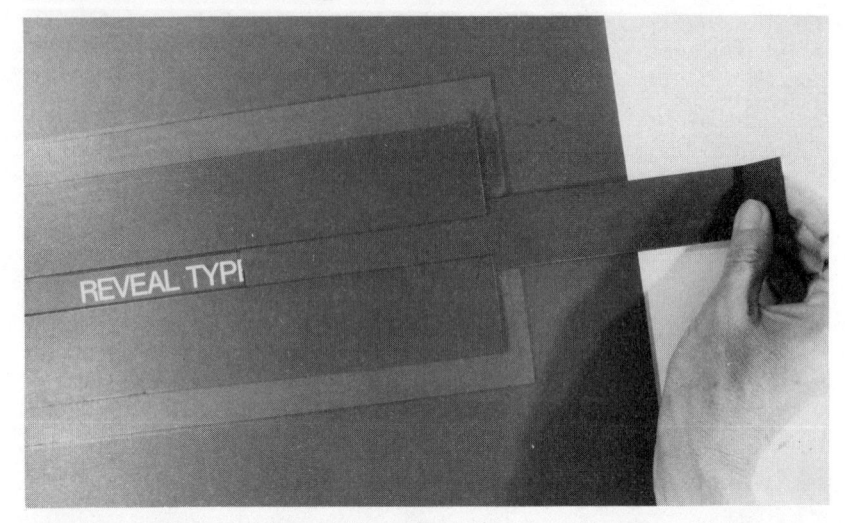

Figure 15-12 Black mat board with tracks.

take away the paper and paint, the type, and the color papers and replace them all with a keyboard (and an operating manual about 1 inch thick). Color background? Press a button; you can have 24 of them. Typeface? Another button or two, and there's a choice of 100 (but only 8 at one time). Color for the type? Another button, another 24 choices. Edges, outlines, borders, rolls, crawls? With a skilled operator, all these elements are possible.

Logos and symbols can be converted into digital format and then stored, to be called up at a later date. Limited animation can also be done on a character generator. In an area approximately one-fifteenth the size of the screen, a figure can walk, turn on its axis, move back and forth in space, and bounce around the screen.

These character generators have been around for only a few years, but they are being updated all the time and their potential is seemingly limitless. While sales literature will say that anyone can create graphics (which may be a truthful statement, in and of itself), do not assume that you will get quality graphics. If you plan to purchase a better-grade character generator, make certain that you have a designer on staff (or a free-lancer you can call upon) to make the thousand decisions involving fonts, colors, spacing, composition, and layout.

CONCLUSION

Being a graphics artist in a private television facility can be a rewarding and often creative outlet. Graphic techniques, formulas, and concepts are

Figure 15-13 Character generator in control room. (*Merrill Lynch Video Network.*)

as limitless as the artist's imagination. The freedom to experiment is more common in private television than in most other forms of production. The artist must keep one thing in mind: Even the most beautiful design is useless if it is delivered when the set is being torn down. The artist's deadline must have number 1 priority. Therefore, it is important to develop a standard of quality that can be met under pressing time restrictions. The ability to produce quality in a given time period becomes the artist's most valuable commodity.

16 Lighting

JOEL WILLIS

OVERVIEW

The modern television camera will produce a picture with any reasonable amount of illumination. It is the control of that illumination, its direction, placement, level, and intensity that will enhance the pictorial effect. This appropriate control is the job of the lighting director. This person works closely with engineering, production, talent, costume, set design, makeup, and direction to help achieve the "look" of a program, as well as to deal with the vicissitudes of the television system.

The professional lighting director understands the limitations of the television system, the terminology used in describing light and lighting equipment, the mechanics of controlling light, and the techniques used in lighting the set. This expert also comes to the production prepared with both the knowledge of how to overcome commonly encountered problems and the tools necessary to do the job. This chapter will examine four key areas:

Limitations of the television system

Lighting instruments

Production procedures

Problem solving

LIMITATIONS OF THE SYSTEM

Electronic Conditions

To understand lighting, it is essential to know the limitations of the television system. Your eye is very forgiving when confronted with extremes of dark and light, but the television camera is not. The contrast range in television is limited to about 30:1. Therefore, in terms of composition, set, costume, makeup, and lighting, it is not advisable to have any object in a scene reflect more than 30 times the light reflected by the darkest object. More than a 30:1 contrast ratio will result in washed-out pictures with very little detail in the dark areas. This effect occurs when engineers are forced to reduce the amount of light striking the camera tube (in order not to burn it out) by closing down the iris in the lens. With less light striking the camera tube, darker sections lose all detail. The picture is then described as "being in the mud."

16-1

16-2 PRODUCTION WITHIN PRIVATE TV

Through control of the quantity and quality of light, we will achieve a pleasing contrast range for the camera. But it cannot be done by the manipulation of light alone. Television is a team effort and you *must* collaborate with your colleagues in makeup, set design, artwork, and engineering to achieve the best results.

Private Television Studios

Another limitation of lighting as it relates to private television is the physical setting of studios in office buildings. In contrast to broadcast operations that operate in studios with high ceilings and extensive dimmer boards (a device that controls the voltage to the lamps, and as a result controls intensity), private television studios usually do not have that luxury.

If a private television studio can set its lighting grid at 12 feet, it is considered fortunate. Even at 12 feet, the throw of light is short, and consequently, too much light may strike the scene. Any height below 12 feet is a problem because the instrument will be too close to the subject.

There is also the additional problem that most private TV studios do not have full-sized lighting consoles which allow the lighting director individual control for each instrument. Often four or five lamps are plugged into one control. This arrangement does not afford flexibility for the lighting director.

Finally, most private television studios do not have adequate air conditioning for the amount of heat generated by the lamps. This limitation can make the studio unbearable for both talent and crew.

These problems and limitations are more difficult and costly to correct once a studio is complete; therefore, a qualified lighting consultant should be brought in before a studio is constructed.

Color Temperature

Cameras are adjusted or balanced to a single color temperature. Color temperature is a measurement of how "hot" a color is as expressed in degrees Kelvin (K), which is the Celsius scale starting at absolute zero (−273.16° Celsius).

White light, or the light we think of as white, is actually composed of different wavelengths of light—the same ones that are in the spectrum: red, orange, yellow, green, blue, indigo, violet (in that order). Each color of the spectrum has its own color temperature, with red being a low figure, approximately 1200 K; green at approximately 4700 K; and violet near 8000 K.

Different light sources will have different mixtures of color light. For example, a tungsten-filament lamp will measure approximately 2500 K (and have a yellowish tint to it). A warm white fluorescent tube will measure about 3500 K and have a bit of green in it. A clear blue sky will measure at 10,000 K.

The problem arises when two or more different color temperatures are mixed. Studio and portable quartz instruments are generally 3200 K. When daylight and fluorescent light are mixed in the same shot, color problems can arise. Some solutions will be covered later in the chapter, but in essence, advanced planning, careful selection of lighting instruments, and proper use of filters, gels, and scrims can overcome most color balance problems.

LIGHTING

As with an orchestra, large or small, lighting is made up of a combination of instruments. As strings, woodwinds, percussion, and brass round out the full sound of the orchestra, so do *key, fill, back,* and *set* lights round out the spectrum of lighting. Within each musical grouping, there are the bass and treble sections; in light, we have *brutes* and *inkys* in both hard and soft sources. And, as musical instruments can be modified in their register with capos and mutes, so can *gobos, flags, scrims,* and *gels* shape, define, and soften the tones of what we see.

The orchestration of light requires the same sensitivity as conducting an orchestra. The blend of soft and hard sources, a mixing of color temperatures, and experimentation with new tonal qualities are all a part of the lighting director's symphony. But first, we must know what each instrument is capable of doing.

Fill. Fill instruments, usually soft sources of light, provide a relatively shadowless, flat, broad pattern of illumination. Their main function is to set the base level of illumination while maintaining the shadow areas of a scene at a slightly transparent level. They also serve to balance the hard effects and shadows created by key lights.

Key. Key instruments are the principle sources of light within a scene. They define the center of attention, and they are generally a hard source of light with a focusable control that ranges from flood (wide throw of light) to spot (narrow throw of light). The intensity of the key instruments will determine the proportional mix of other sources of light within a scene. Their placement in terms of height and direction will give the strongest shadow areas within a scene. Light from these sources requires the most control and leaves little margin for error.

Back. The back instrument provides separation of an object from its background. Placement, intensity, and control are all important in creating a well-rounded, defined subject. It also is usually a hard, focusable light source.

Set. As the name suggests, the set lights accent and provide highlights to the overall scene.

Figure 16-1 A 10-inch 2000-watt fresnel light with barn doors. (*Strand Century, Inc.*)

EQUIPMENT

The lighting instruments and accessories used to provide fill, key, back, and set lights will vary according to the complexity of your production and the size of your budget. However, all instruments fall into two general categories: *spotlights* and *floodlights*. In most situations, the focusable beam of the spotlight will be used for key and back lighting, while the softer beam of the floodlights may be used for fill and set illumination.

Spotlights

Fresnel Spotlights. Available in wattages from the 100-watt inky to the massive 10,000-watt brute, the fresnel spotlight (Figure 16-1) has been the standard hard-light source in the motion picture and television industries. It is distinguished by its ability to go from flood position, with a reasonably soft beam of light, to a very hard, narrow-focused beam of light. With *barn doors* to control the spill of light, and its ability to focus by means of a lens, it most often acts as a key light.

Leko Spotlight. Also called ellipsoidal spot, the Leko (Figure 16-2) provides a precise, sharp-edged beam controlled by shutters, iris, and lens. It offers complete control of the distribution and direction of light. It is best

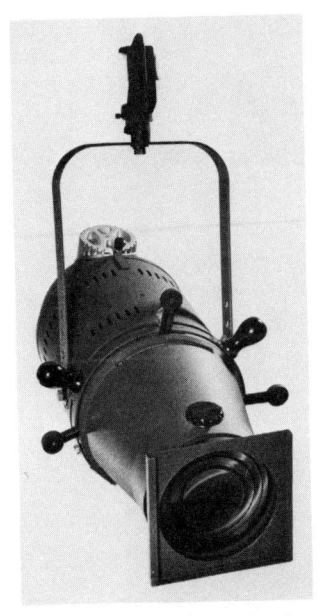

Figure 16-3 Scoop. (*Kliegl Bros.*)

Figure 16-2 Leko light. (*Kliegl Bros.*)

suited for projecting patterns on a scene (and is therefore an indispensable tool in set design and construction) or isolating an object or subject from its surroundings. It is the best instrument to use for a limbo effect because of the high degree of control provided by its internal shutter system that prevents unwanted spill.

Floodlights

Scoops. It looks like a bucket with a bulb inside, but the lowly scoop is the workhorse of the television studio (Figure 16-3). The beam emanating from this instrument is wide, even, and unfocused, which makes it perfect for fill and set lighting.

THE MEANS OF CONTROL

In conjunction with the sources of light are the means of controlling the intensity, shape, shadow, and color temperature of the light. It is with this control that we achieve the full potential of lighting. There are several major devices that facilitate this control.

Barn Door. Mounted on most focusable instruments, the barn door serves to direct and mask the beam of light emanating from that fixture.

16-6 PRODUCTION WITHIN PRIVATE TV

Figure 16-4 Gobo used on set to prevent light from spilling onto camera. (*Matthews Studio Equipment, Inc.*)

Gobo. This is an opaque flag made of metal, cardboard, or fabric that is not mounted on the lamp (Figure 16-4). It offers excellent control over spill light and can sharply define light and dark areas.

Cookaloris (Cookie). This is a cutout pattern that allows light to pass through it in a predetermined pattern. It is excellent for creating visual interest on a background set or cyclorama, i.e., venetian blinds, corporate logos, abstract designs. It can be attached to a fixture or mounted separately. It illustrates perfectly the controlled use of shadow areas within a scene.

Scrim. This wire or fabric is placed in front of a light source to reduce intensity (Figure 16-5). It provides control without a change in color temperature, as opposed to dimmers that vary the voltage available to the lamp.

Diffuser. The diffuser is a translucent glass or plastic material placed in front of a light source to soften and reduce the harshness of that source.

Gel. The gel, a plastic, translucent material, is placed in front of a light source to change color in, or add color to, a scene. It can also be used to

Figure 16-5 Scrim placed in front of fresnel light to reduce intensity. (*Matthews Studio Equipment, Inc.*)

change the color temperature of a light source from 5400 K to 3200 K or vice versa.

Intensity Controls

In most studios, instruments are plugged into dimmers (Figure 16-6). Their major advantage is that they allow the lighting director the ability to lower the intensity of the lamps in gradual stages. The major disadvantage is that once a lamp has had its voltage lowered, and therefore is not operating at its designed output, its reduced color temperature creates a warmer, reddish-yellow texture in the light. The result: The color balance of the scene may look wrong.

PRODUCTION PROCEDURES

Whether the production is on location or in the studio, you, as lighting director, should first meet with your program director to determine the tone, scope, and feel the program will have. This knowledge will help to define the number of instruments you will need to do the job.

If the shot is on location, a survey is absolutely necessary, and it should be made as far in advance of production as possible. You will then be able to ascertain the location and source of power distribution. Work closely with a house electrician on that survey so that you will know where breaker boxes are located and what load capacity they are, and can thus avoid

Figure 16-6 Small memory dimmer control console. (*Strand Century, Inc.*)

overloading on production day. Also, ask that the boxes be left unlocked the day of the shoot or that you be given the key. If possible, have the electrician around for the entire shoot.

The survey will also let you find and plan the best places to secure lighting equipment to ceilings, walls, moldings, etc. If the location is outdoors or includes a window, be sure and have gels and reflectors to help control light.

Whenever possible, create a floor plan that shows set pieces in place. It is then possible to plot a lighting plan as an overlay. Graph paper with four spaces to the inch is easy to work with. This lighting plan will save you a great deal of setup time on the day of the shoot; create a minimum of distraction to other people on the production team; and help assure that lights are standing by when tape and talent are ready to shoot. Use stand-ins of the same height and coloration as your on-camera talent. And remember, leave time to adjust for any last-minute corrections.

Before I go on to more specific production situations, let me remind you of safety. All light cables should be gaffer-taped securely to the floor to avoid being tripped over, as well as to prevent an instrument on a floor stand from being knocked over. Sandbags or water weights should be affixed to all stands for lighting or grip equipment to provide a lower center of gravity. Most important are safety chains secured to anything hung overhead.

While you're at it, do not forget your own personal safety; wear gloves when handling instruments or changing lamps even though it is obvious that they may be hot or may accidentally break. Besides, quartz bulbs should never be touched with bare hands, as the oil on your skin will ruin their efficiency.

Three Contrast-Balancing Techniques

One: Eliminate the distraction of color from your eye. For this I use a panchro glass, available at all professional camera supply houses. It is a monoclelike device that, when held to the eye, effectively renders a black-and-white image of the setting, making it far easier to determine glaring contrast errors that will disturb the video signal. Color information only distracts your eye by imparting a false sense of brightness.

Two: Balance the quantity of light on the subject by using a light meter that gives you a reading in footcandles. This meter will aid you in determining a quantitative measure of illumination that can be very difficult for your eye to distinguish. Most meters read either incident or reflective light. The former has a white domelike receptor surface and is read at the subject's position and pointed toward the camera. It measures the light falling on your subject from three sides. Be careful not to block any light with your body, as your shadow will throw off your reading. The reflective meter is used from the camera position and pointed at the scene. Its receptor surface is flat and receives light from only one direction. It will reveal, as its name suggests, the light reflected by your subject.

Three: The ultimate technical aid to lighting for the television system is the video engineer's waveform monitor. One is shown in Figure 16-7. It will show you electronically the peaks and valleys of the scene you have lit. The contrast levels are graphically displayed from every position that the camera is placed in, from wide angle to close-up—any field that the camera can view. There will be no question as to what is hot or cool, or the relation between foreground and background.

The waveform is available only after engineering has had a chance to set up the cameras, and it can be completed only after you have established the base illumination of the scene. Work closely with your video engineer before the two of you turn the set and cameras over to the director. Of course, the final result must look right on the control room monitor. It is the final test of success for the image that has been lit. (This assumes, of course, that the monitor has been properly adjusted.)

PROBLEM SOLVING

Lighting for television often seems to be composed of 10 percent artistry and 90 percent problem solving. Visual techniques such as chroma keying and rear projection limit the placement and intensity of the lighting instru-

16-10 PRODUCTION WITHIN PRIVATE TV

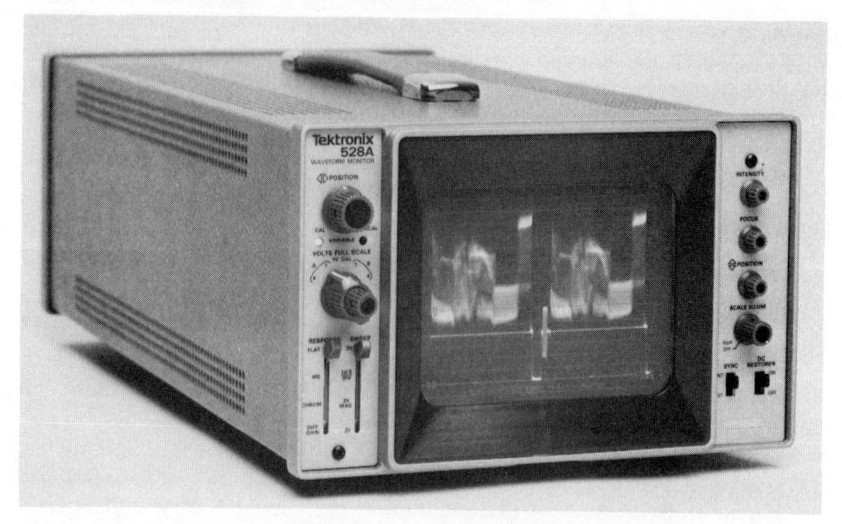

Figure 16-7 Tektronix 528A Waveform Monitor measures brightness of image received by camera. (*Tektronix, Inc.*)

ments. Remote productions sometimes require the mixing of sunlight with artificial light, while nonprofessional talent poses its own problems. Outside the studio, rooms are either too large or too small to allow textbook answers to lighting questions, and movement on the set creates a multitude of difficulties. All these problems are common to all lighting directors, so, over time, generally accepted answers have been developed. New problems require new answers, but the experience of others can at least provide a starting point.

The following are major problem areas and potential solutions that a lighting director can employ.

Chroma Key

This technique requires the control of light so that a foreground subject can be isolated from its background, and a different background inserted electronically.

- Generally, we key against a blue background. Therefore, no blue should be present in the foreground subject, such as a blue shirt.
- The color blue should be as pure as possible and have no elements of green or yellow. They will only make it more difficult to achieve a clean key.
- From a lighting standpoint, we must keep in mind the new background scene and light the foreground subject to fit the appropriate mood, location, and time of day.

LIGHTING 16-11

- Isolate your subject in the foreground pool of light, perhaps even using side "kicker" lights.
- Keep the subject as far away as possible from the blue background to avoid shadows.
- You will want to light the blue background with as flat an overall wash as possible. There must be no highlights on the background at all.
- Be careful of the background's reflecting blue luminance back at your subject. Any blue at all on or in the foreground will deteriorate the keyed image.
- When the floor is in the shot, shadows falling on it will give you difficulty in keying. Keep shadows as soft and transparent and soft-edged as possible. This will make it easier for the electronics to key the background over the blue.

Rear Projection

Similar to chroma key in the need for control of light, we again must relate background to foreground. Of prime importance is balance.

- First, consider the intensity of the projection. It will determine the overall level of illumination. Think in terms of the projection being your key source, the intensity of which should never be exceeded by the foreground subject.
- Second, key, fill, and backlight are essential on the foreground subject in order to give it definition.
- The most important aspect of achieving an effective rear projection is to control all spill light. You will need more than barn doors to do the job right. You will require cards, flags, gobos, more focusable instruments, a higher angle of light, and, most important, the space to separate your foreground from the projection screen.
- If the above equipment is not available to you, try lighting the foreground subject from the side, rather than using conventional front lighting. This will help to isolate your subject in a pool of light and make it easier to control spill on the projection.
- Keep the intensity of these sidelight keys at the same level. The only objection you will run into is a nose shadow that might be too prominent. Experiment with the situation.

Mixing Daylight with Artificial Light

If you are lighting a person in front of a window, the daylight will do two things to your picture. First, it will dominate in terms of being brighter than your key light source. This problem can be easily dealt with by closing the shades or blinds. If, however, you need a daytime setting, you can cut down the intensity of the daylight by adding neutral-density gel to the window, by adding more light in front of the subject to compensate, or by combining both solutions. The problem is the mix of color temperatures. The solution is achieved with filters.

- Use an 85 gel on windows, with a neutral density (ND) factor of 1 to 6 depending on time of day, exposure, weather conditions, obstructions, etc.
- Normal quartz instruments can be used on the foreground subject. Because the 85 filter converts daylight to tungsten color temperature, you can be looking directly outdoors with a very pleasing contrast range, as well as having a far more interesting background than most interior situations offer. Another way is to convert your tungsten source to daylight with a blue dichroic filter.
- Keep in mind that most modern cameras, both studio and portable, have a filter wheel that should be utilized to balance the camera to the predominant color temperature of light within a scene.

Eyeglasses

An often-recurring lighting problem is caused by the talent's glasses.

The problem can manifest itself in two distracting ways. One is the kick and highlight created by light reflecting on the lens or frame of the glasses; the second is the annoying shadow often cast upon the subject's face. Both these conditions are extremely difficult to deal with and require precise control of lighting, talent movement, and camera placement.

- Soft key-lighting instruments help reduce the annoyance of hard shadows and tend to give less harsh kicks.
- Light at a lower angle than usual, thereby keeping the shadow cast by the light close to the position of the eyeglass frame on the face. This is far less annoying than having a shadow cast down onto the cheeks.
- Light closer to the center, that is, closer to the camera angle, in order to help minimize objectionable shadows. Be careful not to reflect the light directly back at the camera.
- Frames that reflect can easily be covered with strips of black masking tape at the points of kick, or you can apply some makeup base.
- Never use dulling spray. It creates a mess, and is not good for the glasses or the person wearing them.
- If your talent is going to be on camera on a regular basis, he or she should be encouraged to invest in a pair of glasses that work well on television. The best are made with nonreflective glass and with thin, matte-finish frames. Anything else will give you a recurrent lighting problem.

A Bald Head

- Try lowering the height of the backlight so that it does not strike the top of the head, but instead, rims the shoulders with light. Thus, you will separate foreground and background without drawing the eye away from the most important part of the scene.
- The bald head has a sharp definition and will separate naturally from the background. You may find it better to soften the instruments with scrim or to flood them to their widest position.
- Work closely with your makeup artist; he or she can eliminate any shine that the head may have, thereby making your job a bit easier.
- When all else fails, do not use a backlight. The results will be far more pleasing than a hot spot.

Low Ceiling or Small Space

With most interior remote lighting conditions, we encounter cramped quarters.

Ceiling heights of 8 feet are common. Combined with a confined space, we have no place to throw shadows so they will not be objectionable. We must use extreme control, not only in the placement of light, but in the choice of background.

- We will most likely be dealing with one or two people in conversation, standing or sitting. Wherever possible, they should be far removed from walls or furniture that will reveal shadows.
- If a subject is seated, it is most important to have a camera position low enough so you are looking slightly up at him or her. Most tripods or pedestals force you to look down on your subject in confined spaces, thereby emphasizing shadows that fall below and behind the figure.
- In these natural location settings, it is advisable to place your camera as far from your subject as possible. Thus, you will be able to utilize the depth of field of your lens to narrow your focus so that distracting details of a background will be in soft focus, and thereby less distracting to the business at hand.
- Probably the most difficult problem of a low ceiling is the placement of the backlight. Key and fill lights can generally be placed on stands; backlights need to be either hung from the wall or suspended from a ceiling.

Large Rooms

Lighting vast spaces is a problem that can be dealt with in many ways.

- One method is to flood the area to be lit with large amounts of light. This solution requires a large budget and enough electricity and instruments to do the job.

- More practical for most private television situations is to be selective by using pools of lights. Their illumination will give you a sense of the perspective and depth of the space at hand.
- Your foreground subject should be lit according to the sound principles of lighting that have been described throughout this chapter.
- It helps to fill out the setting with small pools of light that highlight key points in the view of the camera. This key point may be a desk, machine, plant, or wall.
- Try to avoid a sharp fall-off into darkness, as well as flat lighting that offers no sense of depth or perspective within a scene.

Hiding from the camera's eye the instruments that create these pools of light is the challenge you must deal with. The wide shot leaves no room for concealing cables and lights. Your skills will be hard pressed to mask and dress cable and light (gaffer tape in all colors can come to the rescue).

Lighting for Movement

The problem of a subject's moving through a scene requires that the subject be evenly illuminated throughout that movement.

- When the movement is parallel to the camera, you will get your best results by flooding your instruments and pooling the cumulative light in the overlap areas so they will strike an even balance across the set.
- When the movement is diagonal or perpendicular to the camera, you have to deal with the increase of intensity of the light as the subject approaches the source of illumination. To diminish this problem, reduce the light so that it trails off by using half scrims. This method enables you to use one instrument to carry a move without the necessity of adding more instruments.

With practice and a broader understanding of the use of these scrims, a single instrument utilizing double, half, and quarter scrims can handle your subject through the entire course of movement toward the camera. The effect of that movement as it relates to lighting must never call attention to itself. Only good pictures stand up in the end.

Outdoors

Lighting outdoors, for most situations, requires that you either harness your key source, the sun, or use it to provide your fill.

- In most cases, you will fill in where the sun doesn't shine by utilizing reflectors to fill in the harsh, directional light source that daylight is.
- Reflectors are available commercially that can provide you with hard or soft fill. Or, you can devise your own with aluminum foil and cardboard.

In either case, the results will be far more pleasing if you try to manipulate the sunlight than if you just let it go as it is.

- The other choice is to diminish the harshness of the sun by building a tent with gel over your subject that will even out the contrast range. This solution is especially advantageous when you are shooting over the course of the day and have to even out the effects of passing time.
- Nature can cooperate by providing you with an overcast day that will let you achieve exactly the same effect.
- If you can run an extension cord from a nearby source of electricity, you can place a lamp near the subject for fill.

The unpredictability of shooting outdoors requires that you be prepared for both extremes. When worse comes to worst, you can always use the tent that you have pitched to protect you from the elements.

Additional information about lighting, particularly as it pertains to location shots, will be found in Chapter 21.

In conclusion, I would like to leave you with one of the most helpful hints in my craft. It is the idea of starting a notebook of lighting setups.

I have used one for years, for both successes and failures. In the book I keep a list of the equipment used on the shoot, the effects achieved, and diagrams that would make the setups easy to transpose to another situation. I suggest that you do the same. Remember to keep track of your lighting failures; you will learn as much from an unsuccessful design as from one that works. From these notes will come your own style and formula for successful lighting.

17 Makeup

LESLIE SHREVE

AN OVERVIEW FOR THE PRODUCER

There are two aspects to makeup for television—the obvious, and the not-so-obvious. In private television, the not-so-obvious often takes on a greater importance.

The *obvious* is to make people look as attractive as they can, and to keep them looking their best throughout the shoot. In the strong lights needed for television, one tends to look washed out, pale, flat, shiny, or all four at once. Makeup can make your talent look more natural—more like themselves—than they would look with no makeup at all.

With the use of makeup you can also hide blemishes, control perspiration, minimize "five o'clock shadows," tone up pallid complexions, and bring those with deep tans into even balance with more pallid people on the program.

Most private television makeup application is corrective in nature. It can minimize poor features and emphasize good ones.

The *not-so-obvious*, especially in the case of nonprofessional talent (corporate executives, teachers, doctors, specialists), is the psychological boost. With makeup you help your client to look and feel as professional as possible.

- Makeup adds to the feeling that this is an important event—that the producers are taking a real interest in the executive's image.
- The makeup procedure (when done correctly) helps the talent relax in the intense moments before the shoot begins. It sets them up; gets them ready to do their best.
- Perhaps most important, it eases them into the television environment (which, as we all know, can be quite terrifying—even once one becomes accustomed to it). Getting the shoot off to a smooth start helps the entire production team greatly.
- During the shoot, refreshing the makeup can provide the talent with "breathing room" if necessary—a private moment to get revved up again.
- Makeup also can be an aid to the producer if the flow of production is not running smoothly. The makeup procedure can be stretched to insulate your client while the technicians tweak out the bugs in the equipment, or the director goes into extra takes with the talent already in the studio. Conversely, makeup can be applied quickly—to show the executive that the producer values his or her time.

OVERVIEW FOR THE MAKEUP ARTIST

Let me speak first to those of you who already do makeup work. The world of private television places some major strains on the art of applying makeup.

To begin with, the crew will often be smaller than in commercial television. You may find that the set is someone's office (you do the best you can with the colors and textures found there); the executive, doctor, or teacher will be wearing his or her own clothes (which have not been camera-tested); and you may be expected to double as hair stylist, floor manager, or both, along with your responsibilities as *the* makeup artist.

Efficiency and speed are often essential in these on-location private television shoots. Many times, the interview will be squeezed in between a breakfast meeting and a flight to Hong Kong.

Most important, for many of your industrial clients, this will be one of their first exposures to having makeup applied. Some will be actually frightened by this intimate art. They are not used to being touched. Often, also, they will be coming to your makeup table full of confusion and fear caused by their impending "performance" in front of that strange and awesome device—the camera.

THE PSYCHOLOGY OF APPLYING MAKEUP

Before you even pick up a sponge, you must begin to build these people's confidence both in themselves and in you as a professional part of the production team. I try to make my clients feel that I am there *only* for them. Once they feel sure that someone is totally devoted to how *they* feel, look, and "come across," they can be eased into the television environment. This emotional support is just as much a part of your job as your makeup skill.

Following along with this, let me stress the importance of an aspect of makeup that becomes second nature to those of us who work in this area a lot: the *touch*. Every time you touch your client, you must do so with that firm but gentle, confident manner which in turn adds to his or her confidence in you. A tentative touch, especially at the beginning of the makeup session, can destroy any trust established between you and your client.

After seating my client in the chair and placing towels or a makeup cape around the neck and shoulders (I prefer towels or those giant-sized Kleenex tissues to capes, which can appear too beauty-parlorish) I engage the talent in conversation: "Who are you going to be talking to on this tape today?" "Have you ever had television makeup before?" "What a terrific tan! Are you just back from vacation or a business trip?" As we are chatting, I'm looking over the individual very deliberately.

While continuing to look directly into the eyes, I put one hand under the chin, and with the other hand touch the face (usually the cheek first, then the nose and forehead). Here I am accomplishing three tasks:

1. Establishing that our work is now beginning
2. Checking for skin quality to see what kind of preparation will be needed before the makeup is applied
3. Getting the client used to my touching him or her (a surprise attack, so to speak)

Usually some comment follows this, such as "You have excellent skin!" "Your beard is so heavy; I'll bet you have to shave a couple of times a day," or "Do you have any allergies?" If you have nothing to say, a knowing smile (indicating "this is going to be easy and fun to do") works almost equally well. Your calm assuredness will already be having a positive effect on your client.

Unlike production days with professional performers where you can slide off on a break as they deliver their lines, your usefulness to the production team and your client remains strong as the cameras begin to roll. If you have created a positive working relationship with your client in the makeup area, it is vital that you continue it on the set. During the shooting day, you are the best person to smooth down a suit jacket, pull down a vest, or even pull up sagging socks. If you do these functions with the same attitude of "this is all a part of making you look your best," your talent will feel catered to rather than foolish.

It is usually a good idea to speak softly with your talent as you go to rearrange the clothing. For example, "Come forward to me—I want to smooth your jacket a little." However, if he or she is engaged in conversation with other on-camera participants, do your work quickly and quietly. If this is your first rearranging trip onto the set, I suggest that you smooth out everyone to some extent. Thus you will prevent one persons' feeling singled out as the most rumpled. You will also keep any other participant from feeling uncared for. Remember, nerves are heightened at the outset of take 1.

During the course of the shoot, you should remain aware of your clients' needs. More important, the clients should know that you are still "with" them. They can therefore think about what is being said rather than how they look. Thus, they are free to do their best. Stay on the floor or in the control room. Be attentive to your clients during breaks.

Finally, we come to makeup removal. As you know, in commercial television productions this is often a function of the actor. That is perfectly acceptable under those circumstances. In private television, makeup removal is very much a part of the makeup artist's job. It is a necessary element of the service you are supplying to your client.

Approximately one in three first-time on-camera talents in this business ask, "Will you be here to get this stuff off me?" It is a very real concern to them at the beginning of the production day. The assurance that you will be with them from the beginning to the end is extremely important at this time.

GETTING INTO MAKEUP

For those of you entering the field of television makeup for the first time—welcome! You are preparing for very interesting and satisfying work. However, there will be no formulas to be followed exactly to do the job. There are so many variables to be considered that you will be facing a brand new challenge every time!

Makeup depends first of all on the individual to be done, and everyone, of course, is different. Here is where you begin to assess the situation. You must make your decisions quickly, basing them on the following conditions.

1. *Facial type.* Is it basically oval, round, square, oblong, heart-shaped, triangular, inverted triangular, or perhaps diamond-shaped? Your makeup should help to correct the facial balance.
2. *Skin type.* Is it oily, average, or dry, or, as is usually the case, some combination of these? What is its texture? Smooth, pock-marked, sunburned? The type will affect the kind of makeup you will select to use.
3. *Outstanding feature.* Does the person have beautiful eyes, cleft chin, or a strong jaw line—something you will want to emphasize? Look for the most attractive feature in each face and make the most of it.
4. *Problem areas.* These you will want to minimize: heavy perspiration, bald head, five o'clock shadow, light eyebrows, bulbous nose. Makeup for television can truly appear to correct these features. But first you must be aware that the problem exists.
5. *The personality of the individual.* For some more timid types, you will want to do a makeup procedure that is as quick as possible. (You are there to cater to, not upset, your client.) Strangely enough, the more beautiful the face, the longer the makeup application takes. Models, for example, feel jilted if they think not enough attention has been paid to them.

The show as a whole also shapes your makeup decision. The six production variables which must affect your makeup style are:

- The clothes to be worn
- The background of the shot
- The lighting techniques to be used
- Others who are appearing in the segment (everyone must be balanced)
- The approach the director is taking
- And, of course, what the producer has asked for

Before discussing problem areas and their solutions, let us quickly touch on the types of makeup that exist.

The Base Coat. There are various types of makeup bases available to you:

- Liquids—either water base or oil base. They can be streaky to apply. They give the skin a glow.
- Whipped—slightly heavier than liquids, they leave the skin slightly moist. Good for very dry skin.
- Creme sticks (also known as pan sticks or velvet sticks)—heavy, creamy coverage.
- Creme cakes—heavier, creamier coverage.

All the above are usually applied with a dry sponge, except in the case of a very bad complexion, when you might prefer to use a "silk" sponge to get heavier coverage.

- Cake makeup—mixes with water to give a heavier matte finish. Must always be applied with a natural silk sponge, which you thoroughly wet, then squeeze out. It should be moist, but not dripping, for best results.

All things being equal, I prefer to use the pancake makeup (Max Factor's colors and texture suit me) because I find it easy to blend colors, and it covers quickly. You should, however, experiment with all types of foundations to see which you work with best and to help you be prepared to choose a correct medium for each client's overall optimum look.

In choosing base colors, you will do best to try to find those as close as possible to your client's own skin tone. However, no one color will ever match a living flesh tone. Therefore, it will be necessary to blend two or three colors.

Blend the colors right on the sponge. Then apply, adjusting as you work on various areas of the face. The use of these slight variants also enables you to begin your contour as you apply your foundations. This same procedure holds true for every skin—white, olive, tan, yellow, brown, pink, red, and black—male and female. Skin, is skin, is skin! Cover the entire skin area that shows. Don't forget the neck, the ears, and if necessary, the hands.

Powder. The entire face should be powered at the end of the makeup procedure to set the makeup and give it a soft matte finish. Use a translucent powder that will not change the color of the makeup you worked so hard to create.

There are many brands available. The no-color powder from the Research Concil of Make-Up Artists is an especially good one. If you are in a pinch, baby powder (which appears white but can cover rather translucently) will do the trick. Powder the face *before* applying dry rouge. Powder the face *after* applying *creme* rouge. Don't powder heavily under the eyes. Doing so would soon point up the laugh lines or crow's-feet when combined with the makeup in the eye area.

The procedure for application of powder is once again a matter of personal preference. You may use a variety of brushes or a powder puff. For television, I prefer the puff because it affords greater control. It is likely that you will be repowdering on the set during the production day, especially with a client who perspires heavily. And you will be less likely to leave a dusting of powder on the clothes or set if you use a puff.

Using a Powder Puff. Work the powder into the puff by rubbing its surfaces together briskly. Apply evenly with a rolling, patting motion. *Never rub.* If too much powder is applied, use an unpowdered section of the puff and the same firm-pressured, rolling, patting motion to even it out.

Using a Brush. Hold the brush like a pencil. Dip the end into the container of powder or into the palm of your hand in which you've poured some powder. Shake the excess powder loose with a quick flick of your wrist. Briskly work the powder onto the skin with the tips of the bristles. This will often cause the skin to tingle.

THE TECHNIQUES OF APPLYING MAKEUP

The following areas are ones you will most likely encounter when applying makeup.

Faces

Earlier, I mentioned that there are a variety of different facial types. People's faces are their own, and for an industrial videotape, you should not plan to do corrective surgery. However, let me suggest points to keep in mind when you encounter certain facial structures.

The Oval Face. This is considered a standard face and one which often has a natural balance of features (Figure 17-1). In general, the eye is considered to find this symmetrical look most attractive. Therefore, not too much work will be needed when you encounter a face of this shape.

Though other face shapes may be much more interesting, the makeup artist will often want to soften or even alter their optical effect by making them look a bit more oval. These changes are created with the use of light and dark shadings (*highlighting* and *contouring*). Just remember that a darker color absorbs the light, making that area seem to recede. Conversely, a light color, and also a bright one, will reflect the light, making that area seem more prominent.

Highlighting and contouring *must* be done subtly enough to look natural. Blending is very important. You are creating an illusion with contouring tones. Do this carefully.

The Round Face. This shape calls for contour (shading) on the outer sides of the jaws—cheeks, temples, and forehead (Figure 17-2). Blend in further to "hollow out" under the cheekbones. Then, highlight the cheekbones and perhaps blend in a thin amount of the highlighter on the very

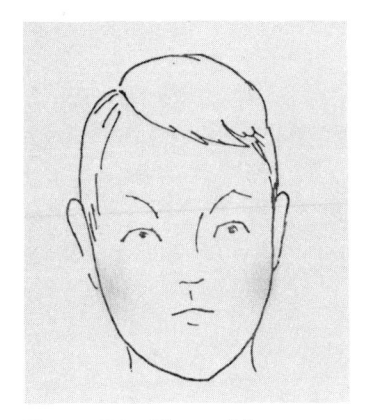

Figure 17-1 The oval face. 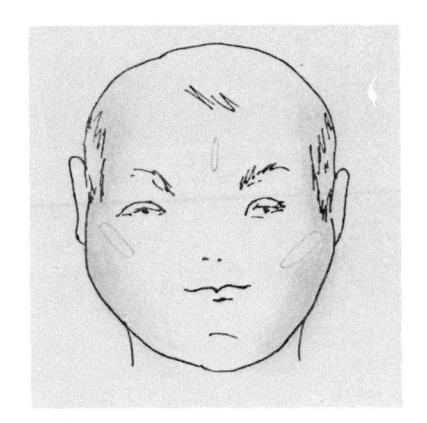   Figure 17-2 The round face.

center of the forehead, adding to the angular look created in this round "canvas."

The Square Face. Such a face can sometimes add to an "executive power" look (Figure 17-3). For certain projects, you may even want to accentuate it. However, it can often also photograph flat, and you will usually choose to contour it to a more rounded appearance. Soften the angular jaws and forehead by contouring. Highlight the cheekbones to diminish the flat look. You may decide to add a small smudge of highlight on the chin to further round out the squareness.

The Oblong Face. This face photographs best when the cheeks are highlighted sufficiently to break up the long look from temple to chin (Figure 17-4). Be sure to add contour under the cheek highlighter to add to this effect. Contour at the temples and hairline can also help this face appear shorter. You may find it necessary to soften the chin area too.

The Heart-Shaped Face. Here, better balance can be achieved by applying contour to the lower chin area, the temples, and the roundness of the forehead (Figure 17-5). Extend the cheekbone highlighter slightly into the hollows. You may also apply a highlighter along the jawline, in a thin line in the center of the forehead, or in both places.

The Triangular Face. This type is characterized by wide jawline and narrow forehead (Figure 17-6). Highlight the cheeks and the side of the forehead. Contour under the cheeks and pay special attention to contouring the jawline. You may need to add a horizontal line of highlight across the forehead in some extreme cases.

The Inverted Triangular Face. This face should be contoured from the sides of the forehead down through the temples (Figure 17-7). Also contour the point of the chin to soften it a bit. Apply the cheek highlight a bit lower than usual to bring emphasis to this portion of the face. You may

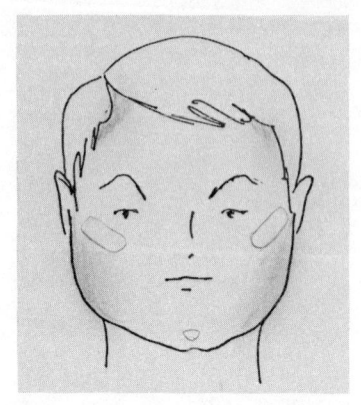

Figure 17-3 The square face. 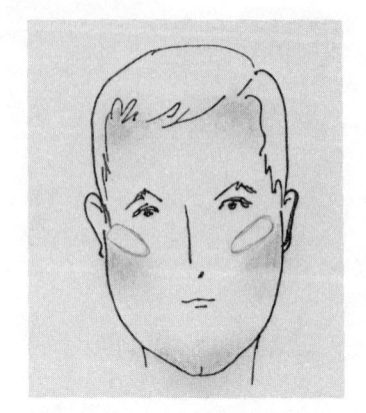   **Figure 17-4** The oblong face.

need to add a thin line of highlight to the jawline above and to the sides of the contour.

The Diamond-Shaped Face. Such faces need contouring at all four "corner" areas: the pointed chin, the top hairline, and along both cheek lines (Figure 17-8). Add highlighter all along the temple hairline—with a little extra at the appropriate places to create the look of wider headbones. Your cheek highlight should be somewhat lower than usual—with the widest point being toward the center of the face. You may also put a thin line of highlight above the contour along the jawline.

With all these tricks of the trade, don't forget to blend, *blend,* and BLEND!

Handling Hands

If the tone of the client's hands closely matches the face, often you are better off to leave them without makeup. This is especially true if your client is wearing a dark color, as the makeup may rub off on the clothing as the hands fiddle with it or are placed in and out of the pockets. A discussion show, for example, is one where hand makeup may not be called for.

However, if the program talent will be using the hands to point in close-up shots, or has the tendency to rest the chin in the hands, do apply makeup to them. Make sure the makeup has thoroughly dried when using a pancake base, or has been well powdered when using a creme base. Caution your clients not to rub their hands on their clothing, but don't make too big a deal about it. You can always brush the makeup off the clothing with a lint brush, if necessary.

Under-Eye Area

This area can pose specific challenges to you, First, you may notice discoloration. Almost everyone has circles under the eyes to some extent. The circle

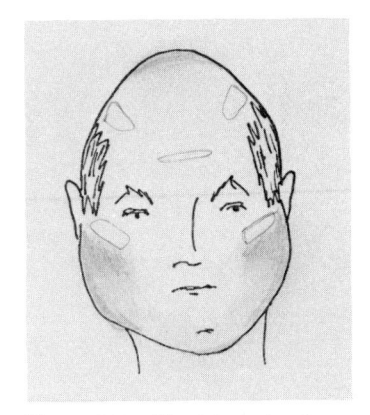

Figure 17-5 The heart-shaped face.   **Figure 17-6** The triangular face.

area is usually several shades darker than the surrounding area, appearing grayish or blueish. This, by the way, has very little to do with how little sleep one had the night before. It is a function of the metabolism, one's heredity, or one's constitution. At any rate, if the circles are there, you will have to deal with them.

Many makeup artists camouflage this area by highlighting (or lighting) it. There are several products on the market specifically designed to conceal this one area alone. You have your choice of powders, pastes, liquids, gels, creams, or cream sticks. (I like the consistency of Filis Forman's creme concealer.)

You may choose to use your concealer over or under the base coat (or a combination of both). Using the concealer under the base gives you greater freedom, as the base will minimize any shading difference. If you need to add concealer over the base, be sure the difference in shading remains minimal. Use the concealer only on the discolored area. Do not let it bleed onto the surrounding areas.

Be extremely careful, when using this procedure, to keep your highlighting looking natural. It can sometimes turn grayish. Remember, on television, lighter colors seem to "travel faster" than darker ones. The last effect you want to produce in your client is that of having owl eyes.

Another camouflaging trick for under the eyes (which I personally prefer) can be used alone or in addition to the highlighting method. Mix a small amount of cream rouge into your foundation. The addition of a warmer color over the discolored area can do wonders. Here, of course, you must be extremely careful with your colors and proportion of rouge to base. Experiment with as many methods and combinations as you can create. Practice until you can easily apply enough concealer thickness to create the effect without going too far into making your client look made up.

In fact, the moment your client begins to *look* made up, you should for all intents and purposes start all over again. Either you've chosen colors too high in intensity, or your application is too heavy. As in many artistic

17-10 PRODUCTION WITHIN PRIVATE TV

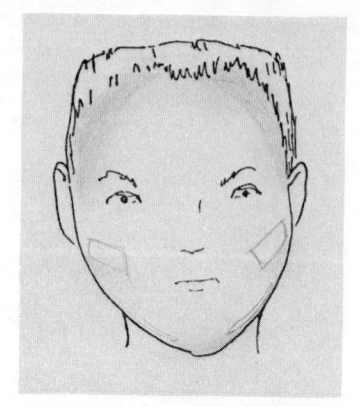

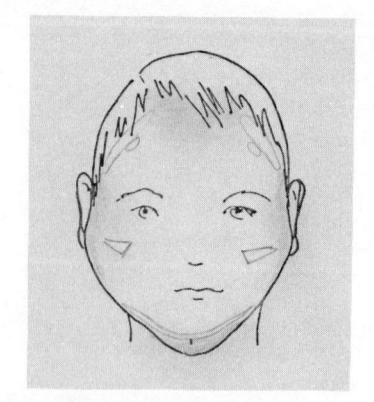

Figure 17-7 The inverted triangular face.   **Figure 17-8** The diamond-shaped face.

endeavors, knowing where to stop is of utmost importance. Remember, less is better—especially in the world of private television.

For clients with under-eye bags, the only real solution is plastic surgery! Since this process is impractical on the set, you should appeal to the artistry of the lighting designer. If soft lights are set below eye level (to hit under the bags), the bags will appear to flatten out.

Natural Discoloration

Occasionally you will encounter a person with a large but natural blemish or discoloration on his or her face or hands. This is sometimes raised, but often is just a concentration of pigmentation in a specific area.

The makeup approach I prefer is to even out and tone down the colors, but not to cover the area so thoroughly that it is hidden. To block out the discolored area completely would be too much of a change. (You wouldn't cover a bald head with a toupée, would you?) This is especially true when the client is an in-house individual whom everyone in the firm knows. Don't ignore this sensitive spot; just make it look as attractive as you are able to.

The same principle holds for extremely heavy beards—five o'clock shadows. Don't create a babyface look. Let the TV picture offer a flattering but realistic impression of the individual.

Receding Hairline

Such a hairline can be accentuated by the bright lights necessary for television pictures. To counteract this combination, you may decide to pencil in the hairs. Do not go beyond the natural hairline. Use a very sharp pencil (stop to resharpen it if necessary). Try to match the actual shade of hair—or use a combination of colors to achieve this effect. Go over the area gently with a brow brush when finished.

Another method is to use a slightly darker base tone as you apply the makeup higher on the face. This shading lessens the impact of the lights and reduces some of the shine. The darker tone, of course, must be blended in with the skin tone applied to the face. There should be an even gradation all the way up.

Eyebrows

Eyebrows may also be too pale to stand out under studio lights, or the hair growth may be so sparse in some places that the shape is not well defined. In these cases, you will definitely want to fill in the brows. As you know, the eyebrows do a lot to convey moods and expressions. They are an important and often neglected facial accent.

Eyebrows may be filled in with the use of a brush-on powder shadow (color-coordinated to the person's own brows). Using a small wedge-shaped brush, apply the powder to the brow in gentle, tiny hairline strokes to create the shape desired. Go over the area gently with a brow brush (stiffer bristles) or a child's toothbrush to soften the effect.

If you prefer to use a pencil (my usual first choice), draw in each little hair as it would naturally grow. This method requires a little more artistic skill, but the result appears more real if you can emulate the surrounding hairs in color and texture. You may desire to use a combination of both brush-on and pencil-on techniques.

The brow should optimally begin directly above the inside corner of the eye. If you pluck the brow in further, the eyes will begin to look closer together.

The outer tail of the brow should end in a direct line with an imaginary straight line (Figure 17-9) from the corner of the nostril on past the outer corner of the eye and straight up. You should pencil (or brush) in the tail ends of the brows so they will reach this imaginary line.

Eyes

Except in extreme cases where some correction is necessary, I believe that less is best in eye makeup for the participants in private television. You want your clients to look as natural and un-made up as possible.

If your client is extremely fair, you may need to add eyeliner, especially on the lower lid line (Figure 17-10). This should be as inconspicuous as you can apply it. On most faces, you'll want to smudge the lower lid liner delicately with a Q-Tip to soften the effect. On other faces, you'll need to leave it more defined so the eye will show up properly.

Never pinch the ends of the liner together. Circling the eye makes it look smaller. Leaving about ¼ inch between the line on the upper lid and that on the lower lid makes the eye appear larger.

If your need to make the eyes look larger is great, you may also add highlighting (a lighter color) in the ¼-inch area between the two liner lines.

17-12 PRODUCTION WITHIN PRIVATE TV

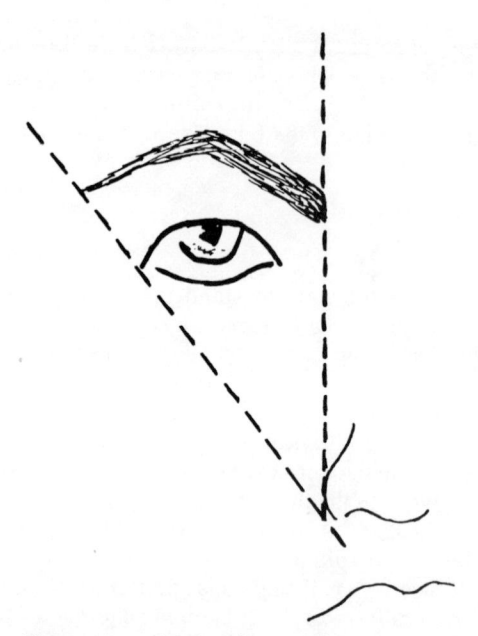

Figure 17-9 Ideal eyebrow alignment.

If you decide to use mascara, *never* use a so-called long-lash type. The little fibers added to the mixture may get in the eyes or cause other irritations. Mascara may be applied to both upper and lower lashes if desired.

For your female clients, in addition to the procedures mentioned above, you may add more eye makeup (according to the person's personality and the look desired for the show).

Highlighting and shading (contouring) in a greater range of colors can be fun and effective. Since it is not unreasonable to expect that a woman wears some makeup on a day-to-day basis, you can often add more makeup and still be true to her character. False lashes may even be used, but only if her natural lashes are sparse. Never let the false lashes be longer than her own lashes; if they are, they will create shadows on her face which will drive your lighting designer crazy!

Lips

Lips need special attention too, when preparing your subject for the camera. Often a nervous performer feels compelled to lick the lips to combat that tense feeling of dryness in the throat and mouth.

If you get a lip-licker, you can:

- Give him or her a soother—a throat lozenge or a sucking candy.
- Do *not* give a lot of water! In this type of nervous reaction, water's effects

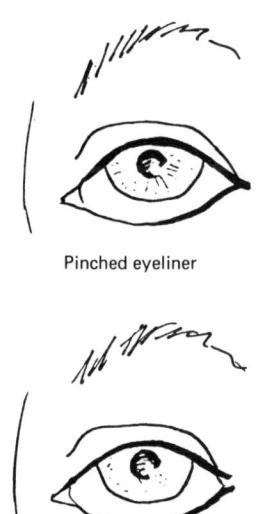

Figure 17-10 Eyeliner for eyes, pinched and open.

are not long-lasting enough. You would not treat your client's real need, and you would take the chance of causing him or her to sweat profusely. This would cause feelings of even more nervousness.
- You can apply Blistex or lip gloss; a light, blotted coat of Mehron clear gloss works well. The lip gloss eliminates the dry feeling and coats the lips, causing a strange feeling when the nervous tongue wants to slide over them (and providing a subtle reminder not to lick).
- Or you can combine the soother and the lip covering. The throat disc promotes salivation.

The look of the lips with some moisturizer on them is pleasing. However, you must remember not to apply the gloss too heavily or it will cause extreme highlights under the studio lights, and the lips' color and shape will disappear in the glare.

Stay away from any lipstick with blue in it. It looks too dark and unnatural on television. The same caution is in order for deep red.

Soft pinks are always a good bet for camera work. You may want to mix the lipstick with the gloss for easier application and extra moisturizing without overwhelming shine.

Finishing the Look

To create a very natural look, it is often nice to "break up" the face with a touch of powder rouge. Apply it lightly, following the bone structure:

the forehead, bridge, and tip of the nose, chin, cheeks, and perhaps the temples. This will give the face its roundness again. It will make it look even less made up! Surprisingly enough, it won't change the color tones that the camera will read.

THE CAMERA AS THE FINAL TEST

After you are satisfied with the makeup designed in the makeup room, try to take your client onto the lit set for a makeup check. Work closely with your video engineer for a few moments to make sure that what you want to see is what he or she is getting. In the case of a film shoot, ask to have a moment to see your client through the camera's eyepiece.

Often you will decide to make some minor changes before the cameras roll (usually something like adding more color in the lips or cheeks, or repowdering the top of a bald head that's reflecting lights). The picture on the tape, film, or television screen is what counts. Don't judge the results by your naked eye alone. Continue to check the control room monitor from time to time during the shoot to be sure that your client's "image" remains constant.

HOW TO REMOVE MAKEUP

There are numerous products available for makeup removal: water-soluble and non-water-soluble creams (which are often perfumed), lotions, gels, oils, astringents such as Sea Breeze, skin toners, fresheners, and soaps. Some performers even use Crisco or olive oil to do the job! After experimenting on various skin types, you will be able to decide for yourself which products and cleansing regimes work best for you and your clients.

On most healthy skin, I prefer to use Albolene Unscented Creme. This product melts down quickly when in contact with body heat and takes off all the makeup in one application. It is kind to the skin, counteracts dryness, and doesn't leave your client smelling of a perfume counter or hospital ward.

You simply apply it to the skin (don't forget the ears, hands, or any other made-up surface); wait until you see the makeup begin to "float"; and then remove it all with a tissue or towel. I then like to hand a tissue or towel to the client to allow him or her to go over his or her face once again to make sure all the remover is removed. The person can feel if there is any residue. This gesture also puts clients back in the position of being in charge of their own faces once again.

If the skin is oily, application of freshener or toner (both with relatively low alcohol content) may then be applied.

A WORD ON CLEANLINESS

It is part of your job to keep your towels, brushes, combs, and hands scrupulously clean. You do not want to take even the smallest chance of passing on a skin irritation to your client. Ideally, each participant should be assigned his or her own sponge, puff, and comb to be used during the production day. At the end of the day, all brushes (those for makeup and for hair as well), combs, and natural silk sponges should be thoroughly cleaned. The latex sponges and powder puffs should be thrown away. Your peace of mind is well worth this extra expense and effort.

PLAN OF ACTION

Entire books have been written on the art of makeup design. If you are new to this field, you should read and reread as many as you can obtain. However, understanding theory alone is not sufficient to prepare you for the work situation. You must practice until you've become proficient enough to handle not only the artistic challenges but the snap decisions and the pressured deadlines as well.

I suggest you begin by working on your own face. Learn which color combinations look the best on you. Feel and see the effect of different types of makeup on your own skin. Be attentive to how the application feels, keeping in mind that you'll want the procedure to be as pleasant for your client as possible. This will involve many hours of research. It is time well spent if you are serious about working in this area of the industry.

Next, set yourself a time limit and race against it with various styles and types of makeup. Ideally, the corrective straight makeup procedure should take about 8 minutes to complete. Unless you are applying hair or prosthetic pieces (highly unlikely in the vast majority of private television makeup jobs), there is no reason the application should ever run longer than 15 minutes. Work at this step until you easily have the speed and the desired results. Develop your timing and proficiency in several makeup media.

Finally, coerce as many friends as possible to be your guinea pigs. Family, friends, lovers—anyone you'll feel comfortable learning on will do. With luck, you'll find many various facial and skin types in this group. Experiment initially to find what makes each individual look best. Then, if your subject will continue to sit for you, work on your speed. If not, continue to use your new knowledge on new "victims." After a while, you'll begin to notice that your snap decisions bring about superior results with greater frequency.

18 Audio in Television

MARK DICHTER

Unfortunately, the element of sound is seldom, if ever, given sufficient attention and creative freedom in video. The dimensions of the audio portion and its quality are usually determined on the basis of delivering a good, clean, understandable narrative track—in itself a feat. Yet it rules out the more substantive and evocative uses of sound. Audio is more than the vehicle to advance the narrative: Sound directs the imagination, shapes the mood, and sets the scene.

Before discussing how to use audio effectively, let us review the basic equipment.

DESCRIPTION OF EQUIPMENT AND TERMINOLOGY

In order to facilitate understanding of audio as presented in this chapter, let's consider a few terms.

Lavalier Microphone. A very small omnidimensional microphone designed to be worn on the chest about 6 inches from the mouth. It gives high intelligibility and low background noise (Figure 18-1).

Cardioid Microphone. A moderately directional microphone used on a boom or desk stand. It must be aimed at the speaker. It rejects unwanted sounds from the rear and the sides (Figure 18-2).

Hypercardioid Microphone. More directional than the cardioid, the hypercardioid must be mounted on a boom (Figure 18-3). The accuracy in using this microphone is more critical because it is more directional than the cardioid. Therefore, if used improperly, an off-mike condition may exist (i.e., the subject sounds further away because of being off the main axis of the microphone).

Shotgun Microphone. Most directional of all microphones, the shotgun is usually hand-held in documentary situations and boomed in studio situations (Figure 18-4). Aiming the shotgun with extreme accuracy is even more critical than aiming the hypercardioid because the off-mike condition will be even more acute.

Boom. With this device a microphone can be hung and suspended over the action on a set. It can be a simple fixed pole or a very elaborate rig that allows for articulation of the microphone across an entire set.

18-2 PRODUCTION WITHIN PRIVATE TV

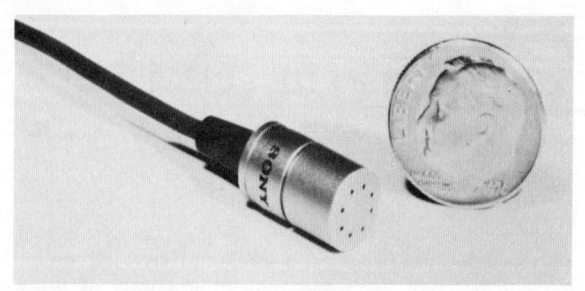

Figure 18-1 The lavalier microphone measured against a dime. (*Sony Corporation of America.*)

Radio Microphone. The so-called radio microphone would be more appropriately called a radio link. The microphone used with it is a lavalier (Figure 18-5). The radio is used instead of the mike cable and allows total mobility of the subject.

Sound Blanket. This is an absorbent blanket used on location to reduce sound reflections off hard walls.

Equalizers. These devices reshape the sound spectrum to compensate for deficiencies in microphones or to correct balance problems.

Filters. These devices selectively eliminate a portion of the frequency spectrum (i.e., low-frequency traffic rumble or high-frequency buzz of a fluorescent lamp).

Mixer. This device combines multiple microphones or multiple recordings to be blended into one output (Figure 18-6).

Limiter Amplifier. With this device, audio peaks are held to a predetermined level so as not to overload tape. It also raises the average level of the whole program.

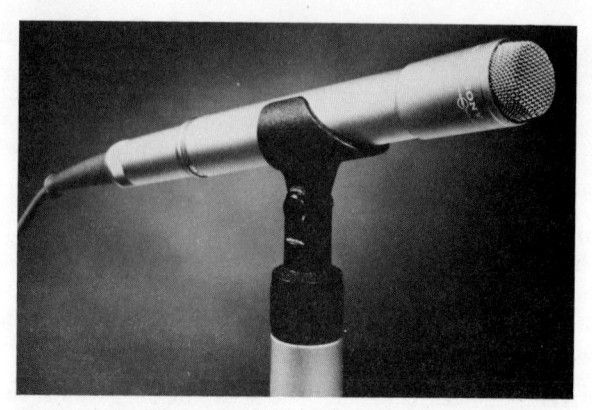

Figure 18-2 The cardioid microphone. (*Sony Corporation of America.*)

AUDIO IN TELEVISION   18-3

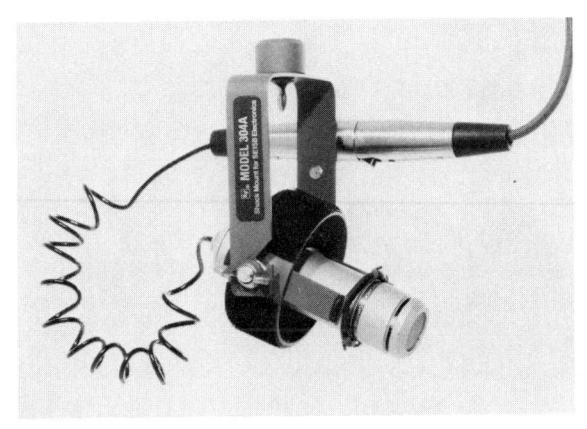

Figure 18-3 Hypercardioid microphone with a shock mount. (*Electro-Voice, Inc.*)

Volume Unit Meter. This is a metering device on a tape machine or console that indicates audio peaks, making it easy to adjust proper recording level. The top scale is measured in decibels; the bottom is in percentage modulation. Good recording practice recommends that the level not exceed the 0 decibel.

Dynamic Range. Difference between the highest level audio and the lowest level. The possible dynamic range of a videotape is considerably

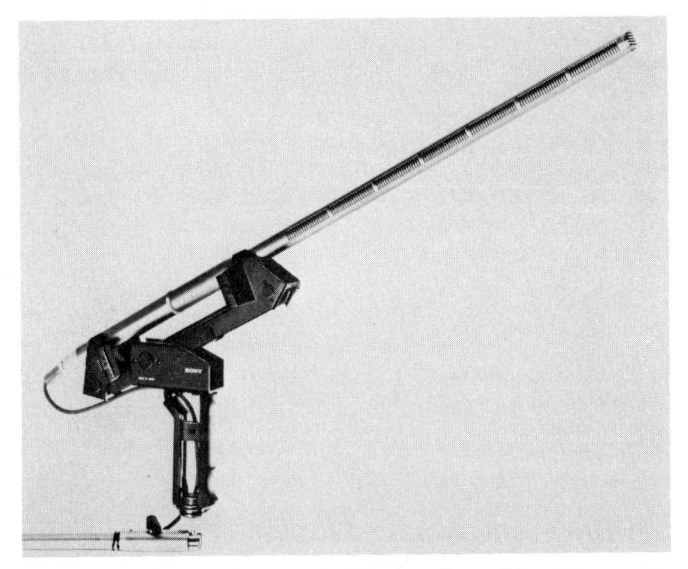

Figure 18-4 Shotgun (hypercardioid) microphone. (*Sony Corporation of America.*)

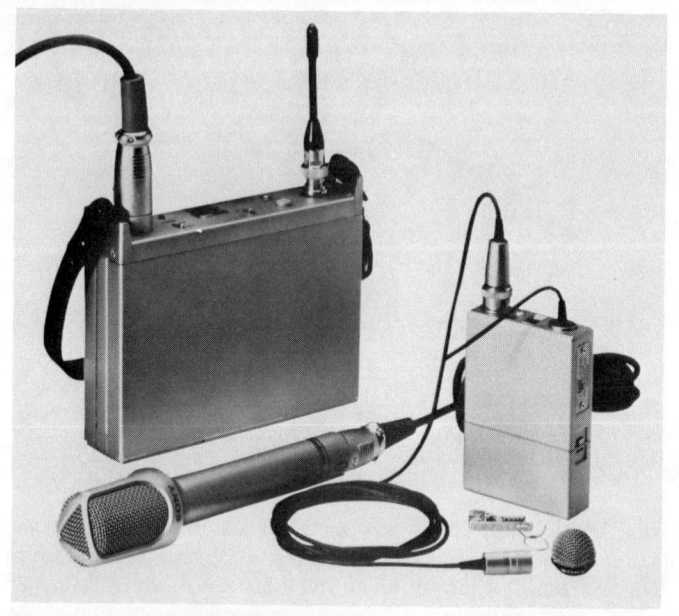

Figure 18-5 A radio microphone consists of a wireless receiver (left) and a pack wireless transmitter (right). (*Sony Corporation of America.*)

less than that of a real-life situation. Real-life sound, therefore, must be manipulated at the point of recording or during postproduction.

Frequency Response. Range of sound frequencies that a system can transmit with fidelity. The specification for television is 50 to 15,000 Hz.

Perspective. The sense of depth. A voice can be placed in any depth, ranging from a close-up to a long shot.

Voice-Over. A narration or voice that is not directly linked to the visual. It doesn't come from within the screen or even from a space just outside the frame.

Fade Up, Fade Down. Directions to increase or decrease the volume of any particular microphone.

Decibel. A measurement of sound power. It expresses a logarithmic relationship. Therefore, a reduction of 3 decibels corresponds to a halving of power.

MICROPHONES AND SOUND PERSPECTIVE

Microphones are the key to sound recording, not only because they change the sound to electricity, but because they alone create the *sound perspective*.

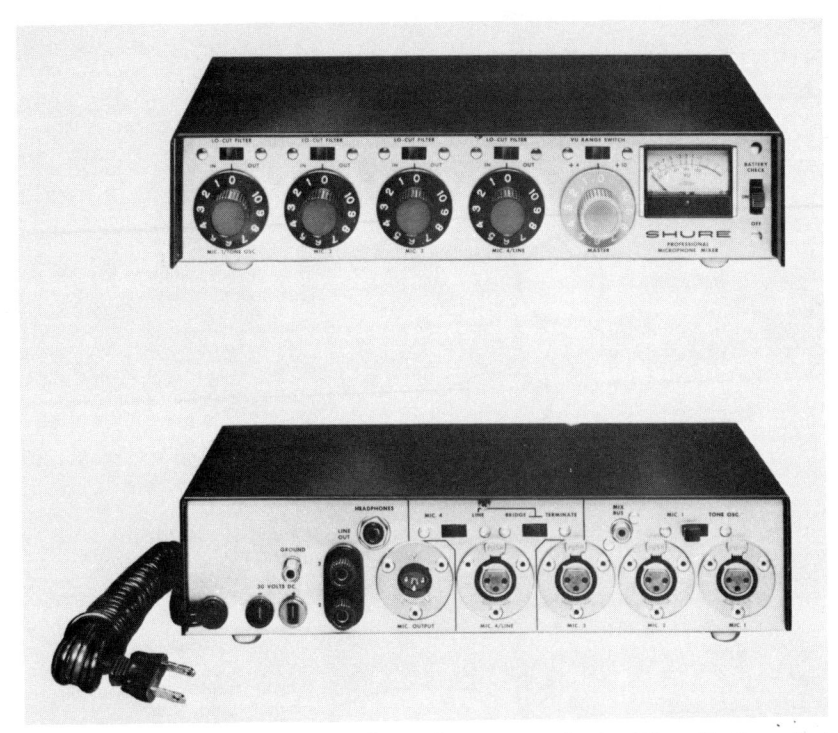

Figure 18-6 Microphone mixer can be used on remote shoots. (*Shure Brothers, Inc.*)

They do this by the manner in which they handle the background sound (all the sounds other than the direct sound of interest—usually the voice) (Figure 18-7). Thus, microphones are classified as to their directionality, ranging from the totally omnidirectional microphone to the ultradirectional varieties.

Let us look at the sound perspective of the lavalier microphone (an omnidirectional microphone). It is placed on the subject's chest, and produces a very distinctive sound. It makes the subject sound close to us and doesn't allow much background sound to be heard. It is good in certain applications (noisy locations), for it truly rejects and suppresses unwanted background noises. However, from an aesthetic vantage point, it is limiting in that it always provides the perspective of a close-up in sound, regardless of the image size.

The more conventional boom microphone (a cardioid, hypercardioid, or shotgun) will always give a sense of space around the voice. It is a tenet of motion picture sound recording that the sound space must match the graphic space. A person in a close-up should have more audio presence than a person in a long shot. By placing the microphone just outside the frame or shot, the sound space will seem to coincide with the visual space.

Television seems to present a special case. More often than not, there is a desire to have a close-up in sound regardless of the graphic information. There are very specific reasons for this.

18-6 PRODUCTION WITHIN PRIVATE TV

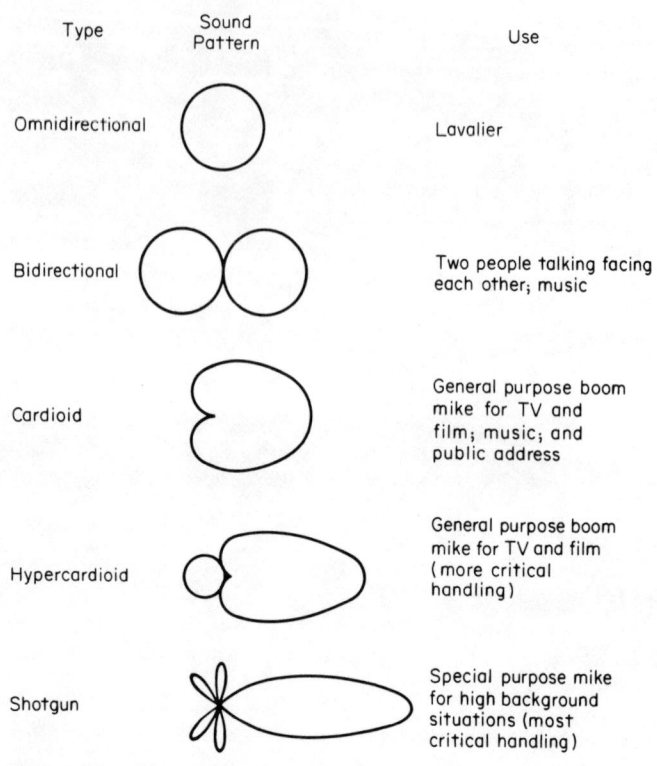

Type	Sound Pattern	Use
Omnidirectional	○	Lavalier
Bidirectional	∞	Two people talking facing each other; music
Cardioid	♡	General purpose boom mike for TV and film; music; and public address
Hypercardioid	⬭	General purpose boom mike for TV and film (more critical handling)
Shotgun	❀	Special purpose mike for high background situations (most critical handling)

Figure 18-7 Chart of microphone sound patterns and their uses.

1. The small speakers of a TV set and low fidelity do not allow for much subtlety in the track.
2. The noise level in the listening situations also demands that the sound track be consistently at full level.
3. Multiple cameras make it impossible to have the proper perspective for each angle. An average perspective must be adopted.

In certain dramatic material made for TV, there is an attempt to maintain some perspective relationships with the visual information. The realistic criteria for private video call for high intelligibility and low background level, with perspective being secondary.

PRODUCTION PROBLEMS AND SOLUTIONS

We have talked about the equipment and sound perspective. Now, let us address the day-to-day applications of sound in production. Specifically, we will examine the common problems that arise and how you can solve them.

The Lavalier

The lavalier has become the microphone of choice for most television use, meeting most of this medium's special requirements—high intelligibility, ease of use (no conflict with lighting), suppression of undesirable background noise, and no extra personnel needed to manipulate it during taping. The lavalier does, however, present some minor problems.

The question as to whether the microphone should be seen or not is totally up to the director. If it is to be buried under a tie or a jacket lapel, certain problems will arise. First of these will be *fabric noise,* which can be caused either by the microphone's rubbing against a fabric or by other fabrics rustling in the immediate vicinity. The condition is aggravated by silks, polyesters, and other stiff fabrics, so, in television wardrobing, cotton and wool are preferred. The caveat is that fabric noise is, of course, a function of movement. A silk ascot will not rustle if the person is sitting rock-steady, but if he or she should so much as move an inch—watch out. Likewise, even a cotton shirt will not help if the subject is involved in some violent physical movement.

The next problem associated with concealing the lavalier may be a marked *loss of high-frequency* sound. If the sound cannot get to the microphone directly, there will be a muffled effect—a distinct loss of high-frequency sound. This loss can be corrected by either rearranging the microphone under the clothing or sweetening the sound in postproduction. Recognize that results will be limited.

Radio Link

If mobility of the subject is a major concern or problem, a radio link can be utilized to replace the microphone cable, and the subject can then move freely without the encumbrances of wires.

Unfortunately, radio mikes leave a great deal to be desired. They are so low-powered that they are open to radio interference of all sorts. It is not uncommon, on location, to have a police radio or taxi-paging system put you out of business for a while.

The Boom Microphone

A boom mike can offer certain advantages over the lavalier and the radio link. There will be no mixing problems and no fabric rustle. But, if more than one person is talking, the boom must be cued (aimed) at the speaker. This maneuver takes a technician of some competence. If you use a boom, the lighting must be taken into account. If the lighting people have not left a slot for the boom, mike shadows are almost inevitable, and a great deal of production time can be lost in trying to eliminate these unwanted shadows.

Certain situations will require more ambitious solutions. For example,

18-8 PRODUCTION WITHIN PRIVATE TV

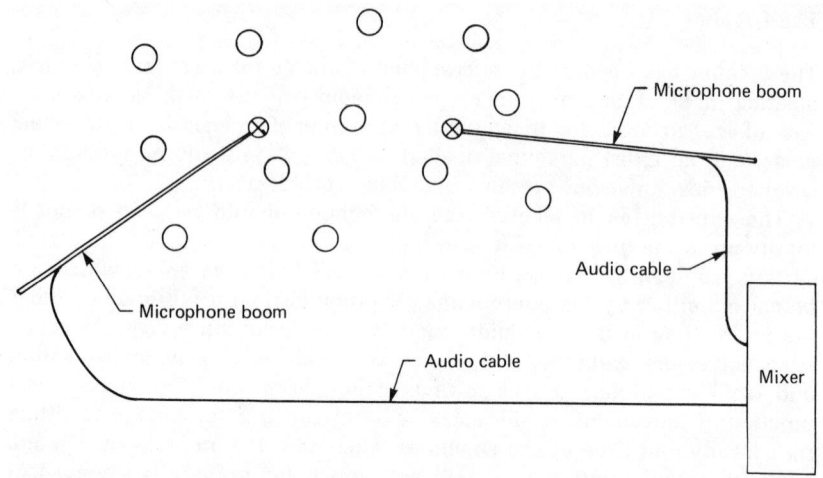

Figure 18-8 Placement of microphones for large discussion group spread throughout room.

a large discussion group, not at tables but spread about the room, takes a solution requiring two booms and mikes, two boom operators, a mixing board, and an audio-mixing technician—a three-person crew (Figure 18-8). Most private video operations do not have that depth and would probably have to hire free-lance operators.

If the group can be located at a round or oval table, then desk mikes can be used, eliminating the need for the two boom operators (Figure 18-9). The lighting will also be simpler in the second case, since the two booms will not be crossing through the set all the time.

Mixing Microphones

Mixing comes into play when there is more than one person speaking in the shot—each person needing his or her own microphone. The balancing of these microphones is not particularly difficult. However, it does require an audio technician to monitor the sound track and balance the level and equalization. A more highly skilled technician will be required for a five-lavalier mix than for a one-microphone setup.

The idea of *balancing* also opens up the question of what to do with the microphone levels of the persons not speaking. If you have three speakers (three microphones) and you set each microphone for full level, every time a person speaks, his or her level will be proper. However, having three open microphones all the time also means you have a tripling of background noise and reverberation. If that is not acceptable, the audio technician should fade down the level of the nonspeaking people. This can be done, however, only with scripted material or with a qualified audio engineer. So, with

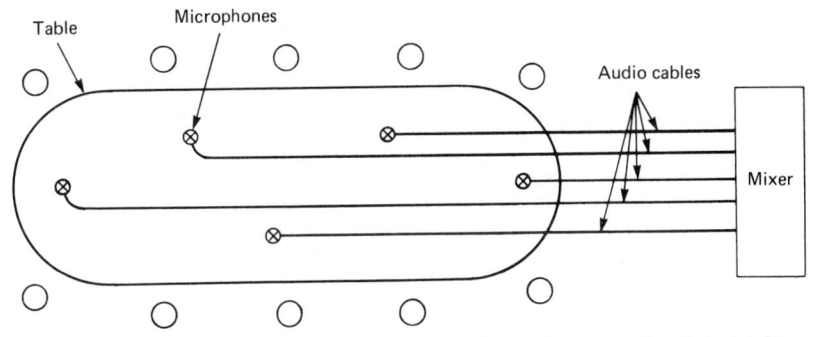

Figure 18-9 Use of desk microphones for large discussion group located at table.

multiple microphones, there are decisions to be made and competent audio engineers to be hired.

Panel Discussion

When participants are at a table, the ideal microphone solution would seem to be a microphone for each person. However, even when the microphone is directly in front of a person in this circumstance, he or she usually leans into it. This fact lets us utilize fewer microphones on occasion—we can let two or three people share a microphone—as long as we don't mind the turning and handling of the microphones.

Large Meeting Situation

Figure 18-10 shows a typical annual meeting with three possible microphone solutions. All three solutions call for microphones on the dais and podium; however, three different audience possibilities should be examined.

Solution A. For a few reasons, hanging microphones from the ceiling is not always a good solution. First, it is quite expensive in terms of equipment and labor requirements. Second, most participants will be too far from the microphones and, in order to get their audio level high enough to match the podium and dais sound, the audio technician may face the possibility of the room going into acoustic feedback, i.e., the sound coming out of the loudspeakers feeding into those hanging microphones.

Solution B. Standing microphones may be placed in the aisles. The participants come up to the microphones to speak. The sound will be good, but the process is time-consuming.

Solution C. Microphones are on booms and are brought to the participants. It is less time-consuming than solution B and also gives better control of the situation, that is, of determining who will speak next.

18-10 PRODUCTION WITHIN PRIVATE TV

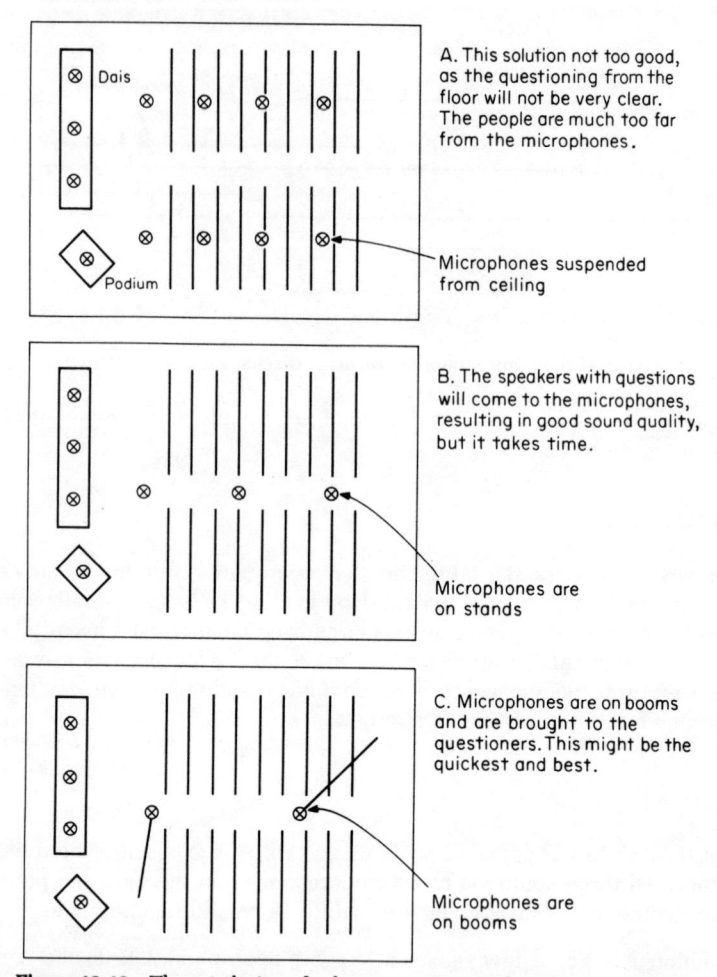

Figure 18-10 Three solutions for large meetings with audience question-and-answer period.

A. This solution not too good, as the questioning from the floor will not be very clear. The people are much too far from the microphones.

Microphones suspended from ceiling

B. The speakers with questions will come to the microphones, resulting in good sound quality, but it takes time.

Microphones are on stands

C. Microphones are on booms and are brought to the questioners. This might be the quickest and best.

Microphones are on booms

Location Shooting

Once you leave the security of the studio, a whole new set of problems arises. Location taping introduces more variables, so the key to good work involves a thorough scouting of the location. The first thing to examine is the *acoustics* of the location. How big is the room? How reverberant is it? What are the surfaces and how "live" or "echoey" will the sound be? If it seems too live, it may be possible to reduce that quality by hanging some sound blankets.

Noise is the major problem in location sound recording. All the sources must be identified: air conditioning, elevators, traffic, airplanes, machinery. Which ones can be shut off for the duration of the taping? If they can't be turned off, can they be tolerated? The crucial point, when you go out to scout a location, is that you must evaluate whether or not usable sound can be obtained. If you feel usable sound cannot be obtained at a given location, quieter, substitute locations should be investigated.

The major tool on location is the lavalier (remember, it's only 6 inches from the mouth). When its use is impractical, the boom mike must be utilized. If you are not in a quiet place, the normal studio cardioid is inadequate and microphones with more severe directional patterns will be needed. These microphones (the hypercardioid and especially the shotgun) do their job of suppressing unwanted background noises very well, but there is a buy-off. Their aiming is very critical—*so* critical that any off-mike sound is likely to be unusable.

The other tool which is indispensable on location is the radio microphone. As indicated before, external crosstalk may be a problem. You may be able to wait it out, or you may have to go to another type of microphone.

Sibilance

A tearing sort of "SSSS" sound quality in a person's speech pattern is defined as sibilance. If the sibilance energy is at too high a level, it can cause distortion on the recording. Reduce this sound by aiming the microphone off axis (a trifle away from the mouth) or using a "duller" microphone. The excessive "S" sound can be reduced in postproduction, but nothing can be done once the tape has been overloaded.

A good check is that old tongue twister, "She sells sea shells down by the sea shore." If the sound sounds raspy and spitty on all the S's, you have sibilance!

Popping and Wind Noises

Microphones are sensitive to high-velocity air. A voice with a lot of energy in the plosives "P" and "B" can also cause problems. Outdoors, the wind always creates severe problems. Even in a studio, fast movement of the boom microphone can produce an extraneous sound. *Pop filters* (really small wind screens), *wind screens,* and bass *roll-off filters* are used to reduce the effects of wind and popping.

Distracting Background Noises

Hums: 60, 120, 180 cycles per second hums from fluorescent fixtures can always be filtered out. Not so with more complex waveforms due to mechanical sounds (a motor) or electrical noises (a buzz). Sometimes other waveforms can be eliminated in the mix, but deleting them is a tricky business.

Traffic: Usually all one can do with traffic noise is to roll off as much bass energy as possible without affecting the voice quality.

Air conditioning and similar machinery: Such sounds usually cannot be filtered out.

Changing Mikes in Mid-Shot

A good rule of thumb is not to change microphones during a scene. A lavalier cannot be closely matched with a boom mike or a different manufacturer's mike. Changing microphones from scene to scene is not so great a problem—a visual change will support even a strong sound change along with it.

Phase Problems

When using multiple microphones, one has to be alert to the problem of *phasing*. Microphones respond to variations in sound pressure, and it is possible that one microphone's electrical output may be "out of phase" with another microphone's output even though they are responding to the same sound. The result of two microphones being connected "out of phase" with each other is characterized by a hollow sound, due to the fact that certain audio frequencies are being canceled. This is most likely caused by a "flipped" or transposed electrical circuit somewhere between the microphone's transducing element and the audio mixer. The simple solution to phase problems is to reverse one set of cables into the audio mixer.

POSTPRODUCTION

The two basic areas of sound postproduction are the corrective or "sweetening" process and the creative mixing process.

The Corrective Process

Many inconsistencies in a sound recording can be balanced out by the equalization of the track. Equalization can eliminate unwanted frequencies such as excessive bass energy that would make voices sound too tubby or boomy, or excessively high frequencies that might make a voice sound too scratchy or tinny. Frequencies can be boosted also to make the sound more intelligible or pleasant to listen to. The equalizer is most important when it comes to balancing various microphones in order to make the voice quality consistent from scene to scene. There are many more sophisticated devices available in postproduction, but most private video facilities will find that a good equalizer, with an audio technician who knows how to use it, will be adequate for most occasions.

Some very sophisticated techniques are available for "sweetening" the sound tracks in postproduction, but two things can't be fixed:

1. Distortion (overmodulation of the tape)
2. High background noise (usually not a problem in the studio).

So the major job in sound recording is to get a full-level audio signal on the tape with as low a background level as possible. The dynamic range of sound, in reality, is much greater than what can come out of a TV speaker. We must always bear that fact in mind when we record and mix sound tracks. Low-level (i.e., amplitude) material that may be fine in real-life conversation will not carry on the television system. It must be raised to an unnatural level to sound realistic.

The Creative Mixing Process

There are some basic thoughts that we must keep in mind when mixing audio tracks. First and most important to remember is that the sound space and graphic space should coincide. The sound should *feel* as though it's coming from within the screen. This effect may not occur if we use a lavalier on an actor walking toward the camera. In the mix, we might raise the level as the performer approaches the camera to give a realistic sound perspective. If the scene were indoors, we would have to add a little reverberation as well.

Realistic sound effects and music must be made to sound as though they are motivated by the action in the frame. Effects and music that are meant to be motivated by activity off camera, yet within the context of the shot, must be equalized to seem that they are "just off camera." It is imperative that the sound space and visual space be positively correlated.

On the other hand, the use of totally off-camera material (e.g., flashback, flash-forward, etc.) presents some interesting problems. Consequently, off-camera material must have *no* sense of space around it to be usable (i.e., no clues about where it was recorded to confuse the audience).

MUSIC AND SOUND EFFECTS SOURCES

Music and effects can greatly enhance the final tape. Where do you get the missing material?

There are three sources of sound effects:

1. *Record them at the time of taping* (water running, door closing, footsteps), either on the videotape or on a separate ¼-inch audiotape machine. This method is the most straightforward but may not be desirable for two reasons: (*a*) it may require too much production time or (*b*) the sound, even though "real," may not be appropriate (blank gun, dummy door).
2. *Record them after editing is completed* on a ¼-inch audiotape recorder. This solution involves going out and amassing all the sound effects needed

to prepare for the mix. This method has the advantage of all the effects being recorded properly and carefully, but usually at a higher expense.

3. *Purchase sound effects as needed* from a sound effects library. If you have access to a large library, this method provides the most flexibility. However, you pay per effect. If the volume of work is large enough, it may pay to purchase a basic sound effects library ($200 to $300) that will cover a high percentage of situations.

Music presents a different problem, that of copyright. If you commission a score or have someone perform a piece in the public domain, the copyright is no problem. But original music for private television isn't practical, primarily for budgetary reasons. The best source is a music library with a broad selection. The library contains a huge catalog of music fitting all types of moods and styles and also meeting certain technical requirements. But the main attraction of using a library is that when you arrange for the use of the music, you are purchasing the physical sound *and* the rights to use it. If, however, the sound design calls for a particular pop number (Frank Sinatra, the Beatles), your lawyers must contact the record company, the publisher, and anyone else who may have the right to license its use. The process is tedious and usually expensive.

The music library can be used in several different ways:

Purchase by the selection, typically $50 per piece of music

Flat rate for program, typically $125 for a 30-minute tape, for as many selections as needed

Lease of the complete library (perhaps 170 discs) to be kept on your premises and used as needed at a flat rate of $600 per year

Another consideration when using background music is that certain kinds of instrumentation lie well under voice. If the selection has many instruments that lie in the same region of the sound spectrum as the human voice (horns and brasses), something has to give when they are mixed together, and, obviously, it can't be the voice. So the music must be played quite low so as not to fight the voice. While Beethoven's Fifth Symphony might seem right, it probably won't have any impact under dialogue. You'd be better off with a much simpler music line, one that can still be played fairly loud under dialogue.

CONCLUSION

In the early days of the talkies, just the addition of sound was a source of wonder. The first sound films were quite content to have a voice track that was loud and clear, plus a little music for mood. Over the years, a much richer and more evocative use of sound has evolved. Today, the film medium finds sound being used by certain directors in such a way that one can no longer say that sound is in a primitive state, compared with the visual.

Similarly, TV began as a medium of reportage (like the early photoplays of the 1930s). Now that it is maturing, it will develop its own aesthetics, taking from film those ideas which work and adding new ideas that are unique to video. The small screen and speaker need not be a limitation. In fact, they can be seen as giving rise to a more intimate situation than exists in a theater and encouraging a different kind of communication. Stereo sound in television has arrived, all the video recorders have two or three audio tracks, and the receivers will be available soon. These developments will open up whole new areas.

The use of sound in private video is still in its early stages. The hardware is there, the experience of fifty years of sound film is there. So I encourage you to use the sound track in a truly creative way.

19 Directing and Rehearsing the Nonprofessional Talent

JUDY ANDERSON

INTRODUCTION

Using "real people" as talent in private television productions can be a source of great pride and pleasure for the company, the producer and director, and the talent.

Conversely, it can also result in painful experiences, unhappy memories, and nicked or truncated careers.

The purpose of this chapter is to present some guidelines that may help producers and directors to experience more pride and less pain when working with nonprofessional talent. First, it will outline the advantages and disadvantages of using in-house (employee) talent; second, it will discuss some elements of the diplomacy required when working with corporate staff members; and third, it will set up some guidelines for preproduction, production day, and postproduction techniques with your in-house talent.

THE ADVANTAGES OF USING COMPANY PEOPLE

Because private television programs are produced for business reasons—to train, to market, to implement policy—rather than for entertainment, it is inevitable that staff people will be used in programs for the organization. This is not a battle to be fought or agonized over. It is simply a fact to be accepted at least graciously, and at best, enthusiastically.

There are six reasons why a media department should use staff personnel:

- To enhance credibility of corporate messages
- To demonstrate the concern of top management
- To attach faces to impersonal memos
- To control cash
- To demonstrate professional competence
- To integrate the unit into the organization

Enhancing Credibility

Employees have indicated that they want to receive information about events occurring within the company directly from the people involved in them. They believe, rightfully so, that the information will be more accurate and more pertinent if the story is told by someone actually involved in it. Therefore, it is a sound idea to tape company people describing what is happening in the most candid and forthright manner possible.

The same is true of management information on marketing results, dividends, appointments to high-level positions, legal actions, and so on. It is far more believable if it is presented by the executive vice president in a candid discussion or interview than when it is described by an actor or coldly outlined in print.

Demonstrating Concern

There isn't an organization that doesn't at one time or other go through a difficult time. It is then that in-house talent can be used to make important communication points and demonstrate to employees that *management cares*. How often, for example, have you said, "I can accept bad news as long as you tell me the extent and severity of the bad news"? In other words, employees don't want to be left with unanswered questions about developments within the company. Management doesn't want it either, because a rumor, fueled by an atmosphere of uncertainty, decreases productivity and increases turnover among valued employees. Thus, the media manager will usually find it simple to line up a high-level company spokesperson to help sort out the facts from the fiction during worrisome periods.

Attaching Faces to Memos

As organizations get larger, it becomes more and more difficult to maintain personal contacts with people in other departments, even in the same building. Decentralization of facilities and the growth of multinational companies throughout the world only dramatize the reality. Video has the capability of bringing these diverse organizations under one metaphorical roof by enabling staff and management people to talk to one another regardless of time and distance.

Controlling Cash

Nonprofessional talent (i.e., company employees who are not trained as actors) are usually volunteers, and therefore cost the audiovisual department nothing. While there may be some downtime as employees leave their regular work assignments and perform on camera, these costs are usually absorbed by the organization or charged back to the client unit in "funny money" charges. For many media operations, the saving of a couple of hundred or even a couple of thousand dollars in cash that accrues from using in-house staff means that they can now make purchases for essential material such as tape stock.

Just because the talent comes "for free" doesn't mean it is worthless. Producers and directors should look upon in-house talent as a natural resource to be treated with all the care and respect you would give a professional.

Demonstrating Professional Competence to Management

Another advantage to using "real people" as talent in private programming is strictly for the good of the media unit itself.

Working with top management enables a video department to show, and to show off, to management its capabilities in a way that legitimately short-circuits a lengthy chain of command. With constant programming that involves top management as talent, the video department can enhance its budgets, promotions, and programming. Remember, the people who make decisions about video budgets are those in top management, and they are going to approve impressive budgets for the media unit only if they are impressed with the work of the people in it.

Integrating the Unit into the Organization

Other chapters in this book stress the importance of a video department's becoming an integral part of the total organization. Using real people is one of the simplest and fastest ways to do this. Every time a company employee appears in a program, a new part of a network is established. Eventually, these crisscrossing communication lines weave themselves into the corporate fabric as they become more and more important to more and more people. This network works for the media department as well. Tapping into the corporate grapevine may not seem to be necessary for a professional media person, but the leads for stories and insights into the corporation it will provide are enormous. You can learn from the electricians that the company is installing a big new computer (or taking one out), and the telephone installer will tell you where and when the marketing department is moving long before it's officially announced.

THE DISADVANTAGES OF USING COMPANY PEOPLE

Together with the advantages of using real people, there are some very real disadvantages that must be considered in the planning of any given program. Three of the reasons relate to the talent itself, and the last one to program objectives:

- Longer rehearsal time
- Inarticulate or unpresentable spokespeople

19-4 PRODUCTION WITHIN PRIVATE TV

- Uncooperative spokespeople
- Inappropriate spokespeople

Longer Rehearsal Time

In-house people are typically chosen for productions because of their expertise in a particular field, rather than for any expertise in front of a camera. Thus, it's not surprising that even very articulate nonprofessional talent will require far longer rehearsal time than a trained actor. Since professional actors can memorize and rehearse scripted programs faster than nonprofessional talent, actors should be considered when there are time limitations to a production.

Inarticulate or Unpresentable Spokespeople

One of the media manager's recurring nightmares is discovering during rehearsal (or worse, during taping), that the nonprofessional talent has some fatal flaw that simply eliminates the possibility of using him or her. It may be that the talent has developed a stutter or laryngitis or uses words so big no one can understand a thing he or she says.

It may be that the talent has two black eyes that won't disappear under makeup, or perhaps even one blue and one brown eye. And there is always the authority who wears a badly fitting wig or toupee or has a gap where two front teeth should be.

Unless you can control for all these possibilities by assiduous preproduction planning, it is best to think very carefully about using nonprofessional talent.

Uncooperative Spokespeople

Some people just want to be coaxed, an exercise for which you may not feel you have the time. Other people genuinely do not want to appear in your program. Unless they are the only people in the world with the expertise this program needs, you are better off not using them. If you must use them, try to be unusually accommodating and understanding.

Inappropriate Spokespeople

Company people are not always the right people for the job. Professional talent can and should be used, for example, to generate greater enthusiasm than might be possible with in-house talent. A professional sports figure or a known comic or a former politician can arouse the enthusiasm of even the most indifferent corporate employee in the right circumstances. Naturally, these people will cost more to use, so you should be sure that the payoff will be worth it.

THE PSYCHOLOGY OF WORKING
WITH IN-HOUSE TALENT

The psychology of working with in-house talent is straightforward. Diplomacy is the key word, as your talent is not an actor but, rather, a business person who expects to be treated in a businesslike fashion.

In-house directors know that they must live and work in the same company as the talent, so they must work with the talent to ensure not only a good performance, but good feelings about the production long after the experience has ended. The talent is doing the company a favor even when things don't go well. For example, when a talent has not memorized lines after having promised to do so, a diplomatic director will make light of the situation and say that the problem is a common one that happens to many individuals. The tactful director asks whether the talent would prefer using cue cards or little crib notes that can be held in one's hand. He or she stops the shoot when the talent is having difficulty and checks to see whether there are some problems with the script. In other words, a diplomatic director accords the nonprofessionals the dignity of their roles as corporate employees rather than responding to their inadequacies as talent.

Being diplomatic, however, must not interfere with the professional standards a director wishes to maintain. These standards include eliciting the best possible performance from the talent. This often means providing plenty of rehearsal time and pushing the talent to their highest levels of performance.

Providing Plenty of Rehearsal Time

Even something that seems so elementary as rehearsing has its complexities. First among these is the necessity to convince people they *should* rehearse.

Ideally, the producer or director has made the requirement for rehearsing clear to the talent at a very early stage, and the talent, in saying yes, has bought the entire package. Nevertheless, from time to time, people have the notion that nonrehearsed presentations will make them appear warmer and more relaxed. They are concerned that a rehearsed presentation will make them appear stiff and unnatural.

Explain to them that the purpose of a rehearsal is in fact just the opposite; it is to make the talent appear as natural and as comfortable as possible. Remember, the in-house talent is in an unfamiliar environment, and only when he or she feels in control of the situation—a feeling that comes with familiarity—will the person be able to relax and project his or her best self.

The second objection that some people have to rehearsing is the time it takes. People will claim they are too busy, that they cannot leave their jobs, or that their supervisor will not give them the time off.

One way of dispelling these objections is to cite your experiences with the president or chairperson of the board. These leaders often appear on broadcast television, understand the importance of being prepared, and rarely resist rehearsing. Explain to your reluctant talent that even the chairperson of the board rehearses, and then ask whether the talent is more busy than that top executive. You will almost always succeed in getting the time you need.

For your part, you should stay flexible enough to accommodate genuine time requirements. You may need to schedule rehearsals before or after business hours or on weekends so that the individual and the supervisor don't feel pressured. Then, too, offering to schedule a rehearsal during off-hours makes people understand how important you consider the production, and often results in an amazing reassessment of the talent's priorities.

A third reason people resist rehearsing is that they do not understand how important the program is. In that case, you need to explain that the videotape will be seen by employees for weeks, months, perhaps even years after they have left the studio. Thus, in exchange for the relatively short time they will spend in rehearsal, they will help create a presentation that will be an effective and forceful one for years to come.

Pushing the Talent

Everyone has a limit beyond which he or she cannot progress. The director may be able to get the talent to improve to a certain level and then no further. For some, that level may be very, very high. For others, it may be just barely acceptable. The director must be able to recognize when the talent has reached that limit. Pushing beyond it will almost certainly backfire and cause resentment that will show up in a poor performance.

THE REHEARSAL STAGE

Rehearsing Prior to Shoot

Whether your program is fully scripted or not, it may be best to start the rehearsal in a nonstudio environment, working one-on-one with the talent. This nonstudio environment may be a conference room or the rehearsal room off the main studio. There are no bright lights and no heavy equipment, and it is an environment in which the talent feels comfortable.

The one-on-one approach also allows you to be more candid with your talent. After all, there is no one else in the room, and what is said there can and should be held in strictest confidence. Start with a talk-through of the material in a casual and relaxed manner. Concentrate on concepts and important facts rather than on style or voice level. Have the talent answer the question, "Am I getting my point across?"

There are any number of ways to introduce video equipment into the rehearsal process, and you should do so at the first opportunity. You don't have to have the talent sit under the studio lights in front of your professional-grade camera. You can start by wheeling into your conference room the inexpensive black-and-white or color portable equipment that is found within private training operations for the purpose of role playing and practice sessions. These minisetups even come complete with a TelePrompter, if you need it.

Record the first rehearsal and, without saying anything, ask the individual to review it. More times than not, the "electronic mirror" will speak for itself and the talent will not need you to point out flaws in the performance: he or she will be able to see them clearly. This use of video greatly speeds up the rehearsal process because you don't have to explain anything: the talent can *see* the problems.

As the rehearsal continues, both you and the talent will see an improvement. Give positive feedback, but be candid. Sometimes people slip into old habits. They will know it and you will know it. Catch them on it, tell them how they slipped and what they must do to move their performance up the scale to the point when you are ready to schedule the shoot.

Using the TelePrompter

If the presentation is fully scripted, it will generally be put on a TelePrompter or, if very short, on cue cards. Unfortunately, talent sometimes think that using a TelePrompter will negate their need to rehearse. The exact opposite is true. Persons unfamiliar with a TelePrompter may stare at it unblinkingly. Because they can read only a few words at a time, they can rarely "scope" the statement with the correct emphasis the first several times through it. Thus, programs in which a TelePrompter is used require as much rehearsal as programs without one.

There are, however, some helpful hints in making the performance more natural, and these should be attended to before the first rehearsal.

1. Use marks on the TelePrompter script to indicate certain movements. For example, use a single slash to mean "Stop here" for a minor pause. Use a double slash to signal a major pause or to cue the talent to look down for a moment and then look up. Use an X in a circle at places you want the talent to stop or to move to a new part of the set. (Figure 19-1.)

2. Let the talent hold some blue 4 x 6 inch cards in their hands. Nothing of significance need be on the cards, but they give the talent something natural to do with their hands and eyes. For example, a person can look down and shuffle the cards as if collecting her or his thoughts, creating a gesture that adds realism to the performance.

3. When shooting dramatizations or role plays with two or more people on the set, have a number of TelePrompter units all around the set—not just on the cameras. This way, each person will feel free to move his or her head. Or have the talent talk to an off-camera TelePrompter

19-8 PRODUCTION WITHIN PRIVATE TV

> SHORTLY, WE WILL DISCUSS THE SPECIFICS OF A COUNTRY'S EXPORT CREDIT PROGRAM. BUT AS WE BOTH KNOW, INFORMATION WITHOUT A PERSPECTIVE IS JUST A COLLECTION OF FACTS. WHAT I AM ABOUT TO DO IS PUT THESE PROGRAMS INTO PERSPECTIVE.// THIS OVERVIEW WILL HAVE THREE PARTS. FIRST, A SHORT PSYCHOLOGICAL PORTRAIT OF THE EXPORTER AND THE IMPORTER./ SECOND, AN IN-DEPTH EXAMINATION OF THE VARIOUS ELEMENTS OF A GOVERNMENT EXPORT PROGRAM./ AND THIRD, THE ROLE OF OUR BANK IN THIS INTERNATIONAL MARKETPLACE. ⊗

Figure 19-1 To make reading the TelePrompter easier, use slashes and stop signs ⊗.

that is situated behind and over the shoulder of the person the talent is addressing. By talking to the TelePrompter instead of the person, the talent avoids rapid eye movement between the TelePrompter and the other talent. This rapid eye movement almost always gives away the fact that the talent is reading from a TelePrompter.

Improving Content

If the presentation is not fully scripted, you can use the initial rehearsal to improve its content. In fact, by getting your nonprofessionals to concentrate on content rather than technique, you help them relax and you eliminate their concerns about the method of presentation. By the time they have finished rehearsing, not only do they have the content down cold, but they also have improved their presentation. They will be far more likely to generate confidence, enthusiasm, and a high degree of knowledge.

THE PRODUCTION DAY

This is the day you, and the talent, make it or break it. This is the day the director and the producer change in the eyes of the talent from their roles as fellow company employees to the roles of professional television director and producer. The performers are in your territory. To keep them from being intimidated and thus giving you a poor performance, you will want to rehearse them again. In the studio, the rehearsal will enable the talent to get comfortable with the lights, the big cameras, the crew, the set, and all the other unfamiliar aspects of the production.

During rehearsals, but most important during the shoot itself, the director should ask the talent to do four things:

- *Maintain high energy levels.* Energy conveys enthusiasm, dynamic range, and desire to succeed, so most in-house talent will agree with the director that they should project it. To do so, they will have to speak in a more animated manner, with less of a monotone than they might normally, and use gestures and facial expressions to convey feelings about the subject. Typical executives may find this uncomfortable at first, but with practice and playback of the tape, they generally pick up the technique fairly easily.

- *Generate warmth.* Smile. Nod approvingly. Create a warm presentation that will be remembered as well as any live presentation.

- *Listen to what you are saying.* The difference between a canned presentation and one that appears to be spontaneous (even though it is well rehearsed) lies in whether the speaker is merely reading words or really listening for the meaning. The minute speakers start reading, they turn off their audience.

- *Believe in what you are saying.* When a speaker believes in what he or she is saying, it shows. The audience will feel confident that it is not just listening to another party line, and therefore, the speaker, the material, and the medium will benefit.

FOLLOW-UP

When the shoot is over, send the person a thank-you note, along with a production photograph of the talent on the set if you have one. Follow up again after the show has been edited and distributed, and invite the talent to a special preview of the show. If the program meets its objectives, let the talent know. During holiday season, you might like to send a greeting card that has been made from a photo taken right off the screen, showing each performer with his or her name superimposed.

In other words, treat in-house talent as valuable friends and business assets, and they will return the courtesy.

CONCLUSION

Directing nonprofessional talent is not an easy task, but for those involved in private television it is a certainty. Remember, before using nonprofessional talent, always weigh the advantages and disadvantages. Be sure it is the most effective method of presenting your program's message.

If you have decided to use nonprofessional talent, be certain to use diplomacy, understand your talent's limitations, and schedule plenty of rehearsal time. This approach will help you attain the high level of quality you desire.

20 Directing

LON MCQUILLIN

In commercial television, the director is usually a free-lance artist who moves from one production to the next. In private television, however, it is quite common for the director to be a staff member, though this by no means indicates that free-lancers are not also used either to supplement the staff directors or exclusively.

The director in private television deals with a different set of needs and responsibilities from those handled by the commercial television director, and often works with more limited resources. This chapter will examine what the job of the private television director is, and how it is done from pre- to postproduction. In addition, some thoughts on the challenges and problems facing private television directors are included.

ZEN AND THE ART OF DIRECTING

One observation on describing the activities of a director is that it is somewhat like describing the step-by-step procedure of performing surgery. It can be done, but sometimes it lacks life.

Directing is a skill that requires years of experience. Even after directing hundreds of programs one has no guarantee that one will be a competent director. There are, however, a few tricks of the trade:

- *Be up.* If a director is excited about a program, the crew and talent will be excited.
- *Set high standards.* Don't be hesitant in telling your crew and talent what you expect.
- *Keep an eye out for detail.* Ultimately, you are responsible for every element in the program. Develop a checkoff system to make certain that items get prepared and executed.
- *Keep a sense of humor.* No production is perfect. Acts of God cannot be controlled. The statement sounds trite, but a happy crew is a productive crew.

PREPRODUCTION

Production Values versus Budget

The key to success in any video endeavor is careful planning during the preproduction phase. Production values, which have a large influence on

20-1

the impact and effect of a program, must be high if the program is to accomplish its task. High production values can be achieved in two ways—the first involving money, and the other involving time and preparation.

Planning a production is much like walking a tightwire: On the one side is that dreaded beast, the budget overrun, while on the other, lurks the most vicious monster of all—ineffectual results. This balancing act starts the moment a production is proposed, and must be maintained all the way through to the completion of the program.

Production Design

One of the first steps will be the creation of the script or storyboard. In private TV, the sequence will normally flow from a department with a communications need, through a subject expert, to a staff or free-lance scriptwriter. Working with the content expert, the writer will put together a script that answers the communications needs, whereupon a producer and a director will be assigned to the project. (See Chapter 13 on scriptwriting.) Working in concert, these three key people will design a production that can be mounted within the budget established for the program. While the producer is off producing, the director will begin to prepare for the task of translating the script into a program.

At this point in the preproduction process, the director must choose the best methods to execute the script. Here decisions must be made on questions of:

Single camera versus multicamera

Special effects

Sets

Studio versus location

Talent and casting

Single Camera versus Multicamera

Single-camera production has become very popular with the increasing availability of good portable equipment, and offers advantages over multicamera production in many instances. The same kind of skilled artistry commonly encountered in film production can now be applied to video, with each scene being carefully set and lit with only a single camera position in mind. Multicamera production, while certainly not limited to the studio, will generally be of greatest advantage there. For such types of programs as interviews and panel discussions, a multicamera arrangement may be preferable, and if the show is to be transmitted live, its use will be mandatory.

Single-camera production is completely dependent on editing to finish the program, whereas multicamera production with live cutting (at the switcher) can allow the show to be finished when tape stops.

There is a method of production, however, that takes advantage, in some ways, of the best of both techniques: isolated recordings (Iso's) of the signal from each camera on a separate recorder. This technique, while retaining the advantages of multiple camera positions, also depends on the edit—a fact that automatically increases production time and costs. This system can be forgiving with regard to errors. In live switching, a mistake is recorded and becomes part of the program. With Iso's you can cut to another shot to cover the error. But remember, it's too late, during the editing of an Isoed show, to call directions to the cameras—a restriction that may take some getting used to.

Perhaps the most flexible technique with multicamera productions is to do a live switch onto a master videotape recorder (VTR), and have a second VTR record Iso shots. This will require an additional switcher, usually operated by an assistant director. The end product on the master VTR will be a complete program. The end product on the Iso VTR will be the "alternate shots," that is, shots not found on the master tape. In this manner, if everything works right during the shooting, the show is done. Furthermore, if there are some rough spots, there is backup material on the Iso VTR that can be edited into the show with a minimum of postproduction. It requires some additional VTRs, but it's as close as one can get to having your cake and eating it, too.

Special Effects

Although special effects are incorporated into the show either in production or during postproduction, they should be planned in the preproduction stage. The best rule of thumb with regard to effects is to keep their use to a minimum, and to make sure there is a good reason for their use when they are employed.

Few things are more distracting than the overuse of all sorts of different wipe transitions just because they're available on the switcher. Somewhere around 80 to 90 percent of all transitions should be cuts, with dissolves making up most of the balance. What few instances that remain may be appropriate places for wipes.

The other major purpose for the effects capabilities of the switcher is what we might call "true special effects." For example, you're doing a takeoff on *Star Trek* for a program on, say, human relations. (Shows such as that are not too uncommon in private television—they can really grab viewer attention and get the message across very effectively.) Then the special effects capabilities of the switcher become very useful in doing such things as "transporting" people around. One note of caution: Creating effects that fall short technically of the viewer's prior viewing experience may do more harm than good.

The increasing availability of digital effects capabilities makes their use also very tempting. Performed by equipment such as Vital's Squeeze-Zoom or a Quantel unit, an image may be frozen, flipped, squeezed, shrunk, rotated, and the like. In using these effects, make doubly sure that you have a valid reason, as they tend to be quite gimmicky.

Sets

Set design begins with the budget, and if there is little money available, then innovative solutions will have to be found. Surrealism may be the answer to a financially beleaguered director's problem. Most people have seen the classic play *Our Town*, which is traditionally staged with minimal sets. It's a good example of how lighting can often take the place of a set. For television, careful attention to the lighting design and video shading will obviously be necessary. Working without a set budget calls for an unorthodox approach that may use odds and ends in designing one. Television is fairly forgiving with regard to the materials and details of sets, and scrounging for set pieces can turn up all sorts of items that can be recycled into set elements. In addition, items normally found around the studio can be used in ways other than those for which they are designed. For example, seamless background paper, cut into shapes or outlines, can be used to create "supergraphics" that break the monotony of a background, and poster blowups from photographs can be used to tie in with the subject matter and provide a visually interesting (though ideally not distracting) backdrop.

Studio versus Location

A number of factors are crucial in getting the show in on time and within budget. Studio production is relatively safe, since most of the elements are under control, and necessary items and people are usually close at hand. Location production—especially exteriors—is much harder to pin down, however. Here, the studio is, in essence, transported to the setting, and anything that has been forgotten or that goes wrong will cause trouble. Outdoor shooting also depends highly on cooperative weather conditions, and this dependency must be taken into account. Contingency plans to provide for the event of a "weather wash" are necessary.

Talent and Casting

Another factor which the director must consider during the preproduction stage is on-camera talent. The word *talent* is a catch-all phrase that can describe anyone from the chairperson of the board presenting a report for stockholders to several actors hired to play various parts. In general, though, talent can be broken down into two rather nebulous types: "real people" (folks appearing as themselves) and actors.

Shooting real people is often an inflexible requirement, while casting a role may include options as to the source of talent. Using someone off the street (or out of the accounting department) as an actor is substantially different from working with a professional, and the added cost of professional performers must be weighed against the greater efficiencies they offer. In general, experience has shown that competent professionals are usually worth the costs they add to a production—all the more when the schedule is tight.

Finding the right talent is of critical importance to the production. Here, a good casting agency can be of immeasurable help, as it is a clearinghouse for actors. One of the worst possible methods for casting is the "cattle call," where an advertisement is taken out announcing auditions. Such an ad usually results in the descent—en masse—of everyone within driving, walking, or limping range who has ever wanted to be a star. A good talent agency, on the other hand, will listen to the client's needs and will go through its files to come up with a handful of potential choices, whom the director can then audition. This method saves time and frustration. Note, though, that in the major metropolitan areas, most of the good talent available will be members of the performing unions, and therefore they will not be available to you directly if you are not a signatory to their contracts. It is at times possible, however, to make other arrangements, such as hiring a freelance producer or director who is a signatory, and therefore is capable of hiring professional talent.

PRODUCTION

This chapter, so far, has covered the preproduction activities with which a director must be concerned. Before discussing directorial techniques, let's look at two production elements in which the director must take an interest: the production schedule, and the staff and crew.

Production Schedule

When production starts, the director's role becomes that of a conductor whose beat everyone on the set will follow. The possibilities for analogy are marvelous—a "well-oiled machine," "team spirit," what have you; however the performance is expressed, a production crew must work in sync if everything is to go smoothly.

A critical part of the operation is the production schedule. It is very easy to build a trap that can swallow up you and your budget without a trace. In particular, estimating the amount of material that can be shot each day requires a mixture of science, art, and just plain guesswork. On remotes, this mixture can require more guesswork than in the studio, since there are so many added variables, many of which are out of the director's control.

As a general rule of thumb, in shows that entail dialogue between two or more characters, roughly five to seven pages a day would be a good middle-ground estimate of what can be accomplished on location. Obviously, the nature of the material, the abilities of the talent, and dozens of other factors will affect this estimate. There are just as many situations when a single page is a good day's work as there are when twenty pages is a stroll. In general, though, carefully crafted productions seem to fall near that range of five pages per day. This schedule would translate to around 5 minutes of material, under the general average of a minute per page of dialogue. When there is no dialogue, such as in getting pickup shots of an industrial plant, estimating is a bit harder. Here, the director must total the number of scenes to be shot and the number of setups they will require, and use past experience as the scale on which to balance time factors against production values. It is quite possible to spend an entire day getting material that will run 20 seconds in the finished program.

Staff and Crew

Assisting the director with the running of a shoot are the staff and crew. The number and type of crew positions to be used will vary with both the production and the resources of the organization, and they may run one or two people to twenty or more. Perhaps the best way to handle the staffing, then, is to list the major job functions as they are found on a full-scale production, since each of the jobs will have to be handled one way or another, even if one person will end up doing them all.

The assistant director (AD) is, as the name implies, the director's right hand. The AD will essentially handle the management of the shoot, making sure that the schedules are correct and have been distributed, that everyone who will be needed is on hand, and that directions are relayed to the cast and crew during actual taping. In the broadcast and motion picture environment, this position is in part an apprentice position on the way to becoming a director, though there are many who enjoy the job and have no intent of jumping from the frying pan into the flames.

The function of the technical director (TD) may vary depending on the circumstances. In the studio, the TD is the one who does the actual hands-on operation of the switcher, at the director's bidding. On location, the TD is often the person in charge of the overall technical facilities.

Production assistants (PA's) may be assigned to be on the set to assist in the production. Often they are considered to be go-fers, though large productions will have other warm bodies there just for that function. PAs do things like badgering the talent to fill out the reporting forms and releases, keeping track of petty cash expenses, keeping the coffee hot and the soft drinks cold, and generally making themselves invaluable. Many others round out the crew roster. With many, such as the audio engineer, the title describes the job. Some of the important people are:

- Production manager
- Floor director

- Camera operators
- Lighting director
- Set director
- Script supervisor
- Talent coordinator
- Wardrobe mistress
- Makeup and hairstylists
  . . . and many others

Good crew people will respond well to suggestions, and they will often surprise you with their innovative ideas. An excellent practice, after a shoot, is to have an informal gathering of the crew to critique the show—your performance and theirs—to develop ideas that will make future productions go more smoothly and come out better on the screen. In shows that are taped live with multiple cameras, if production facilities and techniques permit, record the intercom line on a spare audio channel, and use it during the meeting to help analyze the crew's performance. If nothing else, it's great entertainment.

DIRECTIONAL TECHNIQUES

Everything that happens on the set, from the look of the show to the technical standards of the videotape to the physical safety of the actors and crew, is the responsibility of the director. He or she is obliged to be the most prepared, knowledgeable, energetic, and flexible person on the set. On the day of the shoot, the director's attitude shapes the character and tenor of the day's activities.

This upcoming section details the director's responsibility in six areas of directorial technique:

Technical standards
Audio
Lighting
Composition
Talent
Working style

Technical Standards

Despite the fact that expert engineers are on hand, it is still up to the director to verify that things are as they should be. The director must not simply take the word of the engineering people that everything looks good—even the best crew people sometimes tend to exert themselves just enough and no further to get results that will be acceptable.

The director should know how to read a waveform monitor and understand the information it provides. For instance, engineers may try to set camera iris and gain so that they have a full-range signal, starting at 7.5 Institute of Electrical and Electronics Engineers (IEEE) units (black), and going up to a full 100 units, or 1 volt (white). The trouble with this is that not all pictures should be set for true black and white; the picture itself should be interpreted, and levels set accordingly. For example, if the scene is predominantly gray, the waveform should provide a reading somewhere in the middle range, and may not come close to a full 1-volt signal. Something else to watch for is overmodulation. Some engineers try to get additional "snap" in a picture by running levels over 100 units. This may look great on the program monitor during shooting, but will end up being clipped or trimmed down to 100 units during the editing or duplication stage. The end result will be a darker, less defined picture.

Audio

Audio quality is every bit as important as video quality, yet audio is often virtually ignored, or paid only cursory attention. There are a number of things to watch out for in recording top-quality audio.

Sound perspective is one of the most commonly violated aspects. For example, if the talent is seen in an extreme long shot walking toward the camera, shouting to someone in the foreground, the voice should *sound* as if it were coming from a distance. This is closely related to audio continuity. If you're using a wireless mike for a long shot, stay with it for the close-ups, so that there is no drastic change in the response characteristics at the cut.

Ambience is another factor that must be taken into consideration of continuity. Changes in mike positioning and levels can also change the ambient sound level, and mismatches during editing can be very noticeable. It is normally a good idea to record some "wild track" ambience to be available during editing, where it can be mixed into the program audio if needed.

Lighting

One of the major historical differences between television and film has been TV's long reputation for poor lighting. In the early days, this deficiency was caused by the need to blast incredible amounts of light into the scene simply to get a full signal out of the cameras (1 volt). It has also been abetted by the use of multiple cameras, which requires that a scene be lit for several angles, creating a compromise in lighting where no one angle is properly lit.

Single-camera production allows television lighting to be set using many of the practices that are taken for granted in film. Since pictures *are* light, it is one area that can derive immense benefit from the hiring of a professional.

Composition

Television shares with filmmaking and photography the fact that it is a visual art. Though more technically complex than either of the other two forms, the basic principles are the same, and there is no substitute for a sense of composition and form. This sense is something that some people are born with; others will have to learn some rules.

There is an excellent book called *The Five C's of Cinematography* by Joseph Mascelli (Cine/Graphic Publications, Hollywood, Calif., 1975) that is excellent reading for anyone new to television (and even some old hands). The five C's are: composition, continuity, camera angles, cutting, and close-ups. The guidelines offered in this book are quite useful, though it is important to remember that television differs in some substantial ways from film. The size of the screen is the most significant, requiring more emphasis on close-ups than when working in film. Also, the effective frame rate of 30 frames per second, as opposed to film's 24 frames per second, provides an image that is much freer from flicker. Pans and tilts that would be unacceptable in film can work in television.

In preparing to shoot a scene, the director must visualize the shot as it will appear on the screen, ahead of time. One can develop a sequence of what we might call *checkoffs* in composing a scene.

1. Is the subject matter correctly included and displayed?
2. Is the background clear, with nothing showing that shouldn't?
3. Is the lighting OK, with no boom microphones or actors blocking light?
4. Is the picture composition all right?

Composition is held as the final item. Once all the other elements have been checked, and repositioned if necessary, the final adjustments for composition can be made.

The director will make sure that the on-camera talent knows the lines and understands the flow and the purpose of the scene, and will run through the scene, perhaps rolling tape just in case everything goes right.

To the five C's of cinematography already mentioned, I would add a sixth: camera movement. The vast bulk of the private television I've seen has been shot with the camera mounted on a fixed tripod. Though it increases the complexity and often the cost of shooting, the use of a dolly or other mobile support can be of immense benefit to a program. Used judiciously, camera movement can add a three-dimensional feeling to a show. There is, for example, a great difference between zooming into a scene and dollying into the same scene. The difference is that perspective changes with a dolly, and does not with a zoom. In nature's blueprints, our eyes were not designed with zoom lenses of variable focal length, and people are given a unique (and highly effective) means of doing their own dollying—their feet. The brain even features an automatic image stabilization system that is the envy of cinematographers and videographers. Dollying the camera, then, comes closest to the reality of the way people would see a shot if they were actually on the scene.

Dealing with Talent

It's so easy to get wrapped up with the technicalities of television and with directing the taping that one may overlook the fact that one of the director's main jobs is to direct the talent. Dealing with "real people" can be a problem. Just as some executives or other types of people who appear on camera take naturally to video, there are those who may never come off well.

Perhaps the most common problem are the people who are nervous about being in front of the terrifying beast with the single, cold eye. They've heard stories of how television makes you look heavier, or they're certain that their noses will shine, or that they'll freeze up and blow the show. Freezing on camera, or fumbling and flailing about mentally and verbally, is a frightening prospect, and the more the person worries about it, the more likely such behavior becomes. The best thing a director can do is to try to relax the person and make him or her feel comfortable. Depending on the person, various techniques can help.

With some people, a run-through on tape that is then played back to them can show them that their fears are unjustified. This approach assumes that their fears *are* unjustified—otherwise, it can backfire. It can even provide an opportunity to pull one of the oldest tricks in the director's book— taping a "rehearsal" that is actually take 1. With the talent believing that "this is just so the engineers and camera people can get an idea of how it will go," or some such nonsense, their performance is often relaxed and easy, compared with their later performance when the "real" take supposedly starts.

People tend to think of the "millions" of people out there behind the camera, or, in the case of private television, the "thousands." It often helps to explain to them that the normal viewing situation involves one or two people at a time, and that the best presentations are given by those who think of the camera in terms of being a proxy for those one or two. If they imagine that they're simply talking to one or two friends, the terror of making a gaffe in view of the "teeming hordes" may dissipate.

The process of making inexperienced talent comfortable in front of the camera can also be aided by giving them a tour of the facilities, explaining as you go what the people and gizmos are all about. To the neophyte, television is often a mysterious and formidable operation, and demystification will often help. Last, be aware that there are some people who just don't belong on camera—that it's impossible to make a leather jacket out of polyester.

Working Style—Modus Operandi

The director is a combination of godfather, cheerleader, pit boss, administrator, psychologist, technician, and artist. It's very hard to execute an entire production all by oneself, and teamwork is essential. Motivating the crew to perform at peak levels is an immeasurably important part of the director's role. Despite all the gadgetry that encumbers the process of television production, it's still a "people" business—the equipment doesn't run by itself.

Each director will eventually develop an individual style, which may even become a speciality. This style may be the reason the director was hired to begin with. Observing the styles of other directors becomes easier as experience grows, and it's perfectly legal to borrow ideas or reapply concepts to situations where they have not been used before.

The captive director, working in the same environment continuously, as is quite often the case in private television, needs to guard especially against getting into a rut and repeating things over and over. This is where fresh blood can help, and many companies occasionally bring in outside directors to get some fresh ideas going, as well as to help out when the work load exceeds the capacity of the in-house staff.

CONCLUSION

The best directors, as with the best in any endeavor, are rarely, if ever, satisfied with their work. Part of the process of growth, of course, is to make mistakes and learn from them. Thomas Edison, when having trouble finding the right filament material for his light bulb after trying 9000 things, perhaps said it best: "I now know 9000 things that *won't* work."

Television production involves long hours, hard work, frustrations with people and equipment, very little glamour, and a lot of meals eaten standing up or missed altogether. It also, though, involves creativity, challenge, growth, and learning; the camaraderie that grows within a great crew, and a sense of pride and accomplishment that's hard to find in other fields.

It's addictive almost to the point of being immoral, but, all things considered, it's a vice that few people, having been hooked, ever want to shake.

21 Shooting on Location

JEFFREY J. SILVERSTEIN

INTRODUCTION

Any time you take your camera beyond the studio door, you're on location. Locations range from city streets and company buildings (exteriors) to offices, enclosed sporting arenas, and factories (interiors). On location, you lack the control over the elements of weather, light, and sound that are taken for granted in the studio. Your job is to get back the control that you lose by leaving those protective, soundproof walls. But remember, you want the videotape "in the can," and nature, people, equipment, and time are trying their hardest to prevent you from putting it there.

WHY SHOOT ON LOCATION?

Because location shooting can require more logistics and old-fashioned work than studio shooting, it is wise to examine some reasons you might want to be on location.

Cost

It *may* be less expensive to bring the cameras to the location than to try to recreate that location in a studio. With the cost of portable equipment dropping and the cost of maintaining studio overhead rising, economics can be an important consideration. Make sure, however, to look at *all* the costs. Location shooting can be more expensive and there are trade-offs to be made.

Realism

No matter how much you spend, you may not be able, in a studio, to duplicate the realism of the location. Today's viewing audience is used to seeing location shooting on broadcast news, in feature film work, and in commercials. Being "on location" for a corporate program gives the tape a sense of identity and actuality which can add to the effectiveness of the material.

21-2 PRODUCTION WITHIN PRIVATE TV

Convenience

There is a definite trade-off here. What may be more convenient for you when you're shooting in the studio may be less convenient for the people you wish to use. For example, it may be easier to shoot in the president's office than to have the president come to your studio. The main question is, "Is it more important that the camera go to the event, or that the event come to the camera?"

Variety

Again, your audience is very "media-literate." Today, people are used to seeing a variety of visual information in short time intervals. A location provides much more visual variety and detail. This can be used to your advantage when you want to hold the audience's attention.

THE TECHNOLOGY OF LOCATION SHOOTING

In the not-too-distant past, a "remote" meant dragging cameras that weighed hundreds of pounds off their studio pedestals, loading them into a 50-foot van with 2-inch videotape recorders, miles of cable, switchers, and a large crew.

Location shooting can still be just as involved, especially for large-event coverage. But today's video communicator is now able to shoot "film style." Using a smaller crew and much smaller equipment, the whole world is a location.

Technology has reduced the size of cameras, videotape formats, and most of the related hardware. The sensitivity of video cameras is improving, so reduction in lighting needs can make for a more mobile shooting process.

Because the technology is constantly changing, specific pieces of equipment need not be described here. It is far more important to address the basics of location shooting procedures, skills, organization, and preparation. These are "people" things, and not "hardware" things.

PREPRODUCTION SURVEYS AND CHECKLIST

Production on location has been likened to war. (Why do you think it is called "shooting"?) Two of the most important tools a general (producer) has are *intelligence* and *reconnaissance*. The preproduction location scout or survey is the only way to gather intelligence about what your crew can expect when the tape rolls. On location, time is money. Either you find out the problems you're going to have during the location survey, or you discover them when crew, equipment, and talent are waiting around on the shoot.

A location survey has two basic components. One is creative, and one is technical. The creative aspects involve decisions about what will look best on camera, whether the shot will meet objectives, and what options for angles the director has. The technical part involves where you get power, where needed facilities are, and tying down location contacts.

The Survey

Creative Approach. It is usually critical for the director to be in on the creative end of a location survey. The producer and the writer may also be involved. This is the time when shots are planned around the environment. It often pays to have your director of photography or a camera person on the survey. A Polaroid camera is probably the most important piece of equipment to bring along. Plan to use lots of film, and shoot every potential angle, place, and option so that you can put together a location file on each place (Figure 21-1). This way, back in the comfort of the production office, you can go over the location file and match shots to the script. This file can be tremendously helpful in storyboarding. An additional asset is that the snapshots can clue you in to potential lighting and logistic problems.

Conducting the Survey. Each crew specialist that you are able to bring to the survey will have his or her own checklist of problems to look for. The lighting director will focus on lighting instruments, power requirements, and where to hide the cables. The sound person will focus on the acoustics of the area, ambient noise, and how many mikes, cables, and connectors to bring. The director of photography may work with the lighting director to determine camera setups and moves. The production manager has responsibility for the crew and the logistics.

Depending upon the nature and budget of the production, many of these jobs will be combined. That doesn't mean that the questions don't have to be answered.

If preproduction has been well done, most shoots should be relatively uneventful. You can get down to the matter at hand, which is getting that footage on tape. An anticipated difficulty should be a "presolved" problem.

The Production Manger's Checklist

Everyone has a different system for a technical survey. The checklist in Figure 21-2 forces you to *ask the important questions*. You may modify it, attach it to location file folders with Polaroids, expand upon each section, or create a family of forms to support it.

The Contact Sheet

One of the most valuable pieces of information you need on location is how to get in touch with people—fast. The contact sheet should include

Figure 21-1 Survey photographs can help you compose a shot before you arrive with a crew.

names, addresses, and business, home, and alternate phone numbers. This information should be assembled for crew, cast, suppliers of anything you are using, and local location contacts. A good rule of thumb is, "If they're involved in the production, they should be on the contact sheet." The sheet should be copied and placed in the hands of anybody on the crew who may need it. There is absolutely no excuse for not being able to reach a key person, service, or resource on location.

BE PREPARED

Travel, Lodging, Food, Etc.

Regardless of what anyone says, location shooting is not a nine to five o'clock job. There are long hours and, more than likely, a crew may be away from home overnight. Part of good preproduction planning is determining how your people get there, where they stay, and where they eat.

Make friends with your corporate travel agent. Spend time explaining the "craziness" of contingency plans and the probability that you will have to shift gears quickly. Get your corporate travel and cash-advance requests in early enough so that the system can handle them. If you don't have a travel agent, get one. A good one will save your life when you have to get your tired and hungry crew and all that equipment from one place to another.

Planning for the Shoot: Cash

One of the greatest difficulties your company accountants will have is understanding why you need cash on location. They'll want purchase orders, bills, blue forms, expense reports, and your signature in blood. Give them what they need. Just get the cash. Little expenses on location mount up:

FUSION MEDIA INC.

PRODUCTION MANAGER CHECKLIST

Date of Survey ______

Production ______ Production No. ______

Production Manager ______ Director ______

Shooting Dates ______ Client ______

Contact ______ Phone(s) ______

Location ______ No. of Locations ______

Location Contact ______ Phone ______

Location:
Shooting Area ______ ☐ Photos Attached ☐ Sketch Attached
Lighting ______ ☐ Photos Attached ☐ Sketch Attached
Windows ______ ☐ Photos Attached ☐ Sketch Attached
Ceiling Height ______ Sound Level ______ Heat-A/C ______
Storage/Security ______
Loading Dock/Elevator ______
Communications ______

Permits: ☐ Required - Applied For ______ ☐ Not Required

Security Access: ☐ Required - Applied For ______ ☐ Not Required

Power
Outlets Available (120 VAC) ______ Amperage Available ______
Amperage Needed ______ To Be In Place By ______
Electrician ______ Phone ______

Transportation:
Crew ______
Cast ______
Equipment ______

Lodging: ______ ☐ Not Required

Food: ______ ☐ Refrigerator Available

Personnel:
Subject Matter Expert ______ Phone ______
Assistant Grip ______ Phone ______
Cast ______ Phone ______
______ Phone ______

Crew: ______

Equipment Needed: See "Equipment Checklist" ______
Supplied By Client ______

Comments: ______

Figure 21-2 Production manager's checklist. (*Fusion Media, Inc.*)

a parking ticket here, a roll of aluminum foil there, a dozen coffees, some clothespins. Location shoots roll on money. Many of your expenses will be prepaid or billable, but it's those little things that can slow you down. Just remember, "Cheap is expensive."

Crew Comfort

Don't freeze or fry your people. On cold days, plan to provide hot liquids and a warm, windproof area. If necessary, keep the cars running with their heaters on. Hot weather can be just as difficult. Make sure shade is available for your crew. Cold fluids are a must. Remember, the fight against the elements to get that footage is like war. The crew should be dressed for battle.

Allied Supplies

One of the best sources of information on the location planning needs of your crew is a good outdoor equipment store. You probably aren't surprised to hear that some of the best production people I know are avid sailors, campers, scuba divers, mountain climbers, and skydivers. A quick look through a catalog of this kind of sporting gear will clue you in to crew needs. Gear must be light, small, safe, durable, comfortable, warm, flexible—in short, all the things you need on location. Look closely. Isn't that a Swiss army knife strapped to the belt of your gaffer?

Protecting Equipment

Just as people are affected by weather extremes, so is video equipment. Make sure you investigate the temperature and moisture tolerances of your equipment, especially cameras and videotape recorders. Take measures to protect equipment from environmental extremes. Batteries are particularly sensitive to heat and cold. Protect them, or you may end up without juice—out in the cold.

Weather: Alternate Shooting Plans

Check the weather reports when planning an exterior shoot. The local weather bureau's telephone number should be listed on your contact sheet. But no matter how good the weather forecast is, you always need to plan interior "cover" scenes that require the same people you are using outdoors.

When Bad Weather Is Good

When you really require inclement weather for artistic effect, call in a special effects person. Or wait. If weather is bad and you must shoot in good weather, go to your interior cover set. If weather is bad but the scene would look nice with a little snow falling over the factory, take advantage of it.

Safety

Two words: *Safety first.* Know where you can reach medical help. Keep a complete first-aid kit on location.

Backups and More Checklists

Careful planning of your backup systems on location can mean the difference between a successful shoot and having to pack it up without getting the footage. You must plan, on paper, a backup procedure for everything that could possibly go wrong. For example:

What do you do if the camera goes down?

What do you do if you have one too few audio adapters?

What do you do if traffic slows down your crew?

What do you do if someone gets sick?

What do you do if it rains?

What do you do if you miss the out-of-town flight?

What do you do if you can't find . . . ?

You need to sit down and create your own list, gearing it to the specifics of your production, and decide what you do. The checklist of equipment is crucial (see Figure 21-3). Make sure that every little adapter is itemized and accounted for, and that you know where to find it—fast. If you're on location, make sure that your contact list includes a number of places where you can get backup equipment. If you will be working out of town, contact on-location rental houses, video facilities, and television stations during preproduction to find out what's available, and to whom you have to speak. You can imagine how valuable this can be.

PERMITS, CROWD CONTROL, AND RELEASE FORMS

Local Permits

One bad habit that portable video equipment tends to foster is the idea of running out in the street and shooting without getting a permit to shoot. It's been done, it probably won't land you in jail, but I do recommend obtaining permits. Most major cities, and all states, have offices of television and motion pictures. Call Information for their names. Make friends with your local commissioner. When you're shooting on company property, you might not need to let the city know. But how about the sidewalk in front of corporate headquarters? How about the person-in-the-street interviews?

A call to the film office will often reveal good information about locations. This is important for out-of-town shoots. Shooting permits can sometimes get you special parking status, which can be very convenient.

If your shoot is big enough to require police protection, the city office can usually arrange for it. Because city commissions are often tied up handling the problems of high-budget feature films and broadcast projects, try and give them as much advance notice as possible if you plan to be in the street or any public place.

21-8 PRODUCTION WITHIN PRIVATE TV

FUSION MEDIA INC.

PRODUCTION EQUIPMENT CHECKLIST

Project Title _____   Shooting Date _____

Director _____ Producer _____ Location _____

VIDEO
- ___ Camera/Lens _____
- ___ Recorder _____
- ___ Monitor _____
- ___ Cart
- ___ Batteries/Charger
- ___ Blank Tape Type _____ Qty ____
-              Type _____ Qty ____
- ___ Wave Form Monitor
- ___ Processing Amplifier
- ___ Tripod _____
- ___ Tripod dolly _____ Filters ____
- ___ Body Brace
- ___ Test Chart w/Stand
- ___ Headphones
- ___ Cables   BNC - BNC _____
-          mini – mini _____
-          mini – canon _____
-          canon – canon _____
-          UHF –UHF _____
-          phono – phono _____
-          8 – 8 pin _____
-          8 – 10 pin _____
-          Other _____
- ___ Cable Adapter Kit
- ___ Misc Video Equip: _____

LIGHTS
- ___ Instruments _____
- _____
- _____
- _____
- ___ Extra Bulbs _____
- ___ Barn Doors
- ___ Pic Stands
- ___ Gels and Filters

SOUND
- ___ Recorder _____
- ___ AC Adapter/Batteries
- ___ Take up reel
- ___ 1/4 inch tape 5 in _____
-              7 in _____
- ___ Microphones _____
- ___ Cables _____
- COMMENTS: _____

FILM
- ___ Camera/Lens _____
- ___ Magazines
- ___ Batteries/Charger
- ___ Film Stock _____ Qty ____
-       Stock _____ Qty ____
- ___ Empty reels/cores
- ___ Empty film cans
- ___ Light Meter
- ___ Tripod _____
- ___ Tripod spreader _____ Filters ____
- ___ Tripod dolly
- ___ Body Brace

GRIP EQUIPMENT
- ___ Slate
- ___ Changing Bag
- ___ Dust Off
- ___ Lens Tissue
- ___ Lens Cleaning Fluid
- ___ Head Cleaner (video)
- ___ Head Cleaner (sound)
- ___ AC Extension Cords
- ___ Multiple Outlet Boxes
- ___ 3 – 2 prong adapters
- ___ Camera Tape
- ___ Gaffers Tape
- ___ Dulling Spray
- ___ Paper Towels
- ___ Camera/Sound Log
- ___ Paper, pens, markers
- ___ Flashlight
- ___ Release Forms
- ___ Authorizations
- ___ Script/Storyboard/Production Board
- ___ Megaphone
- ___ Tools
- ___ Stop watch
- ___ Tape measure
- ___ Polaroid camera
- ___ Polaroid film Type _____ Qty ____

ADDITIONAL EQUIPMENT NOT LISTED ABOVE:
_____
_____
_____
_____

Figure 21-3 Production equipment checklist. (*Fusion Media, Inc.*)

Corporate Permission

Clearing company locations in advance is critical. You might think people would be falling all over themselves to give you locations. If they are, consider yourself lucky. Remember, *they are not in the business of making videotapes.* A crew in the workplace is an interruption to business-as-usual.

Your contacts should be firmed up during preproduction. Get to know the people you will need when you arrive at their location. They want to help. Teach them how. A carefully planned checklist, assembled during the technical survey, can be invaluable. Also, let people know if you plan to do anything unusual to their workplace. You might have to rearrange furniture. You might have to disconnect a few phones. You might need to remove a few hung ceiling tiles to place that extra light. Keep as low a profile on location as possible. Hold a preproduction meeting with the crew to remind them of the essentials of etiquette with bystanders and clients.

Another cardinal rule for location shooting is, "Always leave the location exactly the way you found it." Cleaner, if possible, with no gaffer's tape stuck to walls or coffee cups strewn around. Before the crew moves a stick of furniture, have a dozen Polaroids shot of the office or location. That way, you can return everything to its original place.

Crowd Control

Let's face it. Every time you see a crew shooting on location, you walk over and join the crowd of bystanders to find out what's going on. "Who is it?" "Any stars?" "What station?" "What camera?" A camera always draws attention. Crowd control on location is critical to timely completion of that shot. Tell onlookers what they want to know, be courteous, but get them out of the way of progress quickly.

Production Assistants

It makes sense to budget for a number of production assistants whose sole job is crowd control. Controlling the flow of people traffic behind the camera (as opposed to in front of it) is simple but crucial. If you need an area cleared, don't expect the sound person or camera person to be able to do it. They have headphones to monitor and lenses to look through. Walkie-talkies are excellent for your crowd control team. They can let one another know who's coming by to ruin that shot, or when the noisy diesel van is rounding the corner to wreck the sound. Most film-rental houses can supply walkie-talkies.

Release Forms

Whenever anyone appears on camera, get a release signed. This protects the producer or company from legal action. Even though most private television programs appear in-house, suppose the show is so good that your company wants to market it to other companies? Whenever you plan to shoot on location, bring release forms. Some forms state that the consideration received is a dollar; therefore, you may have to bring a stack of one-dollar bills to make it a binding contract. Before you start handing out money, meet with your corporate or institutional legal department to draw up a release that is legally acceptable and not too intimidating to sign, and that addresses the "consideration of value received" issue. I've successfully used versions of the release form shown in Figure 21-4.

INSURANCE

I recommend insurance for all location shoots. It pays to set up a conference with your corporate or institutional insurance department to discuss the special needs of a location shoot. You may find that your organization already has coverage or can obtain what you need. Many low-budget productions try to get by without insurance. Do not, repeat *do not*, make this fatal mistake. Insurance is one thing you hope you don't need. But it takes only one production accident to show you how potentially expensive that mistake can be. Most rental houses require the equipment to be insured. Most city film and television offices require a certificate of liability before issuing a permit to shoot in the street. To give you an idea what you should insure, consider the expense of loss or damage to the following production elements:

Crew, cast, or vehicles

Injury to the public or to property on location

Master footage (insured for the full production cost—think how much it would cost you to replace it)

For specialized advice, make the effort to locate an insurance agent who has experience in film and television coverage. A call to local film commissions, production houses, or TV stations can help in finding one. Find out what's required for every special situation you may stage.

TRANSPORTATION

Transportation needs can be as simple as a station wagon for you, your camera person, a camera, videotape recorder, and a few cables, or as complex as a fleet of vehicles and vans for crew, cast, and lots of lighting equipment. Don't neglect to plan for sufficient vehicles for all your people and equipment needs. Company vehicles are fine, but make sure the company knows how long you will need them. Also make certain, for insurance purposes, that nonemployees (such as your free-lance crew) can legally drive them.

Vehicle rentals are fine also. Be very sure you plan and budget for them. Gas, tolls, and parking cost money also. Unmarked, low-profile vehicles are best for location shoots. They don't attract attention. Carefully plan your equipment packing and crew logistics. People and hardware are both fragile. It doesn't pay to overstuff a car with crew and equipment at the expense of safety or comfort.

FUSION MEDIA INC.

AUTHORIZATION AND RELEASE
FOR VIDEO TAPE RECORDING

SUBJECT: _____ DATE: _____

In consideration of value received, I hereby grant _____ permission to copyright and use video tape recordings of me in connection with the subject production in any manner or form for any lawful purpose at any time. I waive any right that I may have to inspect or approve the finished product or the written copy that may be used in conjunction therewith, or the use to which it may be applied. I release and discharge _____ from any liability to me by virtue of any alteration that may occur in the making or editing of said video tape recordings. I agree to hold _____ harmless from any liability to others arising from the use by _____ of anything I may say or do during said video tape recordings.

I have read this agreement before signing below and warrant that I fully understand its contents.

Signed: _____

Address: _____

Parent or Guardian: _____
(if not of legal age)

WITNESS: _____

Figure 21-4 Release form for appearing in a video presentation. (*Fusion Media, Inc.*)

Drivers

If you can swing it, get a production assistant to drive. The director should be free for conferences with key crew and cast members en route to the location. Those few minutes of script run-through or shot blocking can save hours on location.

PRODUCTION

Location shooting is really the art of preproduction. If you've done that well, this part should be a breeze. Your people all know what they have to do, where they have to be; everything works, nothing breaks, nothing goes wrong. Details of lighting, audio, makeup, and the other crafts are covered in other chapters. But here are some keys to solving location problems.

LOCATION LIGHTING

The basic principles of lighting as discussed in Chapter 16 apply here. The main difference is that on location, you don't have a grid already on the studio ceiling. You have to create the lighting support system around the location. Location lighting is really a matter of improvising with stands, clamps, ladders, and cables to give you effective illumination.

Most small locations—those up to about a dozen feet in depth—can be lit from floor stands. When you have floor stands on a set, never neglect to weight down their bases with sandbags. Make sure cables are safely dressed. Safety is paramount when you are dealing with electricity.

Many office ceilings are suspended, and a tile or two can be pushed up to allow a clamp to hold a lighting instrument in position. An experienced gaffer can safely hang lights in the most inconceivable positions.

When you need to shoot a very large area, floor stands probably won't work because you can't get the lights high enough. Ladders, scaffolds, and other structures can be used here. During the technical survey, the lighting director or director of photography can make notes and sketches about where lights have to be hung. Any special stands, clamps, or devices should be planned for. A good gaffer usually has lots of tricks with clamps, pipe supports, beams, and, of course, gaffer's tape.

Power for lights is a main concern. Trust this job to an electrician. Don't try to tie in to the house system yourself. Once you have determined how many instruments you'll probably need, you can estimate the number of amperes you'll require. Even if you are using small lights that you can plug into the wall, you need to have a house electrician available. In your technical survey, you should locate the fuse boxes or circuit breakers.

It goes without saying that backups are critical in your lighting equipment on location. Extra lamps, extension cables, adapters, more extension cables, and more adapters should be the rule when packing.

AUDIO

I will guarantee that if there is a flaw in an otherwise perfect location shoot, it will be a sound problem. The place may look superb, but stop and listen to what it sounds like. If you plan to voice-over the sequence, fine. If you need to crack open a microphone and record a person or a thing, bring someone with audio expertise to the location survey. Background noise, echo, hum, and wind are big problems on location. You don't want to eliminate them completely, because the ambient sound on location can add to your production value. Personally, I'd rather have less background noise than I need, and add it later in the sound studio, than to have to worry about filtering it out and trying to "fix it in the mix." It pays to record "silence" or location ambience tone for each location. Doing so will make editing a lot smoother.

Windscreens are useful outdoors. So are good headphones. Wireless microphones can be used, but make sure to test everything out and ensure that the receiver will pick up in your particular instance. And whatever you do, don't forget those little batteries that fit inside the mikes that need them. Bring extras.

LOGGING

It is especially important to log all takes as recorded. Because you may never have the opportunity to return to a location, make sure you have created a written *shot sheet* during preproduction. This lists all setups, angles, and script copy to be read. Each required shot should be numbered on the script. It's simple. You know what shots you need before you get there. You have the responsibility to make sure you get them before you leave.

A production assistant (script person) should be responsible for recording all takes on a camera log. A slate should be used for visual identification, and a verbal slate should accompany sound takes.

Everybody has different idiosyncracies when it comes to slating. All I can say is, do it. It will save lots of time on location because you will always know where you are, and you will be able to check the shot off on your shot sheet. Careful logging will also make editing much easier and faster. If your VTR lays down time code, it may help to note this on your log sheet. (See Figure 21-5.) You may also want to note which takes are particularly good or bad.

ATTITUDE AND CREATIVITY

On location, *attitude* is a balance of patience, persistence, and concentrated observation. Communication skills are important in the teamwork aspects of production. A director's basic resources are his or her talent and that of the crew. On location, you and the crew must "feel" what's going on and flow with the situation to your advantage. Often a suggestion or circumstance will give you a great opportunity for that "priceless" shot.

FUSION MEDIA INC.

VIDEOTAPE LOG

Production Title ____________ Director ____________

Location ____________ Camera Operator ____________

Date ____________ Tape Operator ____________

Tape Roll No. ____________ Camera ____________

Tape Stock ____________ VTR ____________

Scene	Shot	Take	Sound	Description	Comments

Comments/Notes: ____________

Figure 21-5 Videotape log form. (*Fusion Media, Inc.*)

How to Deal with Downtime

Patience with people and equipment is essential on location. Equipment problems demand a sense of humor, and should be turned into opportunities for rethinking possibilities. Equipment downtime should be the focus of those responsible for getting it back on line, but the director should divert attention from the delay and use the time to catch up on some preparation or new contingency plan. Constant communication with the crew is essential here.

TALENT ON LOCATION

In the controlled chaos of a location shoot, it's often easy to neglect the on-camera talent. Never forget for a second that what goes on in front of the camera is what private television is all about. That's what you're there for. Private television producers use both professional and nonprofessional performers on camera. Their needs are basically the same. During a recent shoot, I talked with former NBC anchorman and "Today Show" host Jim Hartz. He agreed that it's tougher for talent to shoot on location than in a studio. "When you're out on location, the unknown is most distracting. A performer needs concentration. It's critical to reduce the number of distractions to a minimum."

How to Keep Your Talent Happy

1. *Get the script to the talent in advance.* That way, they understand what you're doing and feel comfortable with the sequence.
2. *Try to create and maintain an atmosphere on location that allows everyone to do his or her best.* Maintain a relaxed, comfortable air of professionalism. Never allow crew tensions, scheduling problems, or the inevitable crisis to reach the on-camera performers. They like to be informed, but tensions can be easily transmitted and will show up on camera. If possible, isolate the talent from the setup until required.
3. *Try to minimize a performer's discomfort before, during, and after each take.* If he or she is to stand during the take, make sure there's a chair available for use between takes. If it's chilly but the script doesn't permit an overcoat, a "coat-holder" production assistant should be provided.
4. *Have what you need when you need it.* A production assistant go-fer should be available to make sure the performers, as well as crew members, have coffee, soda, candy bars, etc. This is not "star treatment." Although you do want to make the performers feel important, comfort is your goal. Anyone who accepts being taped up with gaffer's tape on microphone cables, getting lights and reflectors shined in his or her eyes, and having to walk to the same mark and say the same line through a dozen retakes, should have what he or she needs when it is needed.
5. *Any time the performer tells you something, listen.* The nonpros may not know why they're uncomfortable, but you should find out and remedy it. Everything that happens to every performer will show on the tape.
6. *Patience is essential.* A nonpro isn't used to the "repeat, retake, wait for tape, wait for color bars, wait for sound" schedule that the pro knows

well. Also, the nonpro needs to know why you want to retake (even if you have to make up a reason). The person asks, "What am I doing wrong?" The director's skill is important here.

WRAPPING THE LOCATION SHOOT

Thank-you letters are a must. Copies to a person's boss are nice, especially if you have good things to say about his or her cooperation during a shoot. The boss just might reciprocate. Good relations on location make the next shoot better. Bad relations on location make the next shoot nonexistent.

Many producers cover their location shooting with still photographs. These not only are good for internal promotion, but make a memorable enclosure in your thank-you note. Everybody wants to get into the act, so make sure you get a shot of that key executive or subject-matter expert next to the camera. It really goes a long way for good corporate relations.

To be successful, production is really a matter of attitudinal and organizational control. Other than the fact that you're not shooting in a studio, location shooting really isn't so very different. Through careful planning, you've minimized the things that could go wrong, and you've come up with contingency plans for those things that do go wrong. The goal is to capture control over the environment once you leave the studio.

22 Editing

PATRICK B. HOWLEY

The videotape editing process involves dialogue between director, editor, and equipment. In a private television environment, often the director and editor are the same person. The dialogue between editor and equipment can range from simple button pushing to computer keyboard entry and, in the not-too-distant future, it will probably be in spoken words. This chapter will examine the different types of editing equipment available to the private television operation with equipment and dialogue in mind. Editing techniques and preproduction planning will also be covered.

DEFINITIONS

Before we progress to the various types of editing systems available to the private television environment, it is essential that certain basic concepts are fully understood. The following list of definitions will be helpful in the analysis of editing systems and the editing process in general.

Control Track. A reference signal recorded on tape to allow playback servos to maintain stability. In film, analogous to sprocket holes.

Edit Pulse. A magnetic marker added to the control track to indicate the start of a new frame.

Guard Band. The blank areas on tape separating recorded magnetic tracks.

Cue Track. Usually the second audio track on videotape. It is used to record time-code information or cue tones or as an additional audio track. Some videotape recorders (VTRs) contain three channels of audio, and the third channel is used as the cue track.

SMPTE Time Code. A binary electronic signal that can be recorded on a videotape cue track. Every single video frame is electronically identified by its individual hour, minute, second, and frame count. Analogous to edge numbers in film. (SMPTE refers to the time code devised by the Society of Motion Picture and Television Engineers.)

Burned-In Code. Visual time-code numbers for each frame are electronically keyed into the video and recorded on tape. Still-framed video shows exact time-code number. (See Figure 22-1.)

22-2 PRODUCTION WITHIN PRIVATE TV

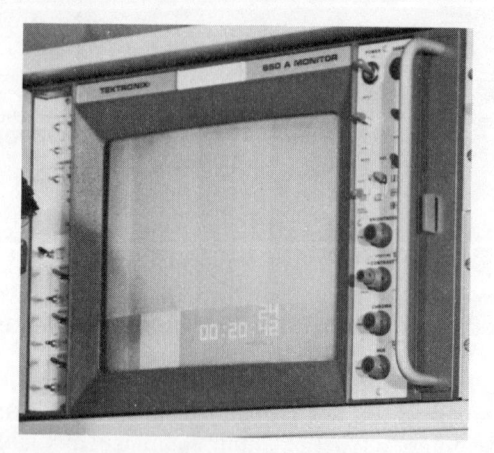

Figure 22-1 Burned-in time code. (*RCA.*)

Assemble Editing. A new control track is laid down along with program material on master tape. This process is satisfactory for sequential butt edits when building the body of a program but cannot be used for dropping in segments in a tape that has been previously recorded. Characterized by clean or stable "in" but poor "out" edits.

Insert Editing. Program material is inserted in the master tape without disturbing previously recorded control track. May be a video-only or audio-only insert or any combination of video and audio. Characterized by both clean "in" and clean "out" edits.

Basic Tape (Black). A videotape that is recorded with black video, time code, and control track. Insert edits are then performed into this tape, which is placed on the master machine.

On-Line Editing. Building a master directly on the same format on which the original footage was shot.

Off-Line Editing. Building a work print and edit-decision list on relatively inexpensive, still-frame, and jogable tape machines. This edit list is used to conform the master on the type of machines on which the original footage was recorded. (See Figure 22-2.)

Edit-Decision List. A list of time-code numbers and associated reel numbers with instructions for transitions such as dissolves or wipes. May be a standard CMX type of paper tape or printout, a floppy disc, or handwritten notes. CMX is a manufacturer of editing equipment. Many of its pioneering contributions, such as edit displays and storage formats, have been adopted by other manufacturers.

Auto Assembly. The conforming process of edits done automatically from a computer edit list.

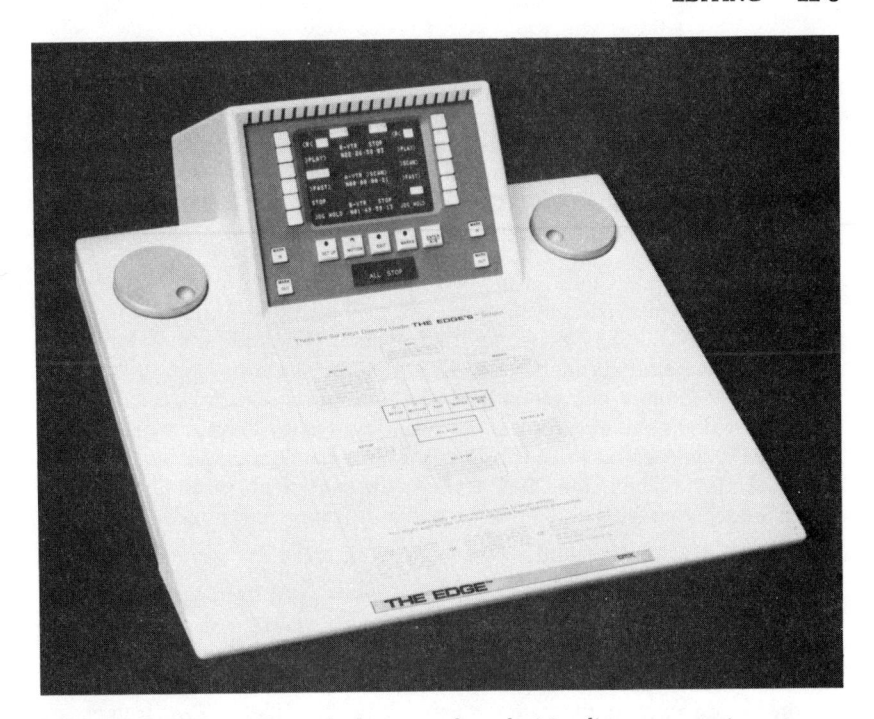

Figure 22-2 Off-line editor which can produce decision lists on paper tape or floppy discs. (*Orrox Corporation.*)

Look Ahead. A feature of some automatic assembly systems. The computer will command the tape machines not involved in the current edit to look ahead and cue up to their next "in" point.

Checkerboard Auto Assembly. The computer will perform all edits involving the reels that are loaded on the tape machines. The program is assembled out of sequence and holes are left to be filled in by later edits involving other reels of tape. This system saves time-consuming reel changes.

List Management. The ability of a computer to rearrange or change an edit-decision list after it is composed.

Time Base Corrector. A device that takes a videotape playback and, by either speeding up or slowing down the signal, makes it synchronous with a reference signal fed into it. Allows videotape playbacks to be passed through a switcher for dissolves or wipes.

User Bits. Thirty-two bits of the SMPTE time-code signal that can be used for any purpose the user desires. Scene or take numbers, date, or film-edge numbers are a few of the more popular uses.

Vertical Interval Time Code (VITC). Time-code information is inserted into two or three lines in the vertical interval of the video being recorded.

When played back at slow speeds or freeze-framed, the time code can be read as long as the tape machine can maintain a locked picture horizontally and vertically. This process achieves all the properties of burned-in code without actually burning the code into the video. Above approximately one-third speed, the VITC reader usually switches back to reading longitudinal time code on the cue track. VITC equipment and on-line 1-inch VTR editing is a very powerful combination.

Drop-Frame Time Code. Converts SMPTE time code to time-of-day type code. The color television system we use in the United States is the NTSC (National Television System Committee) system. (For additional information regarding television standards, see Chapter 25.) The frame rate is approximately 29.97 frames per second, so that for every second of time, 29.97 frames of video have occurred. If our frame rate were 30 frames per second, the SMPTE code could be set to match a time clock and always read time of day. Unfortunately, the 29.97 frame rate causes an error of .03 frame per second. Expressing this error in terms of frames per hour:

Error = .03 frame/second × 60 seconds/minutes × 60 minutes/hour
= (.03) (60) (60) frames/hour
= 108 frames/hour

In other words, the SMPTE clock runs faster than the clock on the wall. Most time-code generators add circuitry to allow dropping frames so that the time code runs at the same speed as the time clock. Two frames are dropped every minute except for the 10th, 20th, 30th, 40th, 50th, and 60th minutes. Thus, 2 × 60 = 120 minus 2 × 6 = 12, or 108 frames, are dropped.

Drop-frame code is useful in editing 1-hour features where the length of program is critical. The program length is calculated by subtracting the time code at the beginning of the show from the time code at the end of the show. Using standard code would give an inaccurate length of show. The 108 frames of error is equivalent to 3.6 seconds of error. Most generators give you the choice of using drop-frame or non-drop-frame SMPTE time code.

Jam Sync. Some time-code generators are capable of slaving to an external source of time code. This external source could be a tape machine time-code playback or another generator. Once locked, they can run in one of two modes of operation: (1) stay in sync with the external code source and follow any change in time-code numbers; or (2) generate a new code in sync with the external source even if the code becomes bad or even disappears.

Mode 1 is called *regenerating time code,* and mode 2 is *jam-syncing time code.* Jam-syncing time code is useful in correcting low-level or marginal-quality time code. Any tape to be corrected (providing it contains continuous code) is fed to the slave sync generator, and once the slave is in jam sync, the new code is rerecorded over the old code. It is worth mentioning that the regeneration of code does not require a slave generator. It can be regenerated by electronic circuitry contained in most time-code readers. Correcting bad code requires time-code generators with the jam-sync option included.

EDITING SYSTEMS AVAILABLE TO THE PRIVATE TELEVISION OPERATION

Two-VTR Systems

In their most basic form, these VTR systems contain a controller and two videotape machines (Figure 22-3). The editor operates the controller which shuttles the tape, searches for in and out points, enters these points on the controller, allows multiple previewing and trimming of the in and out points, and then completes the edit. The accuracies of the edit-points range from plus or minus eight frames to complete frame accuracy.

The controllers can either count control track pulses or read time code to keep track of position on the tape. Some systems use a combination of the two. Time code is read until the tape speed is too slow for accurate reading, and then control track pulses are used to update the time-code display. Once tape speed is fast enough for time-code reading, the display switches back to time code. Good time-code readers should read down to at least one-twentieth speed.

Reading Time Code. Remember that SMPTE time code is a binary serial signal that is recorded on the tape machine cue track. When the tape is not moving, it is impossible to read the code on the cue track. As the tape is slowed down to a stop, the reader will show the last number it was able to read before tape speed was too slow for accurate reading. It is possible that the actual frame on which you have stopped is different from the time code displayed. This is easily verified by using burned-in code along with cue track code and slowly bringing the tape to a stop. If a speed below the reader's capability is reached before stopping, the time-code display will read less than the burned-in code in your video. Entering your in and out points from the burned-in code on the freeze-framed video is the most accurate method of editing. It is, however, time-consuming, as each point has to be manually entered from a key pad. Also, the edit has to be done with burned-in code in the video and therefore must be done off-line.

The vertical interval time code was invented to get around the problems mentioned above. Besides using cue track SMPTE code, two or three lines in the vertical interval of the video are encoded with the same time-code numbers in the cue track. Whenever the video is run at very slow speeds or during freeze frames, the code is read from the vertical interval.

The industry standard for time code is the SMPTE time code. Some editing systems offer alternate versions of time code, and if you are a self-contained in-house facility, you may want to forgo compatibility with SMPTE time-code systems. One advantage you may realize is that some of the non-

22-6 PRODUCTION WITHIN PRIVATE TV

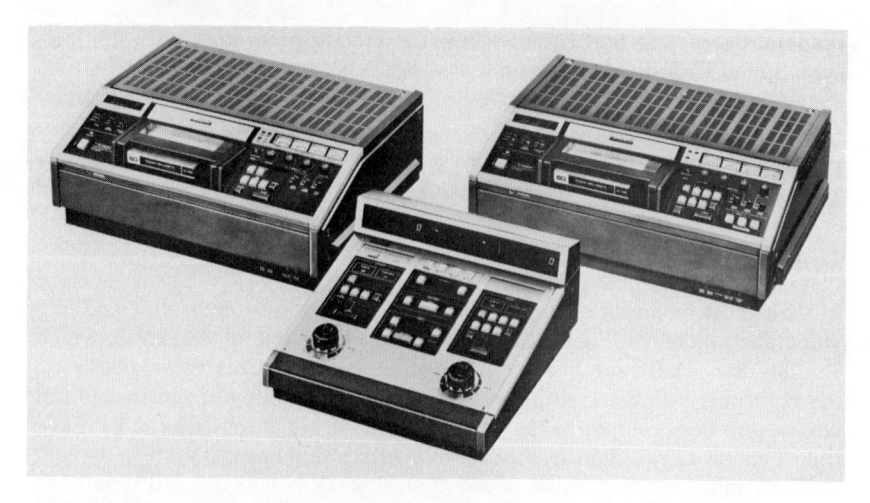

Figure 22-3 Two-VTR editing system. (*Panasonic.*)

SMPTE code systems do not use an audio track, thereby giving you an additional channel for audio.

The more sophisticated editing systems, however, use SMPTE time code, are frame-accurate no matter how many times an edit is previewed, and have elaborate means of speed control and jogging tape. The top-of-the-line systems are expandable to multiple VTR systems and are software-based, allowing changes to be made without extensive hardware modifications. Some systems will remember a number of edit-points; others will generate a CMX compatible edit list and store it in some form: printout, floppy disc, paper tape, or magnetic tape. Some systems will record the edit list on the beginning of the master tape.

Multiple VTR Systems

These systems may range from control of three VTRs (two play, one record) to many playback VTRs and more than one record VTR. Computer control of dissolves or wipes and audio fades is usually standard equipment. Some systems will use a built-in switcher and others will interface with the switcher of your choice. Most systems will generate an edit list and store it. Many users decide to get a printer for edit list printouts and also elect to store the list on either floppy disc, paper tape, or magnetic tape.

Some of the options offered on the top-of-the-line systems are sync roll, auto assembly (standard, look ahead, and checkerboard), control of audiotape machines, film chains, digital special effects, character generator, and intelligent switchers.

Let's start with the basic three VTRs and switcher-audio package. Upon examination, most systems have keyboard data entry and video display of what the computer is doing. It is worth mentioning that other types of systems are being, or shortly will be, offered that don't necessarily use keyboards or buttons for entry of data. Light pens, touch-sensitive cathode-ray tubes (CRTs), digital bit pads and speech recognition may be used. The CMX 600, the first SMPTE time-code computer-assisted editing system that was introduced in 1971, uses a light pen instead of a keyboard and black-and-white video stored on computer discs instead of moving tape machines. I feel that editing systems of the future will be modeled after the CMX 600—but more about that later in the chapter.

Some systems offer light-emitting diodes (LEDs) or liquid crystal displays instead of the video display. This provision eliminates the cost of display monitors. One advantage of the video display is the ability to turn any monitor in the house into a display by connecting a video coaxial cable.

Generally, there are two types of systems: (1) distributed processing and (2) nondistributed processing. The distributed processing systems are usually much more expensive, but for good reason. Each VTR and switcher has its own interface unit that contains a microprocessor. The interface is able to take commands from the central processing unit (CPU), to carry them out, and to report the status back to the CPU. These systems are very powerful, and they are easily expanded by adding additional interfaces. Communication between CPU and interface is usually by a three-wire cable. It is not unusual to see a large distributed processing system with the following devices under computer control at the edit session: four playback VTRs, two record VTRs, a multitrack audiotape machine, a telecine, digital special effects, an intelligent switcher, an audio mixer, and a character generator.

The nondistributed processing systems are not as fast or powerful. The system relies on the CPU for most of its intelligence, and therefore all interfaces must share the CPU to do such tasks as finding time-code numbers and synchronizing tape machines. These systems are usually less expensive.

Most of the systems available today are software-based, which means that modifying or changing the system can be accomplished by computer programmers without involving hardware modifications. One thing to be aware of is that although almost every manufacturer will emphasize how easy it is to update this system by software changes, getting them to do so is quite another story. Software changes are expensive, and most of the manufacturers are not responsive to custom software changes for your operation. Many of the top postproduction houses have their own software people on staff to do these modifications.

THE EDITING PROCESS

How to Shoot Video with the Edit Session in Mind

The editing process can be made easier by thinking ahead when you are shooting. Here are some suggestions.

1. Plan all your transitions between scenes with the edit session in mind. If you plan for these transitions before you shoot, you will not end up with unwanted jump cuts.

2. If you will be editing on a two-VTR edit system, you must build all your transitions other than cuts into the shoot. Live dissolves or wipes are done during the original recording.
3. After the completion of shooting at one location, record some "room tone." This is done by recording, with all the microphones in the same configuration as they were during the previous shooting, with the mixer gain settings the same but with all actors and crew silent. Room tone is necessary when you do audio-only edits and do not want to leave holes in the program sound.
4. Shoot cutaways of all the actors. Also shoot some "wild" footage of the surrounding area if time and tape permit. The cutaways may come in handy to cover unwanted jump cuts, sloppy camera work, or missed switches by the technical director.
5. Record any material sound effects if you are in the field and time permits. Doing so may save you some sound-effect searching before the edit session. Such recording can be done by a single person with an audiotape recorder without tying up the crew.

How to Prepare for an Edit Session

The following guide may be helpful if you work with SMPTE time-code systems:

1. Make sure that all reels have ascending time code. That is, the code is always increasing as the tape goes on. Code out of order will cause the computer to have trouble cuing up tapes to edit points.
2. Always make sure there are at least 15 to 20 seconds of video which contain time code before the edit-point. This provision will allow the computer to be able to cue up the VTR.
3. Make sure that all time code is properly phased and synchronous with the video. Failure to do so will cause repeated aborted edits. The frames bits of the SMPTE code should line up with the vertical interval of the video. All SMPTE generators should be locked to the system sync generator or the remote camera if you are recording in the field. If you are shooting multicamera in the field, each tape machine should receive one common time-code source. The time-code generator should be locked to the sync generator that is driving all the cameras. This is especially important if sync rolls during the edit session are required. Also when shooting on ¾-inch cassette, it is better to record time code after the shooting is complete in order to minimize crosstalk into the program's audio.
4. All reels of tape should be accompanied by a take sheet which contains scene and take numbers, corresponding time-code numbers, description of the scene, and quality of the take. Quality of the take is especially important if, during the edit session, something is found to be wrong

with the intended good take. Finding the next best take will be easy and not require viewing all the remaining takes while the edit session grinds to a halt.

Off Line versus On-Line Editing

Most editing jobs should be done off-line first. With the exception of simple jobs with only a few edits and reel changes, on-line editing should wait until a rough cut has been approved by the client. The on-line edit may then be an auto assembly from the edit list generated during the off-line session.

Some of the advantages of off-line editing first are:

1. **Cost:** It is much cheaper to make edit decisions on a less expensive system.
2. **Creativity:** It is easier to be creative without the added technical pressures of the on-line session.
3. **Speed:** In and out points are easily found because of the nature of the off-line VTRs and the burned-in code. Off-line VTRs are usually selected because they are capable of viewing scenes at various speeds and are able to freeze frame.
4. **Less chance of master tape damage:** Edit decisions are made using burned-in code dupes of master footage. Many previews can be performed without causing tape dropouts or edge damage to the original footage. The master footage is used only to conform the master tape on-line.
5. **Ability to make changes in program easily:** A second rough cut is easy to obtain by taking parts of the first rough cut and going another tape generation.

Reason 5 can't be emphasized enough. Only after the final rough cut has been made and approved is the master assembled. Often the first cut is for content, and later rough cuts are for proper pacing and style. Trying to edit on-line without off-lining first often causes programs to lack the proper pacing necessary to achieve the desired results. Also music and effects can be experimented with during the rough-cut stage and any changes can be made in the rough cut to accommodate them.

Editing Procedure

Here are some guidelines for a successful edit session:

1. Start with preproduction planning; shoot with postproduction techniques in mind. Fixing problems in postproduction doesn't always work.
2. Log scenes with description, quality of performance, time-code numbers, and comments. Accurately label tape reels with the numbers and content of each reel.

3. View all recorded footage, collecting scene and take numbers and time-code numbers. (Dubbing to cassette or reel-to-reel with burned-in time-code numbers can be done at this point.)
4. Record graphics and build any special effects or opticals on a separate reel.
5. Edit either off-line followed by on-line conforming or directly on-line. Editing off-line consists of building a work print and developing a decision list. Conforming on-line from edit list is done either automatically or manually. Art cards and special effects can be done here.

    On-line requires making edit decisions as you build the master. This process is very expensive if there are a lot of critical edits. Art cards and special effects also can be added here. Minor sound editing and mixing may be done at this stage.
6. Add sound track if done separately or strip off audio and lay it back with effects and sound sweetening.

As a footnote to the editing process: This procedure is for a typical program containing most of the elements of production. Some programs will not include all the above steps if the time element is critical.

When designing graphics and special effects, it is important to keep the amount of tape generations down to a minimum. Multilayer graphics can be accomplished by a combination of character generator pages and art cards. Pages may be built on the character generator so that each successive page creates a pop-on effect. Solid-color backgrounds and highly saturated colors deteriorate rapidly when they are tape generations down from the original. Whenever possible, add any solid-color backgrounds in the last step of the edit process. The graphic information may be keyed over the color background as it is being dropped into the master tape to save a generation. Generally 2-inch quad VTRs have the most trouble with solid backgrounds of many generations. This material can be handled somewhat better by 1-inch tape machines that record a single field of information on one head pass.

THE FUTURE

Editing systems of the future will be different in many respects as the nature of videotape machines changes. Soon, digital VTRs will become commonplace, and eventually, digital VTRs with no moving parts (with videotape replaced by memory) will eliminate tape generation and shuttling problems. All scenes will be instantly accessible, chroma keys and matte shots will be flawless, background scenes will be digitally generated, and forty or fifty generations will be commonplace in order to achieve special effects.

A typical production and postproduction facility of the future might operate as follows:

A remote crew goes into the field and shoots with miniature cameras with memory blocks inserted into the recording portion of the camera. Video, multiple channels of audio, and time code are recorded digitally and stored in the memory block.

Upon arrival back at the postproduction facility, all memory blocks are logged into the library computer and the blocks are inserted into the central memory area. The client then sits in the screening booth and views all the footage, adds comments, and picks those scenes needed for the editing session. The video appears on a large, flat screen as it is called up directly from the central memory bank.

Editors will speak to the computer. Voice recognition devices will react to all commands. A digital edited master will be assembled on-line in a real-time mode. The need for off-line editing systems will diminish. The client leaves with an edited master memory block and as many copies as necessary. Backup copies and edit decision lists are kept on file in the library memory bank.

Obviously, we will not see these advances in the immediate future, but all experts state that they are attainable. Dollars, experimentation, and time will be necessary to bring these events to fruition. But let's not lose sight of the fact that equipment does not breed creativity, and that human nature does.

Appendix:
Guidelines for Purchasing Equipment

PATRICK B. HOWLEY

Having purchased over $2 million worth of audiovisual and broadcast equipment, I would like to pass on some general rules for equipment selection.

1. *Stay away from manufacturers with no track record* in the product you are purchasing. Let someone else be their track record.
2. *Do not buy anything that is being exhibited for the first time* at a major trade show (such as the National Association of Broadcasters Convention). The product may be a mock-up or prototype from which the manufacturer hopes to gather information, such as, whether there is a market out there for the product. If you're lucky, the salesperson will not take your order, but might lead you to believe the product will soon be available. "Soon," to me, means a few months; to some manufacturers—a year and a half.

   (Also, some of the new products exhibited require seventeen engineers with tweakers and clip leads hiding behind the curtain to keep the unit running.)

   a. Going one step further, *do not buy the first production run* of a radically new product, even when the manufacturer has a good track record in similar products. Too many manufacturers release a product too early and let their first customers field-test it for them. Then the manufacturer sends teams of field-service engineers to live with their major customers until the bugs are worked out. You must sit back and wait for field-service bulletins to arrive in the mail, telling you how to modify the equipment to bring it up to specifications.
3. *Avoid used equipment* unless it is being overhauled by the original manufacturer or one of the good companies that specialize in refurbishing television equipment. Always get a partial warranty.
4. *Hire a consultant* for major equipment purchases. Then get a second opinion from another trusted engineer.
5. *Check with previous customers* of all your manufacturers. Were they able to get service and parts? How reliable was the equipment?
6. Involve your purchasing department only when you have narrowed your choices down to a few manufacturers. You may be amazed at how many "freebies" the purchasing department may be able to squeeze out of a manufacturer. But do not involve that department too early.

I will surely get hate mail from manufacturers and comments from other engineers, such as "I violated these rules and I have never been burned." I must stress that these are my general rules for private television, and under certain circumstances they may be broken.

If you have an excellent engineering and maintenance department, rules 1 through 4 may be broken. If, however, you have one engineer who does your operating, maintaining, schlepping, and other chores—beware. I might add that I have broken every rule at least once and escaped without major injuries. But, fortunately, I have always had an excellent engineering staff. One final thought. Try not to break rule 5; I have been most sorry when I've done so.

23 Videocassette Duplication and Distribution

WILLIAM B. FOLLETT

The art of duplication has many more factors than most people realize. Whether your facility produces 6 tapes a year to be distributed to 150 locations, or 150 tapes a year to be distributed to 6 locations, there are a number of important considerations which pertain to duplication with which you should become familiar.

Whether you are using an outside duplicator or perform the duplicating in-house, this chapter gives a full description of all the key factors, regardless of the size of the duplicating facility. They are:

Mastering

Duplicating

Quality control

Packaging

Distribution

Recycling

Library and storage

For the purposes of this chapter, all references will be made as they relate to the outside duplicator. Of course, some video facilities will want to do their own duplication. I only suggest that it be managed as a separate entity—with its own designated equipment and time schedule. Otherwise, quality may suffer.

Recognize that an outside duplicator spreads costs over a hundred or more clients. With up-to-date equipment and special staff, the duplicator presents a singlemindedness of purpose that rarely can be found in an in-house duplicating operation.

MASTERING

Most originals arrive at the duplicator in one or more of several formats: tape, 35mm, 16mm, or slides. Usually, original film productions are transferred first to video tape, and then the duplicates are made from the tape. If the original tape is on an outdated format, a transfer is made to a more controllable format (quad or 1-inch type C). This process is called *mastering*.

Film-to-Tape Transfers. The majority of film productions are done in 16mm and 35mm film formats or 35mm slides; only occasionally is 8mm or Super 8 used. The 8mm and Super 8 display have less clarity and stability

23-2 PRODUCTION WITHIN PRIVATE TV

Figure 23-1 Duplicating room. Through the master control panel (in foreground), sound and picture are first enhanced, then reproduced by banks of slave machines (rear) in quantity. (*S/T Videocassette Duplicating Corp.*)

than the larger film formats, but are still usable as master sources for video programming. However, it is always wise to commit film formats to the popular 1-inch type C tape, because it is "broadcast standard" and equipment for this format is widely available. The tape master also permits added flexibility for making changes. Changes, such as repositioning of content, insert titles, holding frames for emphasis, and other video techniques, are all possible after the tape transfer with proper accompanying equipment.

DUPLICATING

The duplicating process entails a controlled environment; proper system design; master tape evaluation; and enhancement and rights protection.

Controlled Environment. The quality of a videotape heavily depends on the conditioned environment in which it is duplicated. Optimum quality is obtained at 68° Fahrenheit, 50 percent humidity. Blank tapes should be stored in the same atmosphere for 24 hours prior to use to assure the same duplicating performance from machine to machine. The air must be filtered, free of dust and lint. A typical duplicating room is illustrated in Figure 23-1.

Figure 23-2 One-inch type C videotape recorder. (*Ampex Corp.*)

System Design. The flexibility of a duplicator's system can be an important part of the service offered. Optimally, all master videotape recorders (VTRs) should be able to feed all duplicating units at a single level of quality. A 2-inch quad master duplicated on a ¾-inch cassette, or a 1-inch C master on a ½-inch cassette can be run at the same time without compromising quality. (Figure 23-2 shows a 1-inch type C VTR.) Every duplication possibility or combination can be switched through the enhancement systems to assure the desired effect.

Master Tape Evaluation. The duplicator, on receiving a master for duplication, should judge it on its technical qualities (edit-points, video levels, noise, color, audio) and record these judgments in a master evaluation report (Figure 23-3). Every master is thus evaluated so that necessary enhancements and special handling can be planned before duplication and a point of comparison established to evaluate cassette quality after duplication. If the program is rerun, the report is at hand.

Enhancement. While a good duplicate is not likely to come from a really bad master, the enhancement skills of a professional duplicator can be used to good effect where minor technical variances in light, color, audio, and video levels are present. Applying enhancement techniques during the

23-4 PRODUCTION WITHIN PRIVATE TV

```
                                            PLAYBACK EVALUATION FORM
    S/T    S/T Videocassette Duplicating Corp.
           500 Willow Tree Road • Leonia, New Jersey 07605    WORK ORDER: _____
                                                              OPERATOR: _____

CLIENT: _____ DATE: _____

TITLE: _____ LENGTH: _____

MASTER FORMAT: _____ MASTER: _____ E E MASTER _____ DUBMASTER _____ COLOR _____
                        MONOCHROME _____ SHB: _____ HB: _____ LB: _____

DUPING FORMAT: _____ S/T STOCK _____ CLIENT STOCK _____ NUM. DUPES _____

VIDEO CHECK LIST
TAPE GUIDE ALIGNMENT _____ VIDEO LEVEL _____ CHROMA LEVEL _____
CHROMA PHASE _____ HORIZONTAL/VERTICAL STABILITY _____ SIGNAL/NOISE RATIO _____

    PROBLEM           LOCATION              PROBLEM           LOCATION
1. BANDING _____   8. MOIRE _____
2. CREASE _____   9. SCALLOPING _____
3. DEVIATION _____  10. SCRATCH _____
4. DROPOUT _____  11. SKEWING _____
5. EDITS _____  12. SPLICES _____
6. HEAD SEPARATION _____  13. TRACKING _____
7. INTERCHANGE _____  14. VELOCITY ERROR _____
                    15. _____

AUDIO CHECK LIST                       ENHANCEMENT EQUIPMENT
    PROBLEM           LOCATION
A. DISTORTION _____   CVS504 _____
B. DROPOUT _____   DELTA _____
C. EXTRANEOUS _____   3M PROC AMP _____
D. LEVEL FLUCTUATION _____   COLOR CORRECTOR _____
E. _____

OPERATOR COMMENTS: _____
_____
_____
_____

AUDIO REQUIREMENTS: NORMAL (2 CHANNEL) _____ TONE REQUIREMENTS: CH1 _____
                    STEREO _____                       CH 2 _____
                    SINGLE _____

VIDEO REQUIREMENTS: AB ROLL _____
                    PREROLL (BLACK) _____

SPECIAL OPERATING INSTRUCTIONS: _____
_____

QUALITY CONTROL COMMENTS: _____
_____

QC'd by: _____ LABELED BY: _____
```

Figure 23-3 Master evaluation report. Every incoming master is evaluated technically, and a master evaluation report is stored within the master as a permanent record of the tape characteristics and performance. (*S/T Videocassette Duplicating Corp.*)

duplication process, the professional duplicator levels out the variances—reduces noise, equalizes the audio, and corrects the time base. Thus, he or she can often improve the overall quality of the program or cassette.

Rights Protection. Barred access to duplicating areas, to master libraries, and to finished product storage should be a part of every duplicator's service, for the duplicator recognizes that he or she is dealing with privileged information. If the client wishes to extend the protection of the property after it has left duplicator premises, the client can request an antipiracy signal on the duplicates. These signals come under various names, such as Stop Copy and Copy Guard. They are designed to thwart thieves from using a duplicate as a master to make other duplicates. This process is encoded on the tapes at the same time that duplicates are being made. The Stop Copy signal on the pirated copy will prevent a viewable picture on a Betamax consumer unit. It will prevent both picture and sound on most industrial and ¾-inch U-Matic units. On a Video Home System (VHS) format, the audio is generally unaffected, but the video is subject to severe distortion, and audio failures may occur.

QUALITY CONTROL

Quality control is vital in three key areas: first, in the blank tape itself; second, in the condition of the duplicator's equipment; and third, in the quality of the recorded signal.

Blank Tape Evaluation. There are critical standards to which every blank must conform if it is to eliminate itself as a variable in the duplication process. Tension in the cassette, dropout ratio, physical thickness, and bonding of oxide for a given cassette are factors that must be known before the duplication process begins so that adjustments can be made for minor variations. Incoming blank tapes, therefore, are inspected batch by batch to be sure that they meet the standards the duplicator has set in order to deliver a final product of predictably high quality.

Equipment Maintenance. Preventive maintenance is an important part of every duplicating service. It is scheduled in the same way an airline schedules its maintenance programs with systematic checkups, regular parts replacements, and constant adjustments. This care is necessary because the mechanical condition of equipment varies on the plus or minus side, and the duplicator's tight preventive maintenance schedule tends to lessen the detrimental skew from one side to the other. Specifically, staff engineers, people capable of assembly and disassembly of the machines, must be on hand with a complete inventory of parts so that it is never necessary to wait for the manufacturer's repair person or for parts to be shipped.

Quality Check. Assuming a good blank tape and equipment which is operating at peak performance, it is still important to check the quality of the recorded signal of every duplicate. (A quality control form is shown in Figure 23-4.) If there is a defect in the recorded signal, it can generally be found in the first and last 2 minutes of the cassette. The cassette is verified for color, dropout, noise, and picture definition. Tapes are tested for tension, stability, and evenness of run. It is not the duplicator's job to

DROPOUTS	Less than 30-per-minute average. A dropout is defined as a signal loss resulting in a horizontal line of greater than ½ the width of the TV monitor screen with the dropout compensator defeated in the tape player. (Most dropouts will not be seen in a normal player because the dropout compensator will correct them.) This spec is the same as Sony tape specs.
INTERCHANGE	A flat RF envelope of maximum amplitude at the detent position of the tracking control (home track) with breathing variations of less than a 25% loss of amplitude. As tracking is adjusted either direction from home track, the envelope shall decrease in amplitude evenly with no mistracking at 50% of initial amplitude measured on an oscilliscope.
AUDIO LEVEL	-5 ± 1.5 db maximum peaks
AUDIO	Clean with good frequency response. No distortion or noise.
TAPE DAMAGE	None
CASSETTE PLASTICS	No physical damage accepted and malfunctions including chattering of reels or parts during fast forward or rewind.
PROGRAM INDEXING	Specified black before first program segment $\pm$ .5 seconds measured by electronically detecting control track pulses and counting from first control track pulse.

VIDEO WAVEFORM	Using a calibrated wave form monitor, the following measurements shall be made with no measurable variations:
	    Sync level—40 IRE units
	    Burst amplitude—40 IRE units
	    Pedestal—7.5 IRE units
	    Peak white video level—100 IRE units

HEAD SWITCHING POINT	6 to 8 lines before vertical sync as measured on WFM and cross pulse monitor.
DIHEDRAL ERROR	Less than ¼ the width of a horizontal sync pulse. Measured on a cross pulse monitor.

Figure 23-4 Quality control checks.

SKEW ERROR	Less than ½ the width of a horizontal sync pulse displayed on a cross pulse monitor.
PICTURE SHAKE OR JITTER	None allowed as viewed on WFM, cross pulse or either type of color monitor.
CHROMAS PHASE CHROMA LEVEL	Correct phase and level viewed on color monitors set up with color bar test pattern.

Figure 23-4 *(Continued)*

criticize the quality of a master delivered for duplication; it is his or her job to get the most out of it. Working from the master evaluation report, the duplicator can generally predict the quality of the duplicates that is likely, and if trouble occurs, he or she can quickly pinpoint its origin.

PACKAGING

The producer of a videotape program for a private TV network is careful to see that this program is a professional presentation. But often the package in which it is distributed is not. The package is a significant part of the total process. Any video document that is likely to be handled by the chairperson of the board should be well packaged. The same should be true of all company videocassettes. The package need not be expensive, but it should be professional in appearance and should reflect the value the company puts on the program. Basically, the label identifies the program with information relating to it; the covers and jackets protect the contents against damage, dust, and lint. But there is more to proper packaging than that.

Labels. Labels are used not only to convey information, but also to enhance the presentation (see Figure 23-5). Your duplicator should be able to supply the type, art, production, and printing facilities on premises so that no time is lost once the duplicates are made. Where labeling facilities are not a part of the duplicator's service, it is important that a good outside printing house be used, and that it be given plenty of lead time to coincide with the project's estimated completion date. More programs have been held up for lack of labels and packaging than for any other reason.

Attention should be given to the use of type and color. Color combinations are frequently used to identify the company division that authored the program. Where the programs are adapted to different audiences, it should be noted boldly on the label, "For Employee Viewing Only," "For Public Viewing." Labels can include information on the care and handling of the cassette, on how to use it in the machine, on when and where to return it, on copyright privileges, and cautions on unauthorized uses.

23-8 PRODUCTION WITHIN PRIVATE TV

> **Merrill Lynch Video Network**
>
> FOR EMPLOYEE VIEWING ONLY
>
> CODE: L 39102
>
> TITLE: MORTGAGE BACKED SECURITIES
> Host: Bob Stefani, MBS Group with Kathy Birk, American Fletcher Bank; Fred Tattersall, First & Merchants Bank; George Bender, United First Mortgage; and Bob Boyd, AMIC
> Discussion examining the mortgage backed securities market, with emphasis on the credit and market concerns of non-government conventional pass-throughs.
>
> LENGTH: 20:00
>
> DATE: March 1980
>
> THE INFORMATION SET FORTH HEREIN WAS OBTAINED FROM SOURCES WHICH WE BELIEVE RELIABLE BUT WE DO NOT GUARANTEE ITS ACCURACY
>
> THIS CASSETTE AND ITS CONTENTS ARE THE SOLE PROPERTY OF:
> MERRILL LYNCH & CO., INC.
> 1 LIBERTY PLAZA, 165 BROADWAY
> NEW YORK, N.Y. 10006
>
> THIS PROGRAM MAY NOT BE TRANSFERRED OR REPRODUCED, BY ANY MEANS WHATSOEVER, IN WHOLE OR IN PART, WITHOUT THE PRIOR EXPRESS WRITTEN CONSENT OF THE MERRILL LYNCH AUDIO VISUAL CENTER.
>
> COPYRIGHT © 1979 MLPF&S INC. ALL RIGHTS RESERVED
>
> PROPER HANDLING
> BOTTOM VIEW
>
> 1. Before inserting, take up slack in the tape by turning reels in the direction of the arrows.
> 2. Avoid repeated insertion and removal of the videocassette without operating the machine.
> 3. Keep this videocassette away from excessive heat, humidity, or strong magnetic fields.
> 4. Rewind the videocassette before storing or shipping.
>
> Printed in U.S.A.

Figure 23-5 Label. An attractive label lends importance to the tape as a presentation. It can also include instruction on the care and use of the cassette. (*Merrill Lynch Video Network.*)

Jacketing. Jackets both protect the cassette and ancillary materials they enclose and enhance the presentation. They may be see-through or opaque plastic, printed or unprinted. If they are not see-through, generally they carry the same information outside that is inside. Frequently, jackets are coordinated to carry corporate or division colors with a corporate logo. Room for collateral material is provided for in some types of jackets.

Jacketing takes place in an assembly area where the jackets, ancillary and collateral materials, and cassettes are gathered. Once the duplicates have passed their quality check, the labels are affixed, ancillary and collateral materials inserted, jackets closed, and the final packages readied for distribution or holding. Where a video presentation carries ancillary materials, they should be arranged for well in advance. They might include a series of instructions, a lesson, a reading list. One accounting firm produces educational video programs that require the use of a printed workbook. These workbooks are kept by the students after the course has been completed and the tape has been returned to the firm. The important thing is that they be ready and waiting for the cassette, rather than the other way around.

DISTRIBUTION

There are two aspects to distribution: shipment and fulfillment. They pertain to getting the cassettes out to the end user. Retrieval and recycling (and eventual reuse of the tape) involves inspection, repackaging, storage, and reissuance. This is discussed in greater detail below. The duplicator will handle any or all of these tasks as part of his or her service, but arrangements should be made in the early stages of planning.

Shipment. Cassettes can be shipped either in bulk for distribution or singly from duplicator to individual user. Special handling is required in either case. Bulk shipments are packed in tightly bundled lots of two to thirty or more cassettes to prevent damage. Individual cassettes are specially packed in jiffy or cushioned envelopes not only to protect against impact, but also to maintain environment. Records are kept in the event of lost shipments.

There is no set way to distribute video programs. Each producer can meet specific needs in one or more ways. One of the large commercial banks uses a combination of bulk and drop shipments. In the United States, the bank distributes its programs singly. Worldwide, it employs the duplicator to handle the shipping arrangements as part of the job.

Fulfillment. The fulfillment of orders by the duplicator requires that stock records be kept and that reports be sent at regular intervals. If a charge is made for the use of the cassette, a strict accounting is kept of monies passed.

The duplicator is hired to fulfill orders for several reasons: staff on hand; proper storage with all parts gathered at a single handling point; and one-stop shipping, so to speak.

RECYCLING

One of the benefits of using videotape is its reusability. Unlike film, it is possible, once a tape is no longer needed, to erase the contents and reuse the tape at a great saving. Recycling, as it is called, is a budgetary necessity for a network of any size. A corporate video program has an average viewing life of less than one year. The original blank ¾-inch tape (without duplication) costs about $16. Its second-time-around costs less than $1.50 for erasing, inspection, cleaning, and storage. For one company with twenty-six divisions, the savings run to well over $100,000 every year, and it began recycling its tapes in 1972.

The number of recycles a tape can go through, without loss of quality, varies from tape to tape, depending on the care and handling and the environmental conditions under which the tape is used. But, generally, a ¾-inch tape averages 6 to 8 recycles, a ½-inch tape, 8 to 10 recycles, or a savings of 75 to 80 percent over what would be necessary to pay for all-new tape.

The duplicator is equipped to handle the recycling process in its three stages: retrieval, check and erase, and tape inventory control.

Retrieval. An effective system for retrieving program tapes from their users is the first step in any recycling process. A simple way to keep track of a cassette series is to code it by end-user location; i.e., No. 1 for Cleveland, No. 2 for Pittsburgh, No. 3 for Syracuse, so that the numbers quickly show who has returned the cassettes and who has not. Some companies charge a fee when the cassette is not returned in a specified time. In either event, instructions for returning every tape should be indicated on the label or jacket, along with the address of the private TV network. This is especially important where privileged information is involved.

Check and Erase. Incoming tapes must be checked for physical damage and their video and audio qualities verified. Those that pass are erased immediately and held for reuse. It is important that the erasure process be strong enough to erase all video and audio information, so that residual noise is avoided in the succeeding duplication. A powerful magnetic eraser is necessary to ensure this. Otherwise, a number of passes through a commercial-grade videocassette eraser is needed.

Tape Inventory Control. Once erasure is complete and the client ownership of the tape is established, the labels should be peeled off as quickly as possible, to prevent the label adhesive from hardening. Cassettes should have only one label affixed at a time, as opposed to label over label. Single labels look better and eliminate the danger of an accumulation of labels jamming the transport mechanism of video players. Erased tapes are assembled by client and by length (60-minute tapes in one bin, 40-minute tapes in another) to facilitate their reuse. Date-stamping the body of the cassette (not the label) each time it has been used and returned is the best way to keep track of the number of times each tape has been used.

LIBRARY AND STORAGE

A big portion of the space required to run a duplicating operation has to be devoted to warehousing, librarying, and other support activities which are not part of the actual duplication process. Secure storage areas of controlled atmosphere where tapes are protected from environment damage and pilferage should be part of the service.

Master Library. Every duplicating facility should maintain a temperature- and humidity-controlled library for the masters it catalogs—one that is under lock and key. Careful alphanumeric cataloging is important, especially where titles are of a similar nature. One federal agency's titles begin with the word "Disaster" because dealing with disasters is prevalent in this agency's work. The numbering system is used to prevent mistakes in both duplicating and shipping, especially when a second disaster occurs in the same place as the first, but a year later.

Cassette Overprints. Often a manager of a private video network will anticipate the future needs of a program that is about to be duplicated. It costs less to make overprints of an original program and store them than to go back and start from scratch. And it is more expeditious to store the collateral and ancillary materials, the labels, and the jackets, under one roof, so that everything is at hand for preparation and shipping when needed.

CONCLUSION

The process of duplicating videocassettes can be a complex and demanding operation. But it is important to make the effort to prepare quality copies. The reason is obvious. Your audience knows of your efforts only via the videocassette they receive. If it is of poor quality, if its label is wrong or illegible, if the support material does not arrive with the tape, if the tape misses its deadline, then all your work will be for naught. A quality videocassette presents you and your department in the most favorable light.

24 The Copyright Law and How to Use It

THOMAS VALENTINO, JR.

You at some time may need to copy some sort of television broadcast, or use a picture, graphic, or music that was originated by someone else. Or, you may want to protect material that you have produced. In either case, this chapter is designed to help you understand the current copyright law and how to work within it.

OBTAINING PERMISSION

The first step is finding out from whom you have to get permission. The U.S. Copyright Office in Washington and music organizations such as Broadcast Music, Inc. (BMI), and American Society of Composers, Authors, and Publishers (ASCAP), both in New York, are good sources of information. Obtaining permission to use copyrighted audiovisual material can become very complicated because the rights may fall into three categories: rights for use of copyrighted text; rights for use of visuals; and rights for use of music or voices. Permission must come from all parties who have any sort of copyright on the material.

Specifying the Material

To obtain permission to reproduce copywritten materials, you must provide the copyright holder with information on the following facts:

1. The correct title of the material.
2. An exact description of the material to be used (i.e., text, visuals, or sound track). If requesting rights to use text material, you must provide the page number and exact quote (or several words from the beginning and several from the end of the portion to be used). If requesting use of visuals from a film, filmstrip, or slide program, you should cite frame numbers.
3. The type of reproduction. If the materials are to be transferred into another format, the reason for the change should be supplied.
4. The number of copies to be made.
5. The use to be made of reproduced material. If the material is a videocassette, specify whether the intended use involves single-receiver playback or multiple receivers. If the intended use involves transmission of the

material, specific information should be supplied as to the method of transmission, whether radio or television, open or closed circuit. In such case, many license agreements will also require the number of people in the intended audience (see item 6 below).

6. The distribution of the reproduced material. If transmission is involved, information should be provided on the extent of the reception area and the probable number of viewers.
7. Whether or not the material is to be sold, whether there will be a charge for viewing, or both.

Making the Request

A request for permission to reproduce copyrighted materials should be sent, together with a self-addressed return envelope, to each copyright owner. Make sure you give yourself enough time to receive a reply. In preparing the request, use the following guidelines:

1. Do not seek permission via telephone. Write and provide the specific information indicated above.
2. Don't seek blanket permission unless absolutely essential. Blanket permission will require a more formal licensing agreement.
3. Allow sufficient lead time for the copyright owner to process your request.
4. Be sure to include your complete return address.
5. Be sure that you are able to comply with all the terms of the permission grant, which may include certifying that videotaped material has been erased following limited use for the specific time period.
6. Most important: Check and double-check to see that all your information is complete. The more complete the request, the more rapid the response.

It may seem easy for you to copy some material—"no one will know"— but you might feel different if the shoe were on the other foot.

Copyrighting a program that encompasses other copywritten material within it is not a problem, so long as permission has been obtained for use of the copywritten material in the newly produced program. Besides payment for the use of copyrighted material, you, as the producer, may be required to give credit in the program credits or some other such identification as to source.

HOW TO REGISTER A COPYRIGHT

Why bother copyrighting private television programs? That seems to be an ever-present question. The answer is simple. If you feel that your program (i.e., demo tape) may be seen and used by many people and you want to protect or control usage, by all means copyright it. The procedure as explained here is quite painless. It is a good idea for private television producers to gain experience with the U.S. Copyright Office and its procedures. So do it! It doesn't hurt to copyright at least one of your programs. The following material, obtained from the Copyright Office, is included for your reference.

The copyright law provides for the registration of "dramatic compositions" in either published or unpublished form. This category (class D) covers the acting versions of dramatic works such as plays, radio and television dramas, musical comedies, motion picture shooting scripts, operas, choreographic works of a dramatic character, and pantomimes. For registration in class D, a work must be more than a story or synopsis that is potentially capable of being dramatized. It should tell its story by means of or through dramatic action rather than through narrative or descriptive material, and it should be complete enough for dramatic performance in its present form.

DURATION OF COPYRIGHT

For works that were created after the effective date of the new statute (January 1, 1978), the basic copyright term will be the life of the author and 50 years after the author's death. For works "made for hire," and for certain anonymous and pseudonymous works, the duration of the copyright will be 75 years from publication or 100 years from creation, whichever is shorter. The same terms of copyright will generally apply to works that were created before 1978 but not published or copyrighted before that date.

WORKS NOT REGISTRABLE FOR COPYRIGHT

Narrative outlines, scenarios, formats, plot summaries of plays and motion pictures, skeletal librettos, and other synopses and outlines cannot be registered for copyright in unpublished form. They are considered "books" under the copyright law, and a "book" cannot be registered unless it has been published with the required copyright notice. As long as they remain unpublished, books are protected by common law against unauthorized use, and no action is required in the Copyright Office.

Copyright protects only the particular manner of expression an author uses in working out his or her ideas or plans in literary, dramatic, musical, or other artistic form. It does not protect any of the ideas or plans on which the author's work is based or which are embodied in it. Therefore, it is not possible to register a claim to copyright an *idea* for a motion picture, television program, story, or any other kind of work.

THE COPYRIGHT NOTICE

To secure and maintain copyright protection when publishing a work, you must make sure that all copies bear the prescribed notice from the time of first publication. It is the responsibility of the copyright owner to place the required notice on his or her work. Notification to the Copyright Office is not needed for this.

The copyright notice generally consists of the word "Copyright," the abbreviation "Copr.," or the international symbol ©, accompanied by the year date of publication and the name of the copyright owner. For example: © 1982 by John Doe.

The specific notice requirements vary from class to class, but for published dramatic works and television and film programs, the notice must be shown, if possible, at the beginning of the program or in a place somewhere in the main credits. In addition, it should appear on the title page of the script, which is also registered with the Copyright Office.

WHO MAY REGISTER A COPYRIGHT?

Unless the work was "made for hire," the individual who actually created the work—a video program, for example—is considered the author-owner. However, most private television programs fall into the category of "made for hire." (Figure 24-1 shows such an agreement.) The term covers (1) a work prepared by an employee as part of his or her job; and (2) a work ordered or commissioned from a "buyer" for a specific use. In these cases, ownership of the video program belongs to the company or employer.

More than one name can be registered on the copyright application. In this case, the work is called a "collective work."

CLASSIFICATION SYSTEM FOR COPYRIGHT REGISTRATION

Class TX: Nondramatic Literary Works

This category is very broad. Except for dramatic works and certain kinds of audiovisual works, class TX includes all types of published and unpublished works written in words (or other verbal or numerical symbols). A few of the many examples of "nondramatic literary works" are fiction, nonfiction, poetry, periodicals, textbooks, reference works, directories, catalogs, advertising copy, and compilations of information.

Class PA: Works of the Performing Arts

This category includes published and unpublished works prepared for the purpose of being performed either directly before an audience or indirectly

```
                    'WORK-MADE-FOR-HIRE' AGREEMENT

_____ ("Company") has commissioned
_____ ("Contributor")
to prepare and supply the "Work"
_____
                    (description )
as a 'work-made-for-hire'* for the in
_____
                    (program)
to be produced or released by the Company.

The Company agrees that the Contributor shall receive upon completion of the Work,
the sum of $_____ as consideration.

The Contributor represents and warrants to the Company that the Work is original
to the best of his or her knowledge except for such excerpts from copyrighted
works as may be included with the written permission of the copyright owners, and
that the Work does not violate or infringe any copyright, trademark, patent,
statutory, common law, or proprietary rights of others, or contain anything
libelous. A copy of each such permission is to be furnished to the Company.

The Contributor acknowledges that the Work was specifically ordered and
commissioned by the Company and was prepared by the Contributor as a
'work-made-for-hire'*. The Contributor understands that the Company owns all of
the exclusive rights of the Work comprised in the copyright under the United
States copyright law and all international copyright convention, including the
right to copyright the Work and all renewals thereof in the name of the Company.


                              _____
                                      Company

        _____        _____
             Date                    Contributor


* 'Work-made-for-hire' is the legal description for contributions specified by the
Copyright Act of 1978, and is used here only as a technical term.
```

Figure 24-1 "Work-made-for-hire" Agreement.

"by means of any device or process." Examples are musical works, including any accompanying words; dramatic works, including any accompanying music; pantomimes and choreographic works; and motion pictures and other audiovisual works.

Class VA: Works of the Visual Arts

This category consists of published and unpublished "pictorial, graphic, and sculptural works," including two-dimensional and three-dimensional works of fine, graphic, and applied art, photographs, prints and art reproductions, maps, globes, charts, technical drawings, diagrams, and models. Within this class are pictorial or graphic labels and advertisements, as well as "works of artistic craftsmanship." The "design of a useful article" may be registrable in class VA, but "only if, and only to the extent that, such design incorporates pictorial, graphic, or sculptural features that can be identified separately from, and are capable of existing independently of, the utilitarian aspects of the article."

Class SR: Sound Recordings

This category is appropriate for registration for both published and unpublished works in two situations: (1) where the copyright claim is limited to the sound recording itself; and (2) where the same copyright claimant is seeking to register not only the sound recording but also the musical, dramatic, or literary work embodied in the sound recording. With one exception, "sound recordings" are works that result from the fixation of a series of musical, spoken, or other sounds. The exception is for the audio portions of audiovisual works, such as a motion picture sound track or an audiocassette accompanying a filmstrip; these are considered an integral part of the audiovisual work as a whole and are registrable in class PA. Sound recordings fixed before February 15, 1972, are not eligible for registration, but may be protected by state law.

Class RE: Renewal Registration

This category covers renewal registrations for copyrights that were already in their first term on January 1, 1978 (that is, renewals for works originally copyrighted between January 1, 1950, and December 31, 1977). Class RE is appropriate for all renewal registrations, regardless of the class in which the original registration was made. (Renewal registration can be made only during the last calendar year of the first 28-year copyright term, and it has the effect of extending copyright protection for an additional 47 years.)

The effective date of copyright registration is the day on which an acceptable application, deposit, and fee have all been received in the U.S. Copyright Office. It is important to keep this fact in mind because delaying the effective date of registration can have serious consequences. In an infringement suit, the court may limit the copyright owner's momentary recovery if the infringement started before the effective date of registration. Make sure that all required forms are filed with the Copyright Office as soon as possible.

DEPOSIT REQUIREMENTS FOR COPYRIGHT

The following information has been furnished by the Copyright Office for deposit requirements for motion pictures, compiled and adapted from Part 202, *Regulations of the Copyright Office*. The material lists definitions and requirements.

I. PUBLISHED MOTION PICTURES
One complete copy of the best edition and a separate description of its contents.

A. The "best edition" is the edition published in the United States at any time before the date of deposit that the Library of Congress determines to be most suitable for its purposes. The formats listed in descending order of acceptability are:
  1. *Film* rather than another medium
     a. Preprint material, by special arrangement
     b. Film gauge in which most widely distributed
     c. 35 mm rather than 16 mm
     d. 16 mm rather than 8 mm
     e. Special formats (e.g., 65 mm) only in exceptional cases
     f. Open reel rather than cartridge or cassette
  2. *Videotape* rather than videodisc
     a. Tape gauge in which most widely distributed
     b. Two-inch tape
     c. One-inch tape
     d. Three-quarter inch tape
     e. One-half-inch tape cassette
B. The separate description may be, for example, a continuity, pressbook, or synopsis. (Inclusion of the running time, publication date, and credits is helpful, but not required.)

II. UNPUBLISHED MOTION PICTURES
One complete copy and a separate description of its contents, such as a continuity, pressbook, or synopsis.

III. ALTERNATIVE DEPOSIT FOR PUBLISHED MOTION PICTURES
Identifying material (instead of one complete copy) and a separate description.
A. Identifying material:
  1. An audiocassette or other phonorecord reproducing the entire sound track or other sound portion of the motion picture, *or*
  2. A set consisting of one frame enlargement or similar visual reproduction from each 10-minute segment of the motion picture
B. Separate description:
  In either case (1 or 2 above), the description may be a continuity, pressbook, or synopsis, but *must* include the following:
  1. The title or continuing title of the work and episode title, if any
  2. The nature and general content of the program
  3. The date when the work was first fixed and whether or not fixation was simultaneous with first transmission
  4. The date of first transmission, if any
  5. The running time
  6. The credits appearing on the work, if any

24-8 PRODUCTION WITHIN PRIVATE TV

FORM PA
UNITED STATES COPYRIGHT OFFICE

REGISTRATION NUMBER

PA   PAU

EFFECTIVE DATE OF REGISTRATION

(Month)   (Day)   (Year)

DO NOT WRITE ABOVE THIS LINE. IF YOU NEED MORE SPACE, USE CONTINUATION SHEET (FORM PA/CON)

① Title
TITLE OF THIS WORK: "FINALLY SOMEBODY DID IT!"
NATURE OF THIS WORK: (See instructions) 3/4" Video Cassette
PREVIOUS OR ALTERNATIVE TITLES:

② Author(s)

IMPORTANT: Under the law, the "author" of a "work made for hire" is generally the employer, not the employee (see instructions). If any part of this work was "made for hire" check "Yes" in the space provided, give the employer (or other person for whom the work was prepared) as "Author" of that part, and leave the space for dates blank.

1. NAME OF AUTHOR: Philip J. Marshall
   Was this author's contribution to the work a "work made for hire"? Yes No X
   DATES OF BIRTH AND DEATH: Born 1/1/58 Died
   AUTHOR'S NATIONALITY OR DOMICILE: Citizen of USA or Domiciled in
   WAS THIS AUTHOR'S CONTRIBUTION TO THE WORK: Anonymous? Yes No X Pseudonymous? Yes No X
   AUTHOR OF: (Briefly describe nature of this author's contribution) Narration

2. NAME OF AUTHOR:
   DATES OF BIRTH AND DEATH: Born Died
   AUTHOR'S NATIONALITY OR DOMICILE:

3. NAME OF AUTHOR:
   DATES OF BIRTH AND DEATH: Born Died
   AUTHOR'S NATIONALITY OR DOMICILE:

③ Creation and Publication
YEAR IN WHICH CREATION OF THIS WORK WAS COMPLETED: Year 1980
DATE AND NATION OF FIRST PUBLICATION: Date (Month) (Day) (Year) Nation (Name of Country)

④ Claimant(s)
NAME(S) AND ADDRESS(ES) OF COPYRIGHT CLAIMANT(S):
Thomas J. Valentino Inc.

TRANSFER: (If the copyright claimant(s) named here in space 4 are different from the author(s) named in space 2, give a brief statement of how the claimant(s) obtained ownership of the copyright.)
Philip Marshall is a staff writer producer for Thomas J. Valentino Inc.

Complete all applicable spaces (numbers 5-9) on the reverse side of this page
Follow detailed instructions attached • Sign the form at line 8

DO NOT WRITE HERE
Page 1 of .... pages

Figure 24-2 Completed form for videotape copyright registration. (*United States Copyright Office.*)

This detailed list may seem confusing at first glance, so a copy of an actual copyright form used to copyright a demo tape is included here. This form (Figure 24-2), which I used for this type of program, is the one that you should be using for most of your productions. Form PA is the general form most production people in private television will be using. The basic fee for filing a registration of copyright is $10. The Copyright Office recommends that you send all remittances in the form of a check, money order,

	EXAMINED BY CHECKED BY	APPLICATION RECEIVED	FOR COPYRIGHT OFFICE USE ONLY
	CORRESPONDENCE ☐ Yes	DEPOSIT RECEIVED	
	DEPOSIT ACCOUNT FUNDS USED ☐	REMITTANCE NUMBER AND DATE	

DO NOT WRITE ABOVE THIS LINE. IF YOU NEED ADDITIONAL SPACE, USE CONTINUATION SHEET (FORM PA/CON)

PREVIOUS REGISTRATION: — **5 Previous Registration**

- Has registration for this work, or for an earlier version of this work, already been made in the Copyright Office? Yes .... No X
- If your answer is "Yes," why is another registration being sought? (Check appropriate box)
  - ☐ This is the first published edition of a work previously registered in unpublished form.
  - ☒ This is the first application submitted by this author as copyright claimant.
  - ☐ This is a changed version of the work, as shown by line 6 of the application.
- If your answer is "Yes," give: Previous Registration Number .... Year of Registration ....

COMPILATION OR DERIVATIVE WORK: (See instructions) — **6 Compilation or Derivative Work**

PREEXISTING MATERIAL: (Identify any preexisting work or works that the work is based on or incorporates.)

MATERIAL ADDED TO THIS WORK: (Give a brief, general statement of the material that has been added to this work and in which copyright is claimed.)

Stock shot material from the Thomas J. Valentino Inc. Videostockshot library.

7 Fee and Correspondence

DEPOSIT ACCOUNT: (If the registration fee is to be charged to a Deposit Account established in the Copyright Office, give name and number of Account.)

Name ....

Account Number ....

CORRESPONDENCE: (Give name and address to which correspondence about this application should be sent.)

Name Thomas J. Valentino Inc.

Address 151 west 46th street (Apt.)

New York (City), New York (State) 10036 (ZIP)

CERTIFICATION: ✱ I, the undersigned, hereby certify that I am the: (Check one) — **8 Certification (Application must be signed)**

☐ author ☐ other copyright claimant ☐ owner of exclusive right(s) ☒ authorized agent of Thomas J. Valentino, Inc. (Name of author or other copyright claimant, or owner of exclusive right(s))

of the work identified in this application and that the statements made by me in this application are correct to the best of my knowledge.

Handwritten signature: (X) ....

Typed or printed name Thomas J. Valentino Jr. Date 4/25/80

MAIL CERTIFICATE TO (Certificate will be mailed in window envelope) — **9 Address For Return of Certificate**

Thomas J. Valentino, Inc. (Video Dept.) (Name)
151 West 46th Street (Number, Street and Apartment Number)
New York, New York 10036 (City) (State) (ZIP code)

✱ 17 U.S.C. §506(e) FALSE REPRESENTATION – Any person who knowingly makes a false representation of a material fact in the application for copyright registration provided for by section 409, or in any written statement filed in connection with the application, shall be fined not more than $2,500.

U.S. GOVERNMENT PRINTING OFFICE: 1978 261 022 17

Figure 24-2 *(Continued)*

or bank draft, payable to Register of Copyrights. If you plan on filing a number of copyrights per year, an account can be set up with the Copyright Office. I suggest that you contact the office for more information. To list the basic submissions, for our demo tape "Finally somebody did it!" we filed Form PA, the $10 fee, and a ¾-inch videocassette copy of the finished program.

CONTINUATION SHEET FOR FORM PA

FORM PA/CON
UNITED STATES COPYRIGHT OFFICE

- If at all possible, try to fit the information called for into the spaces provided on Form PA.
- If you do not have space enough for all of the information you need to give on Form PA, use this continuation sheet and submit it with Form PA.
- If you submit this continuation sheet, leave it attached to Form PA. Or, if it becomes detached, clip (do not tape or staple) and fold the two together before submitting them.
- **PART A** of this sheet is intended to identify the basic application.
- **PART B** is a continuation of Space 2. **PART C** is for the continuation of Spaces 1, 4, or 6. The other spaces on Form PA call for specific items of information, and should not need continuation.

REGISTRATION NUMBER

PA   PAU

EFFECTIVE DATE OF REGISTRATION

(Month)   (Day)   (Year)

CONTINUATION SHEET RECEIVED

Page ___ of ___ pages

DO NOT WRITE ABOVE THIS LINE. FOR COPYRIGHT OFFICE USE ONLY

A Identification of Application

IDENTIFICATION OF CONTINUATION SHEET: This sheet is a continuation of the application for copyright registration on Form PA, submitted for the following work:
- **TITLE:** (Give the title as given under the heading "Title of this Work" in Space 1 of Form PA.)
- **NAME(S) AND ADDRESS(ES) OF COPYRIGHT CLAIMANT(S)** (Give the name and address of at least one copyright claimant as given in Space 4 of Form PA.)

B Continuation of Space 2

NAME OF AUTHOR:
Was this author's contribution to the work a "work made for hire"? Yes ___ No ___
AUTHOR'S NATIONALITY OR DOMICILE:
Citizen of ___ (Name of Country) or Domiciled in ___ (Name of Country)
AUTHOR OF: (Briefly describe nature of this author's contribution)

DATES OF BIRTH AND DEATH:
Born ___ (Year) Died ___ (Year)
WAS THIS AUTHOR'S CONTRIBUTION TO THE WORK:
Anonymous? Yes ___ No ___
Pseudonymous? Yes ___ No ___
If the answer to either of these questions is "Yes," see detailed instructions attached.

NAME OF AUTHOR:
Was this author's contribution to the work a "work made for hire"? Yes ___ No ___
AUTHOR'S NATIONALITY OR DOMICILE:
Citizen of ___ (Name of Country) or Domiciled in ___ (Name of Country)
AUTHOR OF: (Briefly describe nature of this author's contribution)

DATES OF BIRTH AND DEATH:
Born ___ (Year) Died ___ (Year)
WAS THIS AUTHOR'S CONTRIBUTION TO THE WORK:
Anonymous? Yes ___ No ___
Pseudonymous? Yes ___ No ___
If the answer to either of these questions is "Yes," see detailed instructions attached.

NAME OF AUTHOR:
Was this author's contribution to the work a "work made for hire"? Yes ___ No ___
AUTHOR'S NATIONALITY OR DOMICILE:
Citizen of ___ (Name of Country) or Domiciled in ___ (Name of Country)
AUTHOR OF: (Briefly describe nature of this author's contribution)

DATES OF BIRTH AND DEATH:
Born ___ (Year) Died ___ (Year)
WAS THIS AUTHOR'S CONTRIBUTION TO THE WORK:
Anonymous? Yes ___ No ___
Pseudonymous? Yes ___ No ___
If the answer to either of these questions is "Yes," see detailed instructions attached.

C Continuation of Other Spaces

CONTINUATION OF (Check which): ☐ Space 1 ☐ Space 4 ☐ Space 6

Figure 24-2 *(Continued)*

International Video Networking

PART 4

A new development within the field of private television is programming for an international video network. Chapter 25 presents a detailed explanation of the various video standards that are used around the world and describes how programs can be transferred from one standard to another. Chapter 26 discusses how to set up an international video network and how to send video programs abroad, as well as special points to consider when producing a program for a foreign audience.

25 International Video Equipment Standards

GERALD CITRON

The purpose of this chapter is to demonstrate the benefits of an international network; to describe the various standards that exist; and to present methods of how to convert from one standard to another.

A major consideration in exchanging programs internationally is the incompatibility of the various television standards in use throughout the world. Most private video managers have been more concerned with program content than with international television standards. Nevertheless, to successfully exchange videotapes worldwide, one must be aware of the different standards used in each country receiving the material. Since there is no possibility of a single television standard being universally adopted, the private video manager wishing to exchange programs between locations operating on different standards must be prepared to employ one or more of the techniques or specialized equipment that will be described in this chapter.

Most corporations currently use the ¾-inch U-Matic cassette format both domestically and internationally. Some corporations prefer the ½-inch Video Home System (VHS) format, while some European firms have adopted the Philips videocassette recorder (VCR) format. None of these formats are interchangeable. Nor does it seem that video hardware manufacturers will decide to agree on a universal format in the future.

Which is the best format to use in a private video network? The answer depends, to a great extent, on the needs of the individual corporation. Will there be a need for bilingual programs? Is the shipping cost of cassettes a major concern? Where will the tapes be replayed?

Videocassettes: Compatibility and Interchangeability

It is commonly believed that the term *compatibility* refers to the capability of a videotape recorded on one machine to be played on another machine. Essentially, this word is a misnomer since, in practice, a videotape recorded on any machine can be played back on any other machine—*provided* the tape is dubbed from the playing machine onto the other machine, which is then operated in the *record* mode. *Both the playing machine and the recording machine, however, must be operating in the same standard;* for example, in the National Television Systems Committee system (NTSC), phase alternate by line (PAL), or sequential electronic color system with memory (SECAM). Thus, in theory, all machines have a measure of compatibility with one another—as long as the *standard* is maintained.

The question of interchangeability refers to the playing back on one machine of a tape that has been recorded on another machine. For example, a company has an office in New York and one in San Francisco, and a tape is made on the machine in New York. Provided the machine in San Francisco is of the same configuration as the one in New York, the tape will replay on the San Francisco equipment. That is true interchangeability. *Interchangeability*, then, is the ability to replay tapes of a particular format on equipment possessing the same format and operating in the same standard.

COLOR TELEVISION BROADCAST SYSTEMS AND STANDARDS

Researchers in Great Britain and the United States began experimental work on color in the late 1930s—even before a black-and-white system was commercially proven to be practical. The work centered principally on the sequential type of systems employing rotating colored filters on the cameras and in the receivers.

In 1953, the National Television Systems Committee was established by the U.S. Federal Communications Commission (FCC) to recommend a broadcast color system for the United States. The committee's work led to an all-electronic system capable of operating within the then-existing parameters of black-and-white (monochrome) standards and was accepted by the FCC.

The principles of this system, which became known as the NTSC system, formed the basis for all color systems now used throughout the world (Figure 25-1). Public broadcasting in the United States, using the NTSC color system, began in 1954. The same system was adopted by Japan, where service commenced in 1960. Other countries preferred modifications of the American system for technical and political reasons. Germany developed a system known as PAL, or phase alternate by line. This was technically superior to NTSC. SECAM (*s*ystème *É*lectronique *c*ouleur *a*vec *m*emoire), or sequential electronic color system with memory, was a more radical departure. Essentially, both alternatives were designed to minimize the color systems' sensitivity to certain types of distortion inherent in the broadcasting and transmission of two simultaneous images. The variations were particularly designed to cope with the geographic conditions found in Western Europe.

In addition to the three principal broadcast standards, there are subtle variations which exist within each standard. (See Figure 25-2.) Thus, there are in existence NTSC systems working on 525 and 625 lines; there is a PAL standard operating on 525 lines; and there are SECAM systems broadcast with horizontal polarity that are incompatible with those operating on vertical polarity. As a quick reminder, a television picture is composed of a number of horizontal lines. While there are some television sets that have 2000 horizontal lines, generally systems are either 525 lines (most often,

Color encoding system	Principal area of usage	No. of lines/ picture	Voltage and frequency	Remarks
NTSC	U.S.A., Japan, Canada, Mexico, some areas of South America	525	60 hertz 110 volts	1 basic standard, but there are a 525-line PAL (M) and a 625-line NTSC variant
PAL	U.K., Western Europe, some areas of South America	625	50 hertz 220 volts	Variations between systems, resulting in 5 different PAL standards
SECAM	France, Russia, Eastern Europe, some areas of Middle East	625	50 hertz 220 volts	Variations between systems, resulting in 3 different SECAM standards

Figure 25-1 World color TV broadcast systems.

NTSC) or 625 lines (most often, PAL or SECAM). As a rule of thumb, the more lines per picture, the better the picture resolution. Fortunately, however, the videocassette communicator need be concerned with only the three principal standards: NTSC, PAL, and SECAM.

The NTSC System

All color-encoding techniques begin with the three primary colors—Red, Green, and Blue, or "RGB," as they are more commonly referred to by video technicians and engineers. As already noted, both the PAL and SECAM systems utilize the same basic principles as NTSC. PAL and SECAM were developed as an outgrowth of the technical strides made within the decade following the introduction of NTSC and as a basic improvement on the performance of NTSC in one principal area—that of consistency of tint or hue.

Tint information in the NTSC system is carried by a change in the phase angle of the chroma signal, and these phase changes are synchronously detected and recovered within the receiver. In the early days of color transmission in the United States, the phase information was often subject to a great many small errors which resulted from a combination of poor and inept work by technicians in station broadcasting and inter-network connections. Reflections ("ghosts") of broadcast signals from tall buildings and other structures in the vicinity of the receiving antenna also caused errors. These became additive, with the result that phase shifts on the magnitude of 12° or more were often commonplace.

		Current		
Political Unit	Color	Voltage	Frequency	Lines/fields
Aden, Yemen	PAL	230	50	625/50
Afghanistan	PAL	220	50	625/50
Albania	SECAM	220	50	625/50
Algeria	PAL	127–220	50	625/50
Andorra		220	50	625/50
Angola		220	50	625/50
Argentina	PAL	220	50 + direct current	625/50
Australia	PAL	220–230–240	50	625/50
Austria	PAL	220	50	625/50
Azores	PAL	220	50	625/50
Bahamas	NTSC	115	60	525/60
Bahrain	PAL	220	50	625/50
Bangladesh	PAL			625/50
Barbados	NTSC	110–120	50	525/60
Belgium	PAL	220 (110)	50	625/50
Benin				625/50
Bermuda	NTSC	115–120	60	525/60
Bolivia	PAL	110–220–230	50–60	625/50
Brazil	PAL-M	110–127–220	50–60	525/60
Bulgaria	SECAM	220	50	625/50
Burma		230	50	
Burundi		220	50	625/50
Cambodia		120–220	50	
Cameroon		127–220–230	50	625/50
Canada	NTSC	110–115–120–230–240	60	525/60
Canary Islands	PAL	127	50	625/50
Central African Republic		220	50	625/50
Chad		220	50	625/50
Chile	NTSC	220	50	525/60
China (Peoples' Republic)	PAL	220	50	625/50
Colombia	NTSC	110–115–120–150–240	60	525/60
Congo, Brazzaville	SECAM	220	50	625/50
Costa Rica	NTSC	110	60	525/60
Cuba	NTSC	115–120	60	525/60
Cyprus	PAL	240	50	625/50
Czechoslovakia	SECAM	220	50	625/50
Dahomey		220	50	625/50
Denmark	PAL	220	50 + direct current	625/50
Djibouti		220	50	
Dominican Republic	NTSC	115	60	525/60
Ecuador	NTSC	110–120–127	60	525/60
Egypt	SECAM	110–115–220	50 + direct current	625/50
El Salvador	NTSC	110	60	525/60

Figure 25-2 TV broadcast systems and electrical current standards, as reported for nations and selected communities by their national governments.

		Current		
Political Unit	Color	Voltage	Frequency	Lines/fields
Ethiopia		127–220	50	625/50
Eritrea		127–220	50	
Fiji		240	50	625/50
Finland	PAL	220	50	625/50
France	SECAM	110–127–220–115–230	50 + direct current	625/50
French Guiana	SECAM	127	50	625/50
Gabon	SECAM	127–220	50	625/50
Gambia				625/50
Germany, Federal Republic of (West)	PAL	110–220–127–200	50 + direct current	625/50
Germany, Democratic Republic of (East)	SECAM	220	50	625/50
Ghana		230	50	625/50
Gibraltar	PAL	240	50	625/50
Greece		127–220–110	50 + direct current	625/50
Greenland		220	50	525/60
Guadeloupe		127	50	
Guam	NTSC	110	60	525/60
Guatemala	NTSC	110–120–220	60	525/60
Guinea		127–220	50	625/50
Guyana	SECAM	127	50	625/50
Haiti	SECAM	115–220	50 + 60	625/50
Hawaii	NTSC	115	60	525/60
Honduras	NTSC	110–220	60	525/60
Hong Kong	PAL	200	50	625/50
Hungary	SECAM	220	50	625/50
Iceland	PAL	220	50	625/50
India		220–230–250–300	50 + direct current	625/50
Indonesia	PAL	110–127	50	
Iran	SECAM	220	50	625/50
Iraq	SECAM	220	50	625/50
Ireland	PAL	220	50	625/50
Israel	PAL	230	50	625/50
Italy	PAL	60–110–120–127–150–160–220	50	625/50
Ivory Coast	SECAM	220	50	625/50
Jamaica	PAL	110	50 + 60	625/50
Japan	NTSC	100	50 + 60	525/60
Jordan	PAL	220	50	625/50
Kenya	PAL	240	50	625/50
Korea (North)				625/50
Korea (South)	NTSC	100	60	525/50
Kuwait	PAL	240	50	625/50
Laos		127–220	50	
Lebanon	SECAM	110–220	50	625/50

Figure 25-2 *(Continued)*

25-7

		Current		
Political Unit	Color	Voltage	Frequency	Lines/fields
Liberia	PAL	120	60	625/50
Libya	SECAM	125–230	50	625/50
Luxembourg	SECAM	110–220	50	625/50
Macao		110	50	
Malagasy		127–220	50	625/50
Madeira		220	50 + direct current	
Malaya, Federation of		230–240	50	
Malaysia	PAL	230–240	50	625/50
Malawi		220–230	50	625/50
Mali, Republic of		127–220	50	625/50
Majorca, Balearic Islands		127	50	
Malta		240	50	625/50
Martinique	SECAM	127	50	625/50
Mauretania		220	50	625/50
Mauritius	SECAM	230	50	625/50
Mexico	NTSC	110–115–120–125–127	50 + 60	525/60
Monaco	SECAM	127–220	50	625/50
Mongolia				625/50
Morocco	SECAM	110–115–127–230	50	625/50
Mozambique	PAL	220	50	625/50
Nepal		110	50 + 60	
Netherlands	PAL	125–220–127	50	625/50
Netherlands Antilles	NTSC			525/60
Aruba	NTSC	115–127	60	525/60
Bonaire		127	50	
Curaçao	NTSC	127–220	60	525/60
St. Martin	NTSC	120	60	525/60
New Caledonia	SECAM	220	50	625/50
New Zealand	PAL	230	50	625/50
Nicaragua	NTSC	120	60	525/60
Niger		220	50	625/50
Nigeria	PAL	230	50	625/50
Norway	PAL	130–150–230	50	625/50
Oman	PAL	220	50	625/50
Pakistan, West and East	PAL	220–230	50 + direct current	625/50
Panama	NTSC	110–115–120	60	525/60
Paraguay	PAL	220	50 + direct current	625/50
Peru	NTSC	110–220	60	525/60
Philippines	NTSC	110–115–220	60	525/60
Poland	SECAM	220	50	625/50
Portugal		110–220	50	625/50
Puerto Rico	NTSC	115	60	525/60

Figure 25-2 *(Continued)*

		Current		
Political Unit	Color	Voltage	Frequency	Lines/fields
Rhodesia	PAL	220–230	50	625/50
Rumania		220	50	625/50
Rwanda		220	50	625/50
Ryukyu Islands	NTSC	100–120	60	525/60
Samoa	NTSC	120	60	525/60
Saudi Arabia	SECAM	120–127–220–230	50 + 60	625/50
Senegal	SECAM	127	50	625/50
Sierra Leone	PAL	230	50	625/50
Singapore	PAL	220	50	625/50
Somalia, Republic of		110–220–230	50	625/50
South Africa	PAL	200–220–230–240–250	50 + direct current	625/50
Soviet Union	SECAM	127–220	50	625/50
Spain	PAL	127–150–220	50 + direct current	625/50
Spanish Africa		127	50	625/50
Sri Lanka		230	50	625/50
St. Kitts Island	NTSC	230	60	525/60
St. Lucia	NTSC	115	50	525/60
St. Pierre et Miquelon	NTSC	115	60	525/60
Sudan	PAL	240	50	625/50
Surinam	NTSC	115–127	50 + 60	525/50
Swaziland	PAL			625/50
Sweden	PAL	120–127–220	50 + direct current	625/50
Switzerland	PAL	120–125–220–135	50	625/50
Syria	SECAM	115–220	50	625/50
Tahiti	SECAM	120–220	60	625/50
Taiwan	NTSC	100	60	525/60
Tanzania	PAL	230	50	625/50
Thailand	PAL	110–220	50	625/50
Togo, Republic of		127–220	50	625/50
Trinidad, West Indies	NTSC	115	60	525/60
Tunisia	SECAM	110–115–220	50	625/50
Turkey	PAL	110–220	50	625/50
Uganda	PAL	240	50	625/50
United Kingdom	PAL	110–200–220–240	50 + direct current	625/50
United States	NTSC	127	60	525/60
Upper Volta, Republic of		220	50	625/50
Uruguay	PAL	220	50	625/50
Venezuela	NTSC	110–115–220	60	525/60
Vietnam		120–127–230	50	525/60
Virgin Islands	NTSC	115	60	525/60
Yemen		220	50	625/50
Yugoslavia	PAL	220	50	625/50
Zaire	SECAM	220	50	625/50
Zambia		230	50	625/50

Figure 25-2 *(Continued)*

Since the NTSC system does not adequately tolerate phase shift errors of more than plus or minus 5°, the color signals would often be too light in hue and too dark in intensity, with skin tones edged in green or purple. Happily, tint information in the NTSC system is now more accurate, principally because of tighter engineering standards at the broadcast level and the development of the automatic color or tint control circuit. PAL and SECAM systems are, by design, less susceptible to errors in phase.

The PAL System

PAL is essentially similar to the NTSC system in principle. Phase information, however, is reversed in PAL while scanning successive lines; hence its name, *p*hase *a*lternate by *l*ine. When a phase error occurs in the scanning of a line, a compensating error of equal degree but in the opposite direction is introduced during the succeeding line, thus canceling out the error. Since two lines are required to represent corrected tint information, the vertical detail of the tint information is reduced. As long as the phase errors are not too great, there will not be serious degradation of the picture, since the human eye does not require fine detail to determine different hues. The human brain will average out the compensating errors, thus nullifying the effect of the annoying phase shift which would have seriously plagued the NTSC viewer. If the phase error is more than 20°, however, visible degradation does occur in the PAL system. Large phase errors can be corrected by introducing a delay line and an electronic switch in the receiver, as in the SECAM system. It should be noted, however, that phase errors of this magnitude are quite the exception.

The SECAM System

SECAM, like PAL, is similar in principle to NTSC; however, the chroma signal is modulated in only one way. Since the two types of information needed to make up the color picture do not occur simultaneously, the usual errors associated with concurrent amplitude and phase modulation do not happen. In the SECAM system (SECAM III), alternate line scans carry luminance information and *Red*, while the next line contains luminance and *Blue*. The *Green* information is then derived within the receiver by subtracting the *Red* and *Blue* information from the luminance signal. Since each individual line contains only one-half the color information, two sequential lines are necessary to complete the detail of this information. Again, since the human eye is not sensitive to tint and saturation of small details, no adverse effects are perceived.

Since the Green information is derived by subtracting the Red and Blue data from the luminance component—and since this information is transmitted in time sequence—some method to hold one line of data in a storage or memory is required. This is accomplished by a device called a delay line, which, as its name implies, holds or delays the information of each successive line scan for milliseconds. Successive pairs of lines are then matched with the aid of electronic switching. This holding of a line in storage or memory provides the system with its acronym SECAM, meaning sequential electronic color system with memory.

Adjustments such as hue and saturation are usually provided to allow the SECAM viewer to vary the picture to individual taste.

Analysis of the Three Systems

As noted earlier, except for different color-coding techniques, the single most distinguishing aspect between NTSC and the PAL and SECAM color systems is the difference in the number of lines per picture and the number of pictures per second (see Figure 25-3). This differentiation will be examined in greater detail in the section "Standards Conversion."

Over the years, as a consequence of thousands of hours of usage and testing with the three color systems, the following general conclusions represent the consensus of video engineers familiar with all the systems:

- Under normal conditions and with properly adjusted and maintained equipment, *color reproduction is equally good with all three systems.* In visibility of subcarrier, NTSC is the best, PAL somewhat poorer, and SECAM the poorest. SECAM is also the worst in regard to clear color transition in both horizontal and vertical modes.
- Under less than favorable conditions, when noise and interference disturb the pictures, the three systems do not exhibit any great differences except that SECAM can display a "silver fish" type of effect.
- In the event of signal amplitude distortion as an attenuation of higher frequencies, PAL was found to be somewhat less sensitive than NTSC, with SECAM being more sensitive.
- Differential phase errors are much more tolerable in both PAL and SECAM than in NTSC. Phase errors exceeding 10 to 12° cause unacceptable color shifts in NTSC, while PAL can tolerate phase errors to 20° or greater, with SECAM remaining unaffected up to nearly 40°.
- Multiple images or ghosting in cities and mountainous areas are significantly less in PAL than in either NTSC or SECAM.
- Compatibility (reception of color programming on black-and-white receivers) NTSC is the best, PAL slightly worse, and SECAM the worst.
- Production costs are higher for PAL and SECAM receivers because they are more complex than NTSC.
- In studio production, NTSC equipment is the simplest, with PAL and SECAM being successively more difficult.
- In recording color on a video recorder, SECAM is the simplest, followed by PAL, with NTSC being the most costly because of its inherent sensitivity to imperfections in equipment and their adjustments.

25-12 INTERNATIONAL VIDEO NETWORKING

Features	NTSC	PAL	SECAM
Color reproduction	Good	Good	Good
Visibility of subcarrier	Best	Good	Fair
Noise: under less than favorable conditions	Fair	Fair	Poor
Signal amplitude distortions	Fair	Good	Poor
Tolerability of differential phase errors	Poor	Good	Good
Effects of echoes and multiple receptions in cities and mountainous areas	Fair	Good	Fair
Compatibility	Best	Good	Fair
Cost of equipment	Lowest	Higher	Highest
Studio complexity	Lowest	Higher	Highest
VTR requirements	Highest	Low	Lowest

Figure 25-3 Analysis of NTSC, PAL, and SECAM color systems.

It is apparent from the foregoing that no one system is singularly superior in all respects to the other two. Overall, however, the multitude of disadvantages of SECAM—especially when compared with PAL—makes it certainly less desirable than either NTSC or PAL. PAL systems, on the other hand, because of their inherent resistance to tolerate phase error shifts, coupled with the higher definition afforded by the 625-line scan system, are superior to NTSC in difficult reception areas such as mountainous regions and large urban areas.

EQUIPMENT STANDARDS

Having looked at international broadcast standards, let's briefly examine the situation with respect to equipment or machine standards.

Within each broadcast standard, there are different format variations. In NTSC alone, there presently exist no fewer than thirteen viable formats, such as U-Matic, Beta, VHS, and 1-inch type C. Several manufacturers even offer multiple formats within their own product ranges. In addition to the thirteen offered at present, there are at least five others which, although no longer in production, have been sold in sufficient quantities in the market that postproduction houses must maintain compatible equipment in order to interface with tapes recorded on this older equipment.

All the various video hardware manufacturers have long professed the desirability of establishing standards. Indeed, there are such standardization bodies as the Electronics Industries Association (EIA), the Electronics Industries Association—Japan (EIAJ), the Society of Motion Picture and Television Engineers (SMPTE), the International Tape Association (ITA), and the International Television Association (ITVA), to mention a few—of which most, if not all, video hardware manufacturers are either members, or are at least represented at the meetings. Nevertheless, we are no further along toward standardization than we were ten years ago. In point of fact, we are further away now!

As a practical matter, although manufacturers pay lip service to standardization, it really would be in their own worst interests to standardize on someone else's equipment or design. If they were to do so, there would be licensing fees, but that contingency in itself would not be the most significant factor.

The biggest problem would be the write-off of vast amounts of tooling coupled with the investment in new tooling in order to meet the new "standard." Of course, a maker could always buy equipment from the manufacturer of the winning standard, but then the purchaser would be at a distinct market disadvantage from both a time and a cost standpoint.

Many knowledgeable people will logically argue that standardization of format would inhibit further technological advance. Indeed, if we look at the tremendous strides made by equipment makers in the past few years, this fact becomes very evident. On the other hand, equally valid arguments can be advanced for the benefits to be derived from format standardization, such as economies of scale production by various manufacturers and the assurance of longer utilization by users as a consequence of some degree of protection from early obsolescence.

METHODS OF SOLVING INTERCHANGE

As has already become evident, broadcast television systems throughout the world differ in terms of scan lines per picture, pictures per second, frequency of audio subcarrier, color information and encoding, and alternating current (AC) voltage and frequency requirements. Therefore, in order to exchange programs among locations in different areas and countries around the world, specialized techniques and services must be used. There are several different ways of achieving this solution. Summarized in Figure 25-4 are the four principal techniques.

One Company, One Format

If we ignore the broadcast system of the country in question, there is no reason why one universal videocassette system cannot be used anywhere in the world. A multinational company can create a true video communications network since, in effect, the company will be operating within its own closed-loop environment. It is a relatively simple task to operate an NTSC system on 110, 220, or 240 volts and 50 or 60 hertz (cycles per second). Thus, NTSC can be used from Aden to Zanzibar. Similarly, a PAL or SECAM system can also be adapted to do the same. As long as the foreign offices are merely viewing the tapes and not recording their own, this is a feasible system—except for those countries which have established import restrictions on NTSC-only equipment, such as the Soviet Union, Italy, and Spain.

Method	Cost	Remarks
1. Rental or sale of required equipment standard and format, regardless of broadcast standard at location	Usually lowest	Does not allow for two-way compatibility, i.e., local origination. Not always available—depends on location
2. Multistandard playback machines	Relatively expensive	Greater flexibility than #1. Provides limited true two-way compatibility
3. Film	Expensive, but not necessarily as expensive as #2, depending on frequency of use and program lengths	Universal two-way compatibility
4. Standards conversion	Expensive—the most expensive on a per minute of program basis, unless multiple copies are required	Usually required for broadcast purposes. The preferred solution where 10 or more copies are required

Figure 25-4 Methods of solving interchange problems.

Multistandard Equipment

A way to avoid international restrictions as well as to interchange information among foreign locations is to rent or buy multistandard equipment. This is particularly useful if certain locations will be receiving not only NTSC tapes but also PAL and SECAM recordings. Using multistandard video players (Figure 25-5) and multistandard receiver-monitors, these offices will be able to view tapes in all standards.

The only limitation to multistandard equipment is that it cannot play one standard as an input, convert into a second standard, and record the second standard as output. For this reason, if multiple copies in a different standard are required, it is necessary to use standards conversion. For example, if a German subsidiary has produced a videotape in 625 PAL which would be of interest to thirty United States branch offices, it would be quite logical to convert the PAL tape to an NTSC submaster, which could then be duplicated relatively inexpensively onto thirty NTSC copies.

INTERNATIONAL EQUIPMENT STANDARDS   25-15

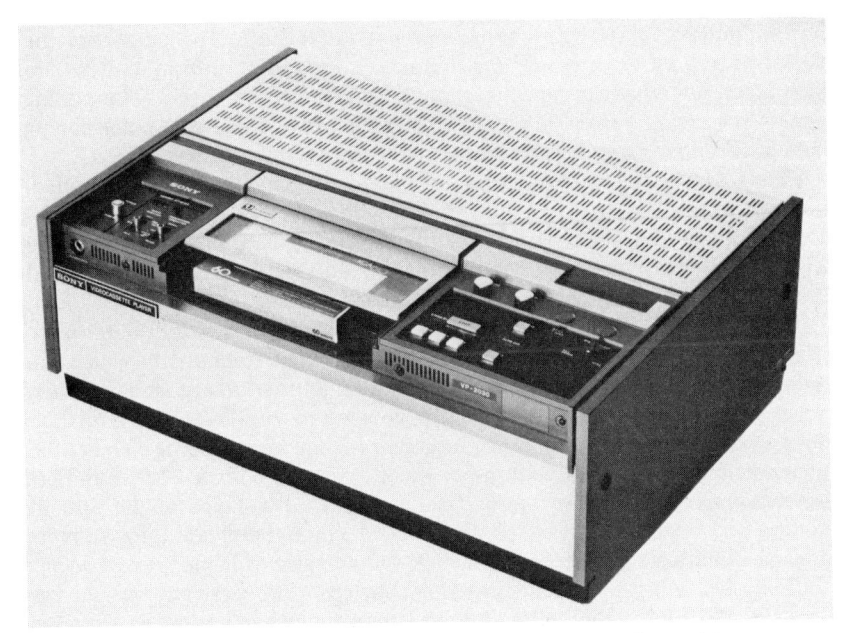

Figure 25-5 NTSC, PAL, and SECAM U-Matic videocassette player. (*Sony Corporation of America.*)

STANDARDS CONVERSION

Standards conversion is the change of one video system into another; in essence, this means going from a 525- to a 625-line standard, or from NTSC to PAL, or to SECAM, or a combination thereof. Color systems can be converted relatively easily, since all color starts as RGB (Red, Green, Blue) and is encoded from there. When the picture is received at a monitor, it is decoded back to RGB. Thus, if the only thing necessary were to convert color from NTSC to PAL, for example, it could be easily done by merely decoding back to RGB and then reencoding in the new system.

The real problem arises from the difference in the number of lines in the different systems. For example, NTSC uses a 525-line system, while PAL uses a 625-line system. When converting, there are two principal difficulties. The first is simply either a loss of some lines or a doubling up of some lines in order to change the number of lines in a picture area. This is further complicated, however, by the fact that 525 lines are, in general, related to 30 pictures per second, while 625 lines are related to 25 pictures per second. Thus, you have to change not only the number of lines in each picture but also the number of pictures in each second. Therefore, you must lose a number of lines or double up on a number of lines. It should be noted that the 30 pictures per second are directly related to 60 hertz, while the 25 pictures relate in a similar manner to 50 hertz. To complicate the matter even further, a recording contains color-coding information within each picture, so that during conversion, one has either to lose 5 color-coding signals when going from 30 pictures to 25 pictures or to find 5 color-coding signals when going from 25 pictures to 30 pictures.

There are basically two types of standard converters in use—optical and electronic. The quality achievable by optical converters is usually inferior to that obtainable with the electronic type—except in cases where the original material contains poor time-base stability. Essentially, an optical converter is akin to the old-fashioned *kinescope* process, wherein a camera is focused on the face of a television monitor displaying the material to be converted. The video camera is operating in the standard to which the material displayed is to be converted. The resultant camera signal is then fed to a video recorder operating in the new standard. Overall, this method loses some color and definition and suffers from an annoying flicker (the result of looking at a 25-picture-per-second source with a 30-picture-per-second camera—or the reverse). The principal advantages of the optical method are lower cost and sometimes speed, since the demand for conversion, especially electronic, is increasing, and sometimes backlogs can occur.

There are two types of electronic converters—the older analog and the newer digital type. The digital standards converter has proven so superior to analog types that most, if not all, those in use have been retired except for those maintained as emergency backup systems. Conversions achieved by digital converters are virtually indistinguishable from the original material.

There are at present three manufacturers. The largest and original manufacturer is Marconi of England, which manufactures DICE, or *d*igital *i*ntercontinental *c*onversion *e*quipment. The second is Quantel, also based in England, which manufactures the DSC 4002 series of converters. Essentially, the two are quite similar, as the developers of Quantel's unit were originally employed by Marconi, and at one point it appeared that Marconi would bring a patent infringement suit against Quantel. This event has not occurred, and the principal differences between the two devices appear to be in the areas of size (the DSC 4002 is much more compact than the DICE) and their respective methods of dealing with movement interpolation when gaining or losing the 5 pictures per second and the 100 lines per picture.

The third manufacturer is McMichael, Ltd., also based in England and a sister company to Marconi. The new digital converter, available in June 1981, was developed by the state-owned British Broadcasting Company (BBC). It is reputed to be the "ultimate in an all digital standards converter," as the manufacturer, with typical British understatement, claims. ACE, or *a*dvanced digital *c*onverter and transcoder *e*quipment, is a four-field converter, in contrast to the two fields offered by DICE and the DSC 4002. The extra two fields are important in improving picture resolution and minimizing the effect of movement interpolation.

Interpolation is a mathematical term referring to the process of inserting intermediate terms in a series, as in a series of numbers. When going from a 625-line 25-frames-per-second format to a 525-line 30-frames-per-second format, the converter must "fill in," or interpolate, 5 additional frames per second in order to complete the picture without leaving an obvious gap. This interpolation process is normally performed without any undue notice except when the video picture contains a series of rapid movements, as when a golf club swings or a motor car turns a corner at high speed. Since the converter must fill in the additional lines, the effect can appear stroboscopic and, hence, disconcerting.

Operation of the converter requires two videotape machines. One unit is used to play the program being converted *into* the standards converter, while the other is used to record the output *from* the converter. The entire operation is performed in real time, that is, a 1-hour program requires 1 hour of conversion time.

Figure 25-6 is a representative rate comparison between the optical and electronic or digital methods of standards conversion.

When to Use Optical or Digital Conversions

When cost is paramount or screening quality is satisfactory, or when both conditions exist, optical conversion is the recommended choice. When the converted material is to be broadcast, when multiple copies are required for distribution, or simply when the best possible quality is desired, digital is the preferred method. Sometimes, however, there are situations or circumstances which preclude the use of the digital method even though cost is secondary and top quality is required. In those instances where the material to be converted is not of the best quality, as in the case of a second- or third-generation master on ½-inch cassette or open reel, or where the original is recorded in black and white with good contrast but extensive video background noise, optical will afford a better conversion than digital.

CONCLUSION

Deciding upon equipment and techniques to be employed in exchanging video programming internationally need not be a harrowing experience. Existing domestic networks can be compatibly expanded in conjunction with various international locations. Equipment currently in use will not necessarily be rendered obsolete. With careful planning and analysis, the inherent difficulties can be surmounted and the benefits highly rewarding, as such corporations as Merrill Lynch, Eaton, Cutler Hammer, Young & Rubicam, IBM, and Xerox, to mention a few, have found.

ELECTRONIC STANDARDS CONVERSION*

35mm/16mm film, 525 NTSC, 625 PAL/SECAM to 625 PAL/SECAM or 525 NTSC

Duration	Cost

2" Quad, 1" Helical 35/16mm to 2" Quad or 1" Helical 5 minutes or less

5 minutes or less	$137.50
15 minutes	$173.25
30 minutes	$275.00
60 minutes	$467.50
90 minutes	$660.00

2" Quad, 1" Helical 35/16mm to U-Matic, VHS, or Beta cassette

5 minutes or less	$137.50
15 minutes	$165.00
30 minutes	$247.50
60 minutes	$385.00
90 minutes	$467.50

U-Matic cassette to U-Matic, VHS, or Beta cassette

5 minutes or less	$110.00
15 minutes	$140.25
30 minutes	$165.00
60 minutes	$247.50
90 minutes	$285.00

Compilation:

Cassette—$10.00 per edit
35/16mm, 2" Quad, 1" Helical—$15.00 per edit or reel charge

Note: 1. Tape stock extra
2. Turn around time approximately 72 hours

*Costs as of April 1, 1981.

Figure 25-6 Representative cost comparisons for electronic digital and optical conversions.

OPTICAL CONVERSION

525 NTSC, 625 PAL/SECAM to 625 PAL/SECAM OR 525 NTSC

Duration	Cost

U-Matic, VHS, or Beta cassette to U-Matic, VHS, or Beta cassette

Duration	Cost
5 minutes or less	$ 82.50
60 minutes	$105.00
90 minutes	$132.50
120 minutes	$157.50
180 minutes	$237.50

16mm film to 625 PAL/SECAM, 525 NTSC U-cassette, VHS, or Beta cassette

Duration	Cost
15 minutes	$ 75.00
30 minutes	$ 93.75
60 minutes	$117.50
90 minutes	$146.50
120 minutes	$185.00

Compilation:

Cassette—$10.00 per edit
16mm film—$10.00 per edit

Note: 1. Tape stock extra
2. Turnaround time approximately 24 hours

Figure 25-2 *(Continued)*

26

International Video Operations

PATRICIA TIERNEY WILSON

Senior managers of many multinational corporations realize that dependable, as well as alternate, ways must be found to communicate with their overseas employees. Such things as the increasing costs of travel and the unreliability of foreign distribution systems for printed material have prompted top management to approve the installation of video networks.

Having been involved in the installation of two such international video networks at Merrill Lynch and Chase Manhattan Bank, I have attempted to include in the following pages my experiences in the implementation of such networks as well as some tips on the smooth operation of the network once it is in place.

RESEARCH FOR AN INTERNATIONAL VIDEO NETWORK

Planning

You have the approval and the capital money budget to install the network. This capital approval, of course, was based on preliminary research done for your international network proposal. The first recommendation is to establish a special international task force, which is different from the task force assigned to set program priorities. The former may have a life span of only a few months. The latter is an ongoing committee. The international committee should be made up of representatives from four areas within your company: the overseas premises installation unit, the purchasing department, the personnel or human resource department, and the audiovisual operation.

There are several reasons for including people from the first three units. First, these people will have an insider's knowledge about individual international offices. The premises person will know, for example, how large the overseas facility is and the appropriate site to place the equipment. The human resource person will know the number of people in the office as well as the attitude of a particular office with regard to head office communications. These representatives will also have firsthand knowledge about local acceptance of video communications. Furthermore, as insiders they will be privy to the long-range plans of office closings. There is nothing so disheartening as installing video equipment in an office only to discover, one year later, that that office is to be shut down.

The inclusion of the company purchasing agent guarantees a smooth

26-1

purchasing phase once that part of the installation begins. True, the agent may be just an observer during the initial workings of the task force, but will be acquiring a working knowledge of video equipment during this investigatory stage.

Selecting Offices

Once the task force is in place, your first task is to determine which offices are to receive the equipment. If every office in the company's branch system is designated to get a unit, you need not go through this exercise. However, if the capital budget is limited, you must carefully scrutinize the branch system. It would be simple to eliminate recipient offices with the fewest employees. But those offices may be just the ones that need to receive messages from the head office. The determining factors, in my opinion, are:

1. The remoteness of location (the premises person supplies the information)
2. The number of employees (human resource supplies the information)
3. The obvious need to communicate to that location (a human resource decision)
4. The total cost, not only of the equipment and its installation, but also of the air freight charges and importation duties

Controlling Costs

You may find out it is just too expensive to install a unit in certain countries. In Korea, for example, the importation duty on video equipment is 300 percent. Furthermore, Korea does not permit any organization except educational institutions and government agencies to import video equipment.

Case Study 1. At Chase Manhattan Bank, the premises task force member supplied this information from his experience in setting up a branch office in Korea. In addition to ordering and supplying typewriters, computer terminals, file cabinets, and the like, he was asked to investigate the importation of a video player for use in the office. Television sets can be imported as household goods, but players cannot. (This particular importation restriction concerned only ¾-inch equipment; ½-inch consumer equipment could be purchased within the country.)

As you can see, it pays to include such staff people in the implementation process.

Host Country's Technical Requirements

The next step in the implementation process is gathering information on individual country TV standards and electrical requirements. Obviously, if the decision is to install NTSC-only equipment, you need not know the local country standard. (The pros and cons of NTSC-only versus tristandard will be discussed later in the chapter.) Finding local TV scanning standards can be difficult, since these standards do change. Peru, for example, changed its standard twice in three years—from PAL to NTSC, then to SECAM; Greece changed from PAL to SECAM. Furthermore, some countries are creating their particular versions of one of the three worldwide standards. Brazil has its own version of PAL called PAL-M.

It is also important to know the electrical voltage and hertz (cycles per second) used in the office. There are three reasons for this.

First, most overseas locations operate on 220 volts/50 hertz. NTSC equipment usually operates on 110–120 volts/60 hertz. Equipment must be modified for 50 hertz and a transformer must be ordered to accommodate the step down from 220 to 110 volts.

Second, tristandard equipment operates on 220 volts/50 hertz. Yet there are areas in the world which operate on 220 volts/60 hertz. There are helpful reference books and pamphlets containing these data but their information is not foolproof. They list the *known* countrywide voltage and hertz. However, there are areas and cities, and even sections within cities, that have different voltage and hertz from the remainder of the country.

Third, despite the fact that certain countries are known to have a standard current and hertz, certain buildings, usually those housing several American multinational companies, have their own generators. These generators usually operate at 110 volts/60 hertz or 220 volts/60 hertz. Why their own generators? Having their own assures the offices of being able to conduct their daily business with American-made electrical and electronic equipment, thus eliminating the need for expensive modifications. Furthermore, that country may have frequent power failures. (Blackouts usually occur during peak business hours.)

The Branch Office Survey

We have found that the best source for accurate local TV standards, voltage, and hertz information is the branch office personnel. To gather this information, we prepared a questionnaire (Figure 26-1) designed to be completed by the overseas office chosen to receive the equipment. The questionnaire asked for the electrical current and hertz used *within* the office by requesting the recipient to check the metal plate on an electric typewriter as well as to double-check with the building electrician or the local electrician servicing the office. To complete the information about the local TV standard, we suggested that the local office call the nearest television station.

We also included several other questions. For example, Who would be the person responsible for the proper usage, storage, and handling of the video equipment and video programming? This query served several purposes. We established contact with one person with whom we could commu-

26-4 INTERNATIONAL VIDEO NETWORKING

> Please complete and return to (your name, department and internal address) no later than (specify a date).
>
> Video coordinator: _____
>   (Name)   (Title)
>
> Telephone number:
> Cable or telex number: _____
> Person to whom equipment will be shipped. This name will appear on all shipping documents.   (Name)   (Title)
>
> Address for equipment delivery (street address, no box numbers). _____
>
> Local import agent (if any), with whom you have a relationship, address, and cable number. _____
>
> Electrical voltage and cycles per second within your office (110/50, 110/60, 220/50, 220/60, 240/50) _____
>
> Do you know of any import restrictions regarding videotape equipment?
>
> _____
> _____
>
> Completed by: _____
>   (Name)
>   (Title)
>   (Date)

Figure 26-1 International video network questionnaire. It is usually accompanied by a covering memorandum announcing the installation of the video network, as well as explaining the purpose of the questionnaire, that is, the need for a video coordinator, the need to know the current and cycles per second, etc.

nicate on a continuing basis. This person's name would be used on all importation documents. By the way, we dubbed this person the video coordinator.

We also asked for the street address of the office. Most offices, at least at Merrill Lynch and Chase, list post office boxes as their addresses. You can imagine the difficulty in delivering a piece of equipment to such a location. Finally, we asked the local office to submit the name and address of a local import agent (if any) with whom it had a relationship. Such a contact speeds up the importation of the equipment because the agent knows the company as well as the people in the office and most likely has serviced the office competently in previous import situations.

This questionnaire method assures you of a way of double-checking the known TV standard, current, and hertz known for that country against what the branch indicates. It also, as we found out, prompts the local office to offer information about local requirements for video equipment. For example, there may have been recent changes in importation requirements, such as an increase in the duty for video from 30 percent to 100 percent. Or, video equipment may cost less when purchased at the local level than when imported. Our Malaysia branch office, when completing the questionnaire, pointed out that it had researched the cost of equipment: Although the purchase price was 40 percent higher in Malaysia than in the United States, when air freight and importation duty were added to that cost, it indeed was cheaper to purchase locally.

EQUIPMENT SELECTION

One-Way versus Two-Way Systems

With this research phase completed, the next decision is, What type of equipment should be purchased? This decision primarily is based on the type of network you want, that is, one-way or two-way. *One-way* means that all information or programming flows in one direction. You produce a program and distribute it to your overseas locations; they view it. A *two-way* video network, as the phrase implies, means that you produce a program and distribute it. The overseas location then records a message and forwards it to the head office. What type of message could that overseas location possibly send that couldn't be covered in a memorandum? If you have a manufacturing facility overseas and if the engineering staff develops a problem, say, in a new process which you recently introduced in a video program, they can record the difficulty and forward their queries to the head office research and development staff. There will be no need for your company to rush a top troubleshooter to the location.

A second reason for two-way video is in the area of training. Generally, the head office conceives, develops, and produces training packages. The programs can range from those totally instructor-led to a combination of instructor and mixed media. An overseas location can record a few training sessions of that program and return the recording to the head office with comments on which areas need to be improved for the overseas location, based on the actual recorded examples that prompted these modifications.

Another reason for two-way video networks is that the overseas location will have the ability to record meetings that are important to senior managers but who, because of travel or time constraints, were unable to attend. Recording the event is the next best thing. Some readers of this chapter may reject the idea of recording meetings with comments of "No production value" or "Who would sit down and view an hour of video?" One thing I've learned is that a viewer sits down in front of a television screen and internally adjusts from his or her "*60 Minutes*-high-production-quality viewing" mode to a lower production-quality "this-is-a-tape-of-a-meeting" mode. Most internal or employee audiences are concerned with content.

Of course, the cost to install a two-way, rather than a one-way, network may exceed your capital budget. One-way networks require players only, while two-way networks require recorder-players. Recorder-players are at least 50 percent more expensive than players. Obviously, the cost alone may determine your choice because importation duties (part of your budget figures) are calculated on equipment cost.

Standards and Formats

Now come the big questions: What standard and format do you wish the network to have? NTSC alone? Tristandard? Three-quarter inch? Half-inch? Videodisc? Here are some considerations you should weigh when deciding which combination to install.

First, determine all the locations where you plan to place a unit. As already discussed, host country requirements may substantially influence your decision. Shipping costs and local duties will also greatly influence your decision on which standard and format to choose.

Second, you must weigh the pros and cons of operating an NTSC-only system or tristandard operation. Chapter 25 gives an explanation of the three worldwide standards: NTSC, PAL, and SECAM. To cite a little background on multistandard units: Equipment manufacturers saw a need to develop multistandard playback equipment, particularly when international private video networks began to mushroom. Initially, they developed two types, the dual standard and tristandard models. The dual standard had the capability of playing back in NTSC and PAL or NTSC and SECAM. The tristandard equipment played back in NTSC, PAL, and SECAM. The dual systems are no longer manufactured. One suspects that the reason is that certain countries change their standards so frequently that the equipment ordered for any of these countries became obsolete. It was more cost-effective to manufacture a tristandard rather than both dual and tristandard playback units.

In making the decision to install NTSC-only or tristandard units, you must determine whether you wish the branch office to use the equipment for your programs only or whether you wish to provide it with the capability of playing back in another standard, so that the local branch office can view programs produced in its local country. There is also the factor of cost. NTSC equipment is less expensive than tristandard equipment. Does your capital budget allow the purchase of the more expensive equipment? And last, does the country restrict the importation of NTSC equipment?

If your choice is to install tristandard units or a combination of NTSC and tristandard, contact several companies that have these types of networks. There are several in operation. Most are ¾-inch formats.

The tristandard ½-inch models, in 1980, were evidently watered-down versions of the consumer models. They are not sturdily built, and the SECAM mode is questionable. Most are Saudi Arabian SECAM. This means that local programming prepared in France, for example, cannot be viewed on this particular ½-inch tristandard player. Furthermore, to my knowledge, there are only two locations in the world that repair the tristandard ½-inch models; London and Jeddah. On the other hand, ¾-inch tristandard models can be repaired in numerous locations around the world. (One of the most agonizing situations is when an irate branch manager wires you every day demanding that his or her equipment be repaired and you, thousands of miles away, are helplessly attempting to get the thing fixed.)

Third, some companies are considering the videodisc for their international network. Several United States companies within the last year or so converted to disc networks for domestic distribution. Videodiscs, even in their infancy, have tremendous advantages over videotape. Among them are their instant access capability, their still frame capacity (54,000 frames), their apparent durability, and their future application of on-line data retrieval with interactive training. However, there are too many unanswered questions. How durable are the players in overseas shipping? Who would service the player overseas? Can I, as a producer, afford a 22- to 75-day turnaround on disc copies versus a 3- to 5-day one for videotape? Will there be enough programming and copies to justify the cost of replication?

Fourth, for local purchases, find out the repair capability within the country. Are there local dealers who can repair the equipment? Remember, there are two types of video dealers; those who are involved strictly in the sale of video equipment and those who handle sales and repair. Are there local dealers who can repair ½-inch equipment? If you are considering a ½-inch NTSC international network, take into account that the local dealer may not know how to repair NTSC or ½-inch equipment.

Of course, a solution to the potential repair problem would be to purchase extra equipment to act as "floaters." When a unit breaks down, the floater is shipped to the location. The defective unit is then packed and returned at which time it is repaired and readied for the next replacement or repair journey. It sounds easy. Unfortunately, it is not. The idea works well in the United States. In the rest of the world, it is a more complicated matter. Export and import licenses must be prepared for the floater as well as for the defective unit. Import fees must be paid on both units. Obviously, meeting these requirements is quite expensive, consumes a great deal of time in shipping, and reduces the effectiveness of the video network.

Finally, the cost of copies is an important consideration. The replication and mailing costs are more expensive for ¾-inch copies than for ½-inch videocassettes.

EQUIPMENT VENDORS

Whichever way the decision goes on the type of video network to be established, it is time to investigate vendors. They are most helpful in this stage of the implementation. I've found that some decisions on hardware made without the advice of a vendor will often change. But where to look for vendors? There are two choices: the specialists and the manufacturers. The specialists are companies whose primary business is the sale and setup of industrial international video networks.

I recommend the specialists. In the two instances in which I was responsible for video networks at Merrill Lynch and the Chase Manhattan Bank, I found the specialists the more informed on the latest import restrictions, hardware availability, and local country standards. Some manufacturers of video equipment, concerned with United States business, may not know the nuances of international video.

FREIGHT FORWARDER

During your information-gathering process, hardware research, and purchase, you should also be investigating a freight forwarder. If your company's daily business is exporting equipment to overseas locations, then all you need to do is contact the unit within your mail department that handles these types of shipments for execution. A note of caution: Your in-house shipping department specializes in one type of equipment exportation. Be certain that the shipping people know that video equipment exportation may require export licenses, import permits, and legalization permits. A country-by-country export-and-import requirement analysis must be done prior to shipment. The specialist-vendor from whom you purchase the equipment can offer guidance to your shipping people.

Those of you who do not have such shipping expertise within your company can seek out the advice and recommendations of the equipment vendor. The vendor may have a working relationship with a freight forwarder. Heed the vendor's counsel. After all, his or her bread and butter depends on efficient, safe, and timely delivery of the product as well as a completely satisfied customer. It doesn't make sense for the vendor to dump the equipment on a customer's doorstep and walk away. Ensuring a complete installation on site at your local overseas office can only lead to more business for the vendor, especially when the network expands or requires updated equipment. However, one of the criteria in your decision on a freight forwarder, aside from the equipment vendor's recommendation, is the number of correspondent agents the shipper has worldwide. Having an agent situated in the city slated to receive equipment assures the presence of a person on the spot to solve any unforeseeable import problems. That local agent, too, will be waiting to receive the shipment with the proper papers.

The freight forwarder will have prepared its correspondent agent with a full list of particulars about the shipment, including a description of the equipment, its purpose, its final destination, and an inquiry as to the agent's opinion and recommended action regarding any known difficulty in the importation process. While the shipper works on the necessary papers needed for exportation, the local agent works toward the same end for the importation.

An alternative to this procedure would be to contract the equipment vendor to handle the shipping entirely. Let the vendor worry about licenses, permits, and legalization approvals. Either way, it is best to have the vendor involved. Do not attempt to do the job yourself.

INVOICES

One thing to remember is that the title of equipment passes from the vendor to your company at the shipping point. So it is necessary that pro forma invoices are prepared on your company's letterhead. Pro forma invoices contain descriptions of the equipment, its weight, serial numbers, and intended purpose. If the equipment is "for educational purposes" (as is the case with most video networks), there sometimes is a lower importation rate imposed, if this phrase is used on the documents (Figure 26-2).

Another thing to remember is that all documentation accompanying the equipment should contain the word *monitor* in the equipment description section. The word *television* has a different connotation in other countries and may cause problems.

KEEPING BRANCH OFFICES INFORMED

While all this research, paperwork, negotiating, and purchasing is taking place, your overseas branches, mind you, are not passive. From your initial inquiry, you will be repeatedly swamped with the one question, *"When?"*

To avoid the ruffling of feathers and to retain your sanity, establish a communications program at the same time as you are proceeding through the implementation process. A series of memoranda outlining your progress to date helps alleviate the pressure exerted by the overseas offices. You must keep the people aware of what is occurring because they are your future audience. Alienating them before the fact can only create a dissatisfied audience at the time of program distribution. If you already have a domestic video network in place and if you have produced programs appropriate for the international network, begin distributing them, even before the equipment arrives. However, send each branch office one cassette at a time, since large packages are likely to be delayed by customs. The benefit of program distribution, coupled with installation news, outweighs the risks of overseas complaints and dissatisfaction. This is simply a program of handholding. You may also develop program and operations manuals as part of your communications program.

26-10 INTERNATIONAL VIDEO NETWORKING

(COMPANY LETTERHEAD)

(ADDRESS OF YOUR BRANCH OFFICE)

PRO FORMA INVOICE
SHIPPED VIA: AIR FREIGHT
TERMS: NO CHARGE INVOICE
FOB AIRPORT (CITY)
SHIP: (PREPAID OR COLLECT)

QUANTITY

DESCRIPTION

MARKS:
AS ADDRESSED

"This material is being shipped on a NO CHARGE BASIS as it will be utilized for EDUCATION PURPOSES. No funds or foreign exchange involved as title of the merchandise remains in the name of (YOUR COMPANY). Value declared only for Customs Purposes."

One Model # of Player/Recorder   (amount)
One Model # *MONITOR*   (amount)

ORIGIN OF MERCHANDISE:
(Country of Manufacture)

PACKING SPECIFICATIONS:
[Here is described the gross weight (carton, equipment packing materials), net weight (the equipment only), and dimensions of the cartons (length, width, and depth).]

We declare the said invoice is true and correct and has been issued for CUSTOMS Purposes only as there is no payment or exchange involved.

(Your company)
(Your name or the name of your company's purchasing agent)
(Title)

Figure 26-2 Sample of pro forma invoice wording.

PURCHASING POLICY

Along with this handholding, ground rules should be established. For example, all purchases will be coordinated through your department. Some offices may state that they can buy locally at lower cost, as did our office in Malaysia. I repeat, at least coordinate the purchase, advising the local office on a recommended dealer, what the local price should be, what format and standard will be used in the network, and model numbers of the equipment. Impress upon the local branch manager the importance of purchasing from the recommended dealer and the necessity of buying the exact models you have suggested.

Other ground rules may be the need to impress the local personnel about the dangers of theft and damage with the movement of equipment in and out of the office, the illegality of duplicating and distributing copyrighted programs among the branch offices, or something as simple as how programs should be mailed back to the head office if you have a recycling system.

PROGRAM DISTRIBUTION

I hope this chapter has covered everything you need to know for a smooth installation process. Now, let's turn to some of the areas you need to know about once the equipment is in place. One of these is the mailing of your programs.

Overseas Mail

Investigate your company's mailing procedures to overseas locations. For different locations, there may be different delivery systems. For example, couriers may be used for major metropolitan locations like London, Hong Kong, and Tokyo. And, of course, regular mail for the rest. You would be well advised to send your programs via regular air mail. Why should you not use the courier system? You may be able to distribute via this system for quite a while, but one day you will receive an irate phone call from the mail department, or perhaps the head of another department, informing you that important documents were held up at customs because of your "infernal" video. An important contract or quotation did not arrive on time. Foreign customs agents, you see, delay everything in a shipment because of a video program. Duty forms must be filled out and a representative of your overseas office must "bail out" all the mail. This can take from one day to several months.

Mailing two or more programs together almost automatically subjects them to customs delay. If you have a three-part program on three separate cassettes or videodiscs, mail them separately, informing the overseas branch either in advance or with an insert in the first package. Conversely, advise the branch, when returning programs for recycling, to use the same procedure. A package containing two or more cassettes subjects it to delay on stateside too. So, use the regular mail for those programs which need not arrive within a few days. Use a customs broker for those programs which must arrive as soon as possible. In fact, there are brokers who specialize in the shipping of media materials. They will prepare all the necessary documents needed for international shipping, including pro forma invoices.

Packaging

A point should be made about how you package your program. Overseas postal authorities define a package as any envelope which opens with a side flap. A letter, despite its size, is any envelope that opens with a top flap (Figure 26-3). An envelope with a top flap will pass through the overseas post office as correspondence or a letter; an envelope with a side flap will be routed to the customs side of the post office. It will be inspected, and duty may be applied. This procedure causes delay in delivery. Furthermore, the customs officials have a right to screen the material for possible political and social statements detrimental to their country. Obviously, this will be difficult for them if your program is in the NTSC format and the country standard is PAL or SECAM. If the authorities do not have the proper playback equipment, they will need to research a facility within their country which can service their need. This can take days, weeks, and even months.

This advice does not guarantee that the program will not be delayed by customs. Fortunately, customs agents are becoming increasingly aware of video, especially in view of the increase in private video networks. If they see that the program material is an electronic message to overseas personnel, they usually permit its entry at little or no cost. One method to assure that you receive this type of treatment is to have the program labels printed with "For educational or internal purposes only."

TRANSLATIONS

Down the road, you will eventually be asked to provide programs in native languages, both for internal and external viewing. With the dual-track capability of players, recording a native language track on the second channel appears simple. Problems arise in the translation itself, however. For one thing, most foreign languages use 20 to 30 percent more words than English. Therefore, any translation cannot be done verbatim. It needs to be done conceptually; that is, it needs to capture the concept of the English and use the foreign language phrase which complements it. We have found that we must time out each scene and then present these timings to the translators, requesting that the foreign language equivalent be said within the number of seconds listed. This stipulation assures that the audio matches the video.

If you have a translation unit within your company, use it. That unit knows the nuances of colloquialisms and the correct way to translate technical terms of your business into the native language. Also have the translation unit check on your choice of narrator. Someone did a Japanese version of a program not too long ago. The only objection to the program was about the use of an American-born Japanese who spoke in phrases somewhat different from the local dialect.

If you don't have a translation unit within your company, choose an outside firm that specializes in film and video foreign language translations.

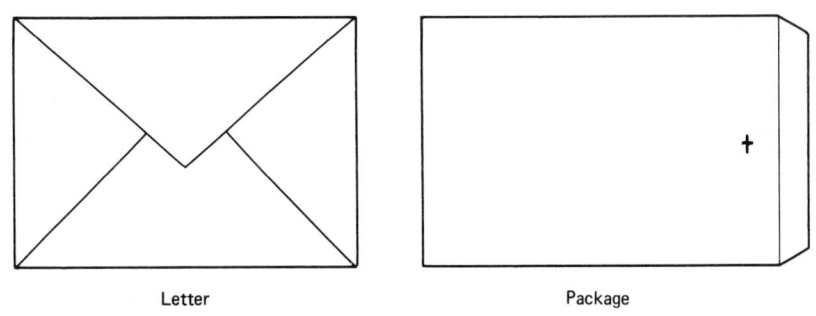

Figure 26-3 Videocassettes sent to international locations should be sent in letters.

To double-check the accuracy of the translation, send it to the country for which the translation is being prepared and have the office there correct any errors. Usually, the errors are in the translation of technical jargon. Furthermore, you can have your local office approve the choice of the narrator by sending an audio demo tape.

One last word: Direct the narrator to do a United Nations type of translation, that is, straight commentary. Attempting to lip-sync is too difficult and becomes expensive. It also drives the foreign language narrator bonkers. The technique is to establish the English, fade under, then begin foreign language translation. Aim for matches of audio and video at cut points.

CONCLUSION

Again, I hope that the previous sections have answered most of the questions you may have about the installation and the smooth operation of international video networks. Certainly, not all questions can be answered, for we in the video business are presently on the cutting edge of new breakthroughs in technology. Such things as international electrodigital teleconferencing, that is, live two-way audio and video meetings, are on the horizon. Interactive video with videodisc, on-line data retrieval and video programs from the same source, and the marriage of word processors and video loom on the horizon as well. Private video networks will play a key role in this telecommunications explosion.

PART 5

Staffing

Those who are employed within private television may be curious about which jobs are assigned which tasks and at what pay scale. For those who are not employed within the field, there is the question of how one gets in. Chapter 27 addresses the former as well as the topic of department organization. Chapter 28 presents a step-by-step procedure for seeking employment within private television.

27

Staffing a Private Television Operation

THOMAS WM. RICHTER

INTRODUCTION

Private television is a creative profession, requiring specialized creative skills in the service of an organization. But private television departments bear no resemblance at all to the production companies that create the programs seen on broadcast television. An overall examination of this young profession, however, does show certain trends in salaries, titles, and organizational structure.

Jobs and salaries in industrial audiovisual production have become a giant numbers game of supply and demand. Overproduction tends to create a "buyer's market." This has impacted greatly the structure and salary grades of audiovisual departments.

Schools and colleges that had never offered courses in photography or television suddenly became aware of the growing need and glamour of communication arts. Students started flocking through the doors to earn degrees in film, television, and all the related disciplines of communication arts and sciences. Each year, over 30,000 new graduates are competing for jobs. It is doubtful that one out of ten of these candidates will find employment in the field of their choice or professional training. They will simply have to do something else. And those who do find jobs will not be in a favorable position to negotiate a commanding starting salary—not, at least, until they have "paid their dues" and gained the necessary industrial experience.

What do people who have worked to gain basic experience have to look forward to in the line of position and salary? The International Television Association (ITVA), through its salary surveys, has been tracking the various staff positions since 1975.

ITVA's first salary survey (Figure 27-1) was a questionnaire. It caused a lot of stir in the industry. Some said it was awful, others liked it, and suggestions for improvements started to come in. One of the suggestions was to start to develop a set of job descriptions that could be used as a standard in the industry. So ITVA put out a call for job descriptions from its members, and the response showed job titles and descriptions that could span the industry. From the meager beginnings of four areas of responsibility, ITVA's job study has evolved to the following descriptions.

27-4 STAFFING

1975 ITVA
Salary Survey

    Company Size (Check one)
- ☐ Large
- ☐ Medium
- ☐ Small
- ☐ Vendor

    Position Type (Check the <u>one</u> area you spend the majority of your time)
- ☐ Management or Supervision
- ☐ Production (Writer, Producer, Director)
- ☐ Engineering
- ☐ Sales Equipment
- ☐ Sales Services

    _____ Years of Professional Experience

<u>Your</u> 1974 Income _____

<u>Your</u> 1975 Projected Income _____

Figure 27-1 The original 1975 International Television Association Survey. (*International Television Association.*)

JOB DESCRIPTIONS

Title: Manager

General functions:

- Is responsible for the planning, organizing, budgeting, staffing, and overall administration of a media department.
- Develops, recommends, and coordinates policies, plans, and programs for application of existing and new media technology on a corporation-wide basis.
- Approves or recommends annual operational and capital budgets.
- Administrates an operational budget.
- Approves or recommends staff and salary changes.
- Has at least two levels of personnel reporting to him or her.
- Determines long-range departmental goals.
- Reports to senior management of his or her major department.

Title: Supervisor

General functions:

- Supervises a media department or section.
- Works with users in planning maximum utilization of audiovisual media.
- Advises users on most effective uses of media, including content and organization of a given program.
- Coordinates scheduling of video productions.
- Directs all company media programs either directly or indirectly through delegation to a qualified staff member.
- Recommends annual operational or capital budgets, or both.
- Develops department or section objectives.
- Recommends staff and salary changes.
- Has at least one person reporting to him or her.
- Is responsible for evaluating and interpreting the users' needs.
- Analyzes and creates program concepts.
- Reports to the manager of his or her department.

Title: Producer

General functions:

- Interprets user needs.
- Establishes program budgets.
- Assigns a writer to the program or performs this function.
- Assigns a director to the program or performs this function.
- Is responsible for the program's being completed on time, within budget, and to the user's satisfaction.
- Coordinates all aspects of assigned videotape productions. This includes developing information from content specialists, working with the assigned writer, graphic designer, studio crew, and participants in the production.
- Directs the actual production, including making decisions on placement of sets, props, lighting, staging, audio timing, technical direction of cameras, and editing.
- Directs the activities performed by the assigned teleproduction crew, on location or in the studio.
- Is responsible to the user for the program.
- Reports to the manager or supervisor described previously.

Title: Writer

General functions:

- Takes suggested program ideas on an assignment basis and develops visualized scripts for television or any other medium.
- Determines program objectives from various technical experts.
- Researches source materials and interviews subject material experts; also prepares script outlines for approval.
- Writes the actual shooting script for the production.
- Assists in staging of shooting the actual program.
- Recommends method or format to achieve the required objectives.
- Develops creative and innovative ideas for further programs.
- Recommends new programs to be produced.
- Reports to either the producer or the manager described previously.

Title: Director

General functions:

- Directs the actual production, including making decisions on placement of sets, props, lighting, staging, audio timing, technical direction of cameras, and editing.
- Directs the activities performed by the assigned teleproduction crew, either on location or in the studio.
- Selects and directs talent for the video productions.
- Reports to the media director, producer, supervisor, or manager described previously.

Title: Media director

General functions:

- Produces, writes, and directs media programs or contracts for outside services for one or all aspects of doing a media program.
- Evaluates and interprets user needs.
- Analyzes the user audience and determines the program objectives.
- Sets program objectives and is responsible for their being achieved.
- Directs the actual production, including making decisions on placement of sets, props, lighting, staging, audio timing, technical direction of cameras, and editing.
- Directs the activities of the crew, either on location or in the studio.
- Writes the proposed program treatment.

- Determines the best medium to use.
- Establishes a program budget.
- Develops a shooting outline.
- Writes the script.
- Auditions and secures talent.
- Evaluates program results and recommends changes.
- Consults with user on the best way to use any media in the presentation.
- May operate and/or maintain the electronic equipment in the department.
- Recommends new equipment to enhance the media programs.
- Experiments with new procedures and techniques that should result in improved efficiency, effectiveness, and quality of the overall television operation.
- Sells the services of the department.
- Reports to either the producer, supervisor, or manager described previously.

Title: Production assistant

General functions:

- Assists in the production; serves as grip, gaffer, camera person, script person, go-fer, etc.
- Knows the operational function of all audiovisual equipment.
- Builds sets or arranges to have them built.
- Secures all props for a program.
- Maintains production outtake files.
- Assists the camera person.
- Acts as floor manager on a studio shoot.
- Maintains an inventory of equipment.
- Assists in editing a program.
- Keeps detailed log sheets on location shooting.
- Reports to the producer or media director described previously.

Title: Engineer

General functions:

- Evaluates and recommends audiovisual and teleproduction equipment.
- Assumes responsibility for technical performance of videotape recording systems, switching equipment, audio recording, and distribution

equipment (with associated controls, installation, preventive maintenance, repair, fabrication, and design of audiovisual systems as required).
- Trains other personnel in proper use of systems and electronic equipment.
- Directs setup of special equipment and remote installations, and ensures that the operations meet the objectives.
- Continually reviews new developments in equipment and techniques.
- Maintains records of maintenance.
- Maintains inventory control of spare and replacement parts.
- Operates all electronic equipment.
- Designs new or modified installation of electronic equipment.
- Assists in studio setup.
- Assists in lighting both in the studio and on location.
- Serves as technical member of both studio and location crew.
- Suggests production techniques within the limits of production equipment.
- Reports to the manager or supervisor described previously.

Title: Technician

General functions:

- Operates all common types of electronic equipment and instruments (cameras, audio recorders, video recorders, projectors, etc.).
- Serves as a technical member of an audio, video, or film crew.
- Performs minor maintenance and troubleshoots electronic equipment.
- Sets up and arranges studios and conference and training rooms.
- Reports to the engineer, supervisor, or manager described previously.

GROWTH WITHIN POSITIONS

Operationally, these job descriptions should be expanded to allow for at least ten years of professional growth and development. For example, let's take the overall classification of a media director. This is a creative person who can write, direct, and produce for film, television, or print media. A person coming out of school may be trained in these areas, but he or she will certainly not be as proficient as the person who has five or more years of practical experience. Yet, the job description for a media director often reads the same for both the experienced professional and the recent graduate.

In a large operation, the job descriptions should encompass multilevels of skill and proficiency. This format would allow a person to enter the field as a junior assistant and move up the line in successive steps to the top. For example, this natural progression could be through the suggested job titles of junior media director, associate media director, media director, senior media director, staff media director, and group leader of media directors.

These six suggested steps allow a continuum for personal growth and development that could span a ten- to fifteen-year career and allow anywhere from eighteen to thirty-six months in each of the subordinate jobs. Although the basic job functions of being a media director would remain the same, the degree of sophistication and responsibility would grow at each step. This kind of career path provides the individual with continual financial growth and personal satisfaction. It sets specific goals for personal achievements, maturity, and responsibility.

With this type of career progression built into the industrial job system, it could generate the kind of experience that goes beyond creativity. Most companies need more than artistic talent. They need people who can work within the politics of a corporate system, satisfy the needs of their corporate clients, and give a return on the investment back to the corporation. In other words, an in-house media employee must prove his or her value to the corporation.

INPUT → PROCESSING → OUTPUT

Now that we have the job descriptions and a career path, how should we position them in a company to function and give a dollar return to the corporation? One way is to set up a department using the simple system of input → processing → output. This means that as an audiovisual manager, you should handle everything from the needs analysis through the production of the program to the control and handling of the program viewing site.

Let's go through a sample organization using this simple system, and see how it can be organized functionally. I must stress *functionally*, because all the little boxes to be looked at are functions and do not necessarily represent people (Figure 27-2). Thus, these charts could represent a department of three, ten, or one hundred people. In fact, there are departments that fall into each of these three categories and follow the organizational structure that will be described here.

Remember, I began talking about a system that has the responsibility for a program from the needs analysis process through production to handling the final meeting site. Knowing that some of the areas that will be

27-10 STAFFING

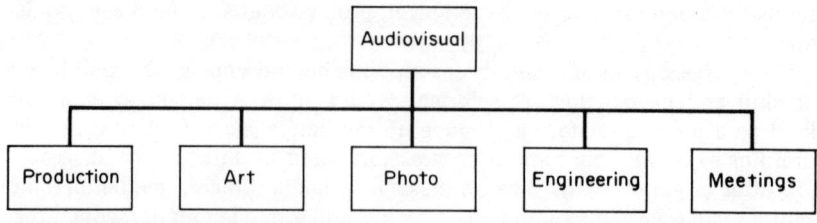

```
                    ┌────────────┐
                    │ Audiovisual│
                    └─────┬──────┘
      ┌───────────┬───────┼───────┬────────────┐
┌─────────┐  ┌────┐  ┌────────┐  ┌──────────┐  ┌────────┐
│Production│  │Art │  │ Photo  │  │Engineering│ │Meetings│
└─────────┘  └────┘  └────────┘  └──────────┘  └────────┘
```

Figure 27-2 Organization chart.

covered will not fit your organizational structure, remember also that none of these lines is a hard line, but each one is a soft line that can be removed with no trouble at all.

INPUT

The first area to cover is the input function. Who decides what programs will be made and how they will be done? The individual whose job it is to see that the department is properly servicing the company (usually the manager, supervisor, or media director) will make these decisions based on the needs and requests of other departments. Sometimes, ideas for programs will come down routinely from the marketing arm or corporate officers. Sometimes, the head of the department will find himself or herself "selling" the idea of using video services to various in-house clients.

The second level of input is directly concerned with the preproduction of a videotape. Engineers, camera operators, artists, and writers, as well as technical specialists from other areas of the company, should be expected to add their suggestions in the early stages of program design.

PROCESSING OR PRODUCTION

Let's take a closer look at the processing section to see how it works. First, we have *project management,* which consists of your producers, writers, directors, graphic designers, and training specialists, who are responsible for conceiving and producing the programs. (See Figure 27-3.) All these functions can be performed by one person as well as by several people performing more than one function. This person—the project manager—gets the production accomplished by either delegating staff personnel or hiring outside free-lancers. The section's functions are production, engineering, camera work (16mm, video and still), editing, technical directing, and production assistance.

As previously mentioned, the delegation of specific job functions will vary from one firm to the next. Several processing activities may be the responsibility of one person. In small departments, many positions, such as that of "writer," will be filled by free-lancers on a job-to-job basis.

STAFFING A PRIVATE OPERATION   27-11

```
                    ┌────────────┐
                    │ Production │
                    └─────┬──────┘
            ┌─────────────┴─────────────┐
   ┌────────┴────────┐         ┌────────┴────────┐
   │    Project      │         │    Project      │
   │    Manager      │         │    Manager      │
   └─────────────────┘         └─────────────────┘
```
- Producers
- Writers
- Directors
- Training Specialist
- Graphic Designer

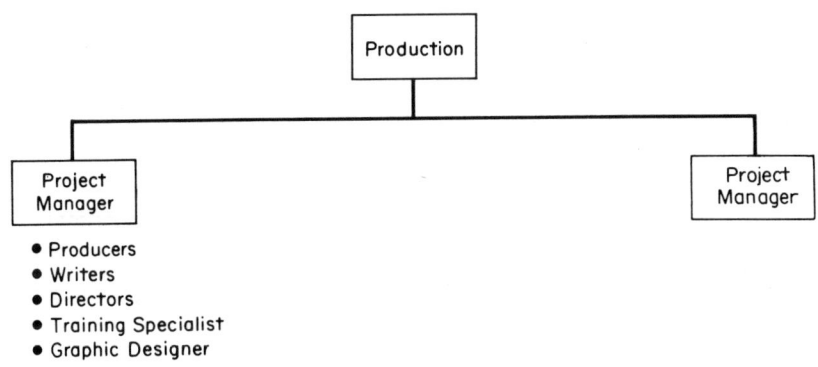

Figure 27-3 Production chart.

OUTPUT

Using our simple model, the section with the responsibility for output may be called "meetings." This section will have the responsibility for administering all the meeting rooms used for training sessions and conferences. This responsibility should not end with the meeting rooms in the general office building, but should also include areas for outside meetings.

Besides scheduling the rooms, arranging for coffee, setting up the equipment, etc., this section should be responsible for all audiovisual equipment used in these facilities. Remember that lights, heat, and a place to seat the audience all play important parts in the successful showing of a program. Meeting services staff members can be your best evaluators because they can see the audience reaction firsthand. Another reason for having meeting facilities and personnel in the communications department is that they can note when a need arises for the use of media in future programs.

SALARIES

Besides refining its job descriptions, ITVA has surveyed its members to determine the pay scale for specific jobs. This survey is published annually. The current annual survey results may be obtained by writing to ITVA, 136 Sherman Ave., Berkeley Heights, NJ 07922.

As you can see from the charts (Figure 27-4), the data gathered in the first three years fluctuated greatly. The reason was that the number of participants surveyed was small, a fact that distorted the final tabulation. Now that ITVA's membership has greatly increased, its data are beginning to show the trend of which all of us would like to be a part, and that is a gradual rising of salaries on an annual basis.

I hope this chapter has been successful in providing you with job descriptions, career paths, departmental organization, and salary trends that will help you organize your staff.

ITVA MEDIAN SALARIES 1975 TO 1980

Position	1975[1]	1976	1977[2]	1978	1979	1980[3]
Manager	19,500	25,000	22,500	25,000	27,000	27,600
Supervisor		17,500	17,500	18,000	20,000	20,500
Producer	15,375	[4]	16,500	17,000	18,500	22,300
Writer		13,000	15,000	12,500	16,200	19,500
Director		13,000	16,000	17,000	17,300	19,000
Media director		16,000	16,500	15,200	16,300	18,000
Production assistant[5]				11,000	11,500	16,100
Engineer	18,000	17,000	17,500	20,000[6]	18,000	20,500
Technician		12,000	12,500	11,000[7]	13,600	14,500

[1] Only three positions were surveyed the first year.

[2] 1977 results were published as ranges. The numbers given here represent the arithmetic means of those ranges.

[3] 1980 was the first year the survey was reported in percentiles. These figures represent the 50th percentile. More important than this figure is the 25th to 75th percentile range, which is available through ITVA.

[4] No response this year.

[5] Was not added to the survey until 1978.

[6] Only a few responses that were on the high side.

[7] Only a few responses that were on the low side.

Figure 27-4 Salaries in private television.

28 Getting the Job You Want

NATHAN J. SAMBUL

This chapter is an expanded version of seminars that I have given to both students and professionals in allied fields. A question-and-answer format is used since, in my opinion, this is the best way to guide persons seeking work in television.

Question: Is the marketplace for positions truly very competitive?

Answer: Unfortunately, yes. While the whole field of private television is growing at a rapid rate, so is the number of people who are trying to enter the field.

At the beginning of the 1980s, 238 colleges and universities had full-fledged communications departments that taught radio, TV, and film. If one uses the conservative figure of 25 graduates per year per department entering the marketplace, an additional 5950 people annually are apparently seeking jobs in the field. This figure does not include students from other departments (theater, journalism, English, art) or those individuals already in the field who are looking for a job change. Finally, there are professionals in allied fields, such as advertising, public relations, and publishing, who are also in the job market. Admittedly, not everyone is entering the field of private television—jobs in broadcast radio, television, and cable are also widely sought. But, in pure economic terms, supply far exceeds demand.

With this overabundance of supply, getting the job you want is extremely difficult, but it can be done. It requires that you first examine your strengths, develop a résumé, create an appropriate cover letter, and consistently follow up all leads. The adage of "getting a job is a full-time job" is true. You can, however, minimize your efforts with a strong plan of attack.

Question: What, then, should be my first stage of attack?

Answer: Polonius stated, "To thine own self be true,/And it must follow, as the night the day,/Thou canst not then be false to any man." That was good advice in Shakespeare's day, and it still holds true.

Of course, Hamlet became a little overzealous in seeking the truth; your intention is to record your strengths and weaknesses and then to present them in the most favorable light.

The first activity you should perform is the creation of a personal record sheet. Recognize that as you travel through the business community, you may change jobs a number of times. By building a file on yourself, you will have a record of the key items that will make you a valuable employee.

The following are some of the facts you should record on your "Personal Record Summary Form." Remember, this is not a résumé; it is a source document that you will add to as you acquire more skills and training.

PERSONAL RECORD SUMMARY FORM

Business activities

Job titles
Job responsibilities
Accomplishments
Promotions
Dates of employment
Salary history
Budget responsibilities
Supervisory responsibilities
References

Education

Dates of attendance
Major and minor concentrations
Honors and recognition awards
Internships
Informal course work
Professional societies
Knowledge of languages

Personal data

Military service
Dates of service
Special training and skills
Social and business organization membership
Community work
Physical health
Marital status
Willingness to relocate or to travel

Introspection

Major personal accomplishments
Definable short-term goals
Definable long-term goals
Salary requirements
Strong motivating factors
Major strengths
Major weaknesses

While there is always an element of risk in taking any new job, it can be minimized by taking stock of yourself and examining closely whether or not your education, background, and expectations match your employer's needs, goals, and corporate philosophy.

Question: Where do I begin?

Answer: After taking stock of yourself, you should line up your prospects. For those seeking positions within a corporate environment, you might want to begin with the Fortune 500 list (published yearly in *Fortune* magazine's May issue). Not every company on the list has a video operation, but these are the companies most likely to start one up, and you may find yourself in on a ground-floor situation.

A more focused approach would be contacting and joining, when possible, major associations in the field. Their membership rosters list the names of people you want to reach.

The major associations are:

Association of Educational Communications and Technology
1126 16th Street, NW
Washington, DC 20036

International Tape Association
10 West 66th Street
New York, NY 10023

International Television Association
136 Sherman Avenue
Berkeley Heights, NJ 07922

National Academy of Television Arts and Sciences
110 West 57th Street
New York, NY 10019

National Association of Broadcasters
1771 N Street, NW
Washington, DC 20036

28-4 STAFFING

National Audio Visual Association, Inc.
3150 Spring Street
Fairfax, VA 22030

National Cable Television Association
1724 Massachusetts Avenue, NW
Washington, DC 20006

Society of Motion Picture and Television Engineers
862 Scarsdale Avenue
Scarsdale, NY 10583

An equally good source of leads are directories of private television operations and production companies. The best known are:

"Corporate Communications Centers Guide" (published yearly in the February issue)
Audio-Visual Communications Magazine
475 Park Avenue South
New York, NY 10016

The Creative Black Book
Strauss Publications, Inc.
109 East 36th Street
New York, NY 10016

The Madison Avenue Handbook
Peter Glenn Publications
17 East 48th Street
New York, NY 10017

Motion Picture, TV and Theater Directory (published semiannually)
Motion Picture Enterprises Publications, Inc.
Tarrytown, NY 10591

Because names are so important, you should read the trade journals. There you'll find out who has moved, who is expanding, and who is using free-lance help. The major trade publications are:

Audio-Visual Communications
475 Park Avenue South
New York, NY 10016

Biomedical Communications
475 Park Avenue South
New York, NY 10016

Educational & Industrial Television
51 Sugar Hollow Road
Danbury, CT 06810

Filmmakers Monthly
P.O. Box 607
Andover, MA 08180 (editorial)
P.O. Box 115
Ward Hill, MA 01830 (subscription)

Millimeter
12 East 46th Street
New York, NY 10017

On Location
6464 Sunset Boulevard
Hollywood, CA 90028

Training
731 Hennepin Avenue
Minneapolis, MN 55403

Video Systems
P.O. Box 12901
Overland Park, KS 66212

Video User
701 Westchester Avenue
White Plains, NY 10604

Videography
475 Park Avenue South
New York, NY 10016

By utilizing all these names, you will have thousands of leads. It is extremely important to remember to correspond with a workable number of addresses (twenty to fifty per week) rather than to do a mass mailing. Follow-up is one of the keys to success, and it can be accomplished only when you have a handle on your correspondence.

Question: What are the experiences or qualifications that managers look for?

Answer: For a job in production, you should have a working knowledge of set design, lighting, creative writing, directing, graphics, photography, cinematography and videography, with an expertise in any four of them. That's rather demanding, but the competition is tough, and you should try to position yourself ahead of all the rest.

If you don't already have a working knowledge of some of these areas, there are a number of ways of acquiring them. The same local college or university that is pumping out your competition may have an adult education course or a course for nonmatriculated students that will increase your qualifications.

If you prefer a more concentrated dose of video, there are excellent

28-6 STAFFING

workshop sessions conducted at the various video expositions held in cities such as New York, San Francisco, and Toronto (for details, contact Knowledge Industry Publications, Inc., 701 Westchester Avenue, White Plains, NY 10604). Also, find out about the Visual Communications Congress in New York (United Business Publications, 475 Park Avenue South, New York, NY 10016); the Los Angeles Professional Video Show (C. S. Tepper Publishing Co., 51 Sugar Hollow Road, Danbury, CT 06810); or the annual convention of the International Television Association (136 Sherman Avenue, Berkeley Heights, NJ 07922), which rotates on an annual basis. It has been in such cities as Los Angeles, Kansas City, Dallas, Atlanta, Las Vegas, and Washington, D.C. Also, Smith-Mattingly Productions, Ltd. (92560 Huntington Avenue, Alexandria, VA 22303) conducts a number of videotape training sessions at its home location and various workshops around the country.

If your style is to learn it on your own, a self-study program is available through Knowledge Industry Publications, Inc. (in cooperation with Media Works of Arizona) on either videocassette or sound-filmstrip format. The package covers such topics as portable and simple camera systems, graphics for television, lighting and special effects, and audio for television. Imero Fiorentino Associates (44 West 63d Street, New York, NY 10023) also has a series of videocassettes covering lighting and staging techniques, lighting for news sets, and basic makeup techniques.

Question: All right. I've lined up my prospects, I've taken the courses, I've listed my strengths and weaknesses and now I'm all set to send out my résumé. Which résumés get read?

Answer: I'm glad you asked that question, because considering all the mail an audiovisual manager gets on a daily basis, you will have only a few seconds to make a positive first impression. Your résumé has to be clear, well designed, and understandable.

There are many different formats that your résumé can take. Since it is rare to see your competition's work, Figures 28-1, 28-2, and 28-3 show three real résumés that were well designed—not perfect, but ahead of the pack. Also, you may want to consider twelve specific suggestions when preparing your résumé.

1. Try to keep your résumé to two pages and at the same time have as much "white space" as possible; it allows room for notes and looks more professional.
2. Your résumé should contain at least the following sections: Name, Personal data, Work history, Education, Honors, and Other activities.
3. Place your work history before education. An M.A. in communications will probably not hold up as well as two years as a production assistant at a cable operation.
4. If you have no paid work experience, reexamine your educational background. Were you an intern for an outside corporation? Did you do

PATRICIA DUKE

Current Address
315B Woodcreek Road
Apartment 304
Bolingbrook, Illinois 60439
(312) 759-5349 or 983-2396

Permanent Address
14 Brook Street
Union, New Jersey 07083
201-687-8888

MEDIA EXPERIENCE

Bell System Center for Technical Education
Media Development Group
6200 Route 53, Lisle, Illinois 60532

Assistant Director, Video Production
Duties include pre-production planning, coordinating studio and location production (equipment, rentals, crew assignments and clearances), editing, lighting and set design for educational, motivational and employee information programming.
1979-present

Scott Ferguson, Associate Manager, Magnetic Media
(312) 983-2528

Media Production Specialist
Free-lance services provided to several New York metropolitan area in-house corporate video facilities — EFP and studio camera, lighting design, set design and construction, assistant director, technical director. Clients included Bell Laboratories, Chase Manhattan Bank, Chemical Bank, New Jersey Bell Telephone, New York Bell Telephone, Prudential Insurance.
1978-1979

Merrill Lynch, Pierce, Fenner and Smith, Inc.
Audio Visual Center
165 Broadway, New York, New York 10080

Free-lance Production Assistant with responsibilities in all areas of production, including set design and construction, lighting, EFP and studio camera, floor director, teleprompter
1977-1978

Jeanette Lerman, Director
(212) 766-4759

Ithaca College School of Communications
Technical Facilities
Ithaca, New York 14850

Administrative Assistant in charge of audio/visual equipment loan and reservation. Duties included routine maintenance and repair.
1977-1978

E. Paul Abert, Director
(607) 274-3214

Ithaca College School of Communications
Department of Television-Radio
Ithaca, New York 14850

Lab Assistant for classes in broadcast operations. Individual and group instruction.
1978

Donald Woodman, Chairman
(607) 274-3214

WICB-TV
Hanna Broadcast Center
Ithaca, New York 14850

Producer/Director, On-Air Promotion Director, Unit Manager. Experience with all areas of production and station management.
1975-1978

Paul Smith, Director of Technical Operations
(607) 274-3542

EDUCATION

Ithaca College, Ithaca, New York 14850
Bachelor of Science, Magna Cum Laude, Television-Radio
September 1975 - December 1978

Ongoing training through industry meetings, seminars and workshops

RELATED INFORMATION

Member, International Television Association, 1978-present

Member, Alpha Epsilon Rho, National Honorary Broadcasting Society, 1977-present

Member, Phi Kappa Phi, National Honor Society, 1978-present

Elected to Who's Who Among Students in American Universities and Colleges, 1978-79

Recipient, Alpha Epsilon Rho National Scholarship Award, 1978

Holder of F.C.C. Third Class Radio-Telephone Operator's Permit with Broadcast Endorsement

TAPES AND REFERENCES AVAILABLE UPON REQUEST

Figure 28-1 Résumé of Patricia Duke.

free-lance work (even as a volunteer)? Was your final project shown over a local cable operation? All these activities can and should be repositioned as work experience.

5. A statement of your objective is optional. Unless you are seeking a very unusual position within a media operation, you may need the space normally used for this statement.
6. List items in your résumé in an inverse chronological order.

28-8 STAFFING

THEODORE E. SMITH
108 Main Drive
Yonkers, N. Y. 10710
914-779-8888

PROFESSIONAL EXPERIENCE

1/79 - 5/79   *Production Intern*, WIXT, Channel 9, Syracuse, N. Y.
Observed and assisted in various weekly productions; concentrating mostly on commercial productions.

8/78 - 5/79   *Advance Television Lab Assistant*, S.U.N.Y., College at Oswego
Instructed and assisted in the technical aspects of television production. Aided students with their productions and provided possible alternatives to different problems that arose.

1/79 - 5/79   *Teleprompter Intern*, S.U.N.Y., College at Oswego
Produced and directed a series of shows for a local cable company with a major emphasis on a Saturday Night Live type format show.

8/78 - 12/78   *Audio Engineer*, S.U.N.Y., College at Oswego
Ran Audio Board for a weekly news show.

9/78 - 5/78   *Producer/Director*, Produced and directed a series of shows for Station WTOP (Campus Television) of Oswego.

9/76 - 5/77   *Newscaster*, WOCR, (Campus Radio) of Oswego.

GENERAL WORK EXPERIENCE

8/78 - 5/79   *Resident Assistant*, S.U.N.Y., College at Oswego. Liaison between students and directors in a dormitory setting. The position involved meeting and working with people, organizational and administrative skills.

11/74 - 4/79   *Ski Instructor*, Karl Plattner Ski School, Hunter, N. Y.

12/75 - 3/75   *Manager/Tour Leader*, Ski-Conomy, Scarsdale, N. Y.
Organized and ran daily and weekend ski trips. Handled all financial responsibilities.

3/72 - 8/72   *President*, Wix'N'Wax Corp., Yonkers, N. Y.
Accountable for the ordering, processing and distribution of candles in a Junior Achievement type organization. Also responsible for direction of sales staff, financial matters and public relations for the Corporation.

EDUCATION   B.A. May 1979, State University of New York, College at Oswego
Major: Communications   *Minor:* Biology
Concentration: Broadcasting; emphasis on Television Production
Relevant Coursework: Television-Production ... Radio Production ... Advance Television Production ... Directing ... Cinematography ... Graphic Production ... Photography ... Slide Production ... Audio Technology and Problems.

COLLEGE ACTIVITIES   Treasurer, Cayuga Hall Dormitory
Orientation Guide, Cayuga Hall Dormitory
Photographer
Video Visions — Television Club

REFERENCES   Available on request from the Career Planning and Placement Office, State University of New York, College at Oswego.

Figure 28-2   Résumé of Theodore E. Smith.

7. Today, a number of women use their initials in order to make the résumé as unisex as possible. It is my feeling that communications is one of the most open of all fields, and sexual discrimination is rare. Unless your friends call you J. R. or T. K., I recommend that you spell out your name.

8. If your major was not related to media, list all subjects that may apply in your situation, e.g., creative writing, journalism, photography, or mechanical drafting.

```
                                          400 West 80th St.
                                          Apt. 4F
Resume of ROBERT BOND                     New York, N. Y. 10010
                                          212-679-8888


Professional Objective: ART DIRECTOR/PRODUCTION MANAGER, a creative
management position involving film/video, multi-media production.

Recent Work: Directed animation for General Motors, Rolls Royce, U.S.
Navy, Air Safety Foundation. Created multi-media shows for ARCO
and Oxirane. Directed opticals (titles, travelling mattes) for
commercials by Pepsi, Chevrolet, Marlboro, Eastern Airlines, Care.

Employment Experience:

    ART DIRECTOR Gray Studios, N.Y.,N.Y. July 78 - Present. Responsible
    for creation and production of technical animation for educational
    and industrial films/video tapes and multi-media programs. Daily
    management of a four man art department. Titles, mattes and
    travelling mattes for films and commercials. Comprehensive
    knowledge of film production, operation of typesetting equipment
    and the Oxberry animation stand.

    ANIMATOR/CAMERAMAN Spectacolor Inc. N.Y.,N.Y. November 76 - June 78.
    Created and produced computer-animated advertising for NBC Saturday
    Night Live, WNEW-FM,Columbia Pictures, Macy's,Atlantic Records.
    Responsible for all in-house still and motion picture photography
    for promotion, publication and clients. Produced and directed
    commercials for KERN, WBTV, WNET Auction. Computer operation and
    scheduling. Training of staff. Consulting with clients.

    ASSISTANT CARTOGRAPHER Syracuse University, Summer 76. Preparing
    illustrations, maps and charts for publication. Design, layout,
    paste-up and photography of graphics.

    PRODUCER/MUSICIAN February 74 - Present. Freelance work in radio
    (WOUR-FM, WAER-FM), the music business and the Cable TV market.
    Producing radio advertising. Preparing promotional packages,
    artwork and posters. Producing recording sessions for musicians.
    3rd Class FCC Radiotelephone license with broadcast endorsement.

Education:

    M.S. in Television-Radio, Syracuse University Newhouse School of
    Public Communication. Graduate training in film/video production,
    computer graphics, animation.

    B.A. cum laude, Lafayette College

Personal Data:

    Age: 30                   Marital Status: Single
    Health: Excellent         Interests: Antiques, Film-making, Music

References will be furnished upon request.
```

Figure 28-3 Résumé of Robert Bond.

9. You should not list every job you have ever had. Having a paper route when you were twelve may indicate that you are industrious, but it also clutters up your résumé. It takes space away from your media-related activities. Of course, if you have performed an unusual task, such as working one summer as a carpenter's apprentice (very useful experience for set construction), list it.

10. Keep the names and addresses of your references back and supply them upon request. (If desired, you can indicate on your résumé that references are available.)

11. Your résumé should be reproduced by offset rather than by photocopy. It may be typeset, but it is not necessary. Colored paper stock is fine, but preferably in a muted shade. What you want to avoid is the "canned" look. Furthermore, unconventional résumés may do more than just draw attention to themselves. They can easily backfire by suggesting lack of sincerity or stability.
12. Use action words that indicate competency or initiative. A partial list of such words includes:

Developed	Supervised	Expanded
Constructed	Initiated	Planned
Established	Implemented	Directed
Designed	Executed	Recruited
Organized	Instructed	Conceived
Authored	Budgeted	Simplified
Researched	Controlled	Negotiated
Documented	Contracted	Accelerated
Managed		

Question: Is there anything that I should send along with the résumé?

Answer: By all means, prepare a cover letter for every résumé you send out. Often, such letters are as important as, if not more so than, the résumé. Cover letters are sometimes viewed as psychological snapshots of the applicant, indicating his or her style and professionalism.

Two examples of cover letters (Figures 28-4 and 28-5) and a few guidelines for preparing them may be helpful.

1. Always type your cover letter (with a single-strike carbon ribbon if possible).
2. A cover letter can help the interviewer read between the lines. If a résumé states "Produced management information tapes," a cover letter can indicate that you worked directly with your chief executive officer and ultimately wrote speeches for him or her on special occasions.
3. Keep the cover letter as brief as possible.
4. Whenever possible, take the initiative. Don't rely on the simple closing statement, "I would like to thank you for your time and cooperation and I anxiously await hearing from you." You can wait forever—that phone is not going to ring!

    Instead, close with, "I shall call you next week to discuss your department's needs and my qualifications." Then proceed to follow up. Getting a job is not just *being* at the right place at the right time, but rather, *making certain that you are* at the right place at the right time.

Question: Are there any helpful hints about the interview process?

> 173 West 78th Street
> New York, New York 10024
> 15 July 1982

Mr. Nathan J. Sambul
P. O. Box 968
FDR Station
New York, New York 10150

Dear Mr. Sambul:

I'm a producer/writer of audio visual materials. In the six years I've been in the field I've produced award winning programs ranging from Gun Control to Nuclear Energy on both pinched and flushed budgets.

I've made filmstrips, multi-media shows and tapes for audiences ranging from students and teachers to government officials and sales people.

In short, I'm interested in a position that would make good use of my creative and managerial skills. The attached resume will outline the scope and range of my experience.

I will call you next week to set up an appointment.

> Sincerely,
>
> *Pamela Murphy*
>
> Pamela Murphy

Figure 28-4 Cover letter from Pamela Murphy.

Answer: There is one overriding hint for all interviews: Be yourself. If you have taken the time to complete a personal record sheet, you know very well who you are, and you should be able to answer questions about your activities and goals in a straightforward and confident manner.

Being yourself, of course, does not mean letting it "all hang out." There are certain conventions that make sense and should be followed.

Use action words in your conversation—the same words that you incorporated into your résumé.

28-12 STAFFING

<blockquote>
ELLEN R. JONES
6 Thomas Hill Lane
West Hilburn, Michigan

April 5, 1982

Mr. Nathan J. Sambul
P. O. Box 968
FDR Station
New York, New York 10150

Dear Mr. Sambul:

Your name has come to my attention through the **Videography** list of "Who's Who in Corporate TV." I am writing to seek your advice, based on your experience and expertise, as to how I may explore directions for a career in video with your firm.

To highlight what I have done so far:

- Taken courses at the University of Hartford — Introduction to Video and Portable Video Systems. I plan to enroll in a studio course this fall
- Produced and edited documentary and educational tapes for a local voluntary agency
- Trained 12 people in use and care of video equipment
- Worked with elementary school students in the use of the school's video equipment
- Bought a Sony Portapak; my experience has been with ½ inch black and white video systems
- Planned and organized a workshop in TV talk show skills for 50 participants, which I will lead in June '82

Some other accomplishments include the use of my skills in organizing, planning, management, writing, and teaching in the following positions:

- Director of Volunteers, Meals on Wheels
- Docent, Wadsworth Atheneum
- Board of Directors, Junior League of Hartford, Inc.
- Chairman, Vassar College Capital Fund Drive

I have a B. A. with a major in English and have taken continuing education courses in public speaking, management by objectives, voluntarism, the aging, and video.

I hope it will be possible for us to meet at your convenience to discuss employment. I shall call your office in the next week or so to see whether we can arrange an appointment.

Sincerely yours,

Ellen R. Jones
</blockquote>

Figure 28-5 Cover letter from Ellen R. Jones. This is a combination cover letter and résumé which can be used to explore possible job openings. If you get the appointment, bring a résumé.

In addition, factors such as team cooperation, ability to work well under stress, and flexibility are buzz words that many interviewers want to hear.

Know the organization. Read its promotion material, annual reports, press releases, etc., in advance of your meeting. They are easily obtainable from the public affairs officer.

Present yourself well. Dress as though you already have the job or slightly more conservatively. Answer the questions in a direct, warm, and relatively brief manner.

Be prepared to discuss your strengths and your weaknesses. There is no reason to expose the fact that you hate your mother or are allergic to apple pie. A legitimate weakness is that you work too hard or that you're a perfectionist.

Have questions ready to ask your interviewer. They will indicate that you have thought about the position and are interested.

Bring a portfolio. You may not be asked to show it, but it often enables an interviewer to focus quickly on your talents.

Question: What should be in my portfolio?

Answer: Portfolios should be like résumés—short, professional, and representative of the work accomplished. I have been asked to sit through tapes that were over 30 minutes in length and that showed modern dance recitals choreographed to asynchronous music or group encounters with people vomiting off screen. Certainly I didn't need the experience. A short 4- to 7-minute tape with three to nine examples would suit everyone much better. If the interviewer wishes to see more, you will have an opportunity to come back for a second interview.

If you are, or recently have been, a student, don't be ashamed to show your work. Oftentimes, schools will let you make a duplication of your work if you will just bring in your own tape stock.

A portfolio may also include scripts, photographs, slides, floor plans, lighting schemes, etc. Remember, a portfolio shows competency and actual accomplishments. In my experience, the individuals who have good portfolios have a leg up over their competition.

Question: Is that all?

Answer: No, but that's enough for a start.

There are a number of general books which discuss the art of getting a job. If you are serious about employment, it may be to your advantage to look at these publications.

How to Get a Better Job Quicker, Richard A. Payne, Taplinger Publishing Co., New York, 1972.

What Color Is Your Parachute? A Practical Manual for Job-Hunters & Career Changers, Richard Nelson Bolles, Ten Speed Press, Berkeley, CA, 1981.

Writing Résumés, Locating Jobs, and Handling Job Interviews, Michael J. Freedman, Ph.D., Richard D. Irwin, Inc., Homewood, IL 60430, 1976.

These publications can provide helpful hints on working with the boss's secretary or the importance of writing follow-up thank you notes.

Seeking a job is a full-time job. But you shouldn't lose sight of the fact that what you are actually after is a rewarding, challenging, and productive position. Getting it may take a while, but it will be worth your effort.

PART 6

Tomorrow's Private Video Communications

Parts 1 through 5 showed how to develop an ongoing video operation with sound management, quality production standards, an international outlet, and a well-defined staff. In the final chapter of the book (Chapter 29), we take a look at what lies ahead and how private television may expand its scope.

The Future of Private Television

JOHN P. WILISZOWSKI, JR.

THE PERILS OF PREDICTING THE FUTURE

Predicting the future is a hazardous assignment. It's like trying to guess what will happen in a situation where there are seemingly infinite variables, all interacting and all contributing to the final results. This is the task one faces when attempting to look ahead at private television. Its future is affected by many factors: developing technology, the economy, management's directives, the cost of fuel and travel, the services offered by broadcasters and common carriers, and even by the personalities of the people involved.

No one can predict for certain what the future will hold. However, anticipating the future is easier when one realizes that all decisions affecting private television are predicated upon a fundamental principle:

Private television must contribute, in a cost-effective manner, to the task of accomplishing the objectives of the organization it serves.

The factors that will most likely affect the development of private television will be discussed in light of this principle.

MARKETS MAKE TECHNOLOGY POSSIBLE

Private television must be involved in the development of new technologies in order to achieve its goal of cost-effective communications. Equipment and facilities that make the job easier and more efficient are becoming increasingly more available as dynamic market possibilities are realized by equipment manufacturers. The responsibility of the private TV manager is to stay informed about the tools developed for these markets that may find application in his or her particular organization.

Approximately 1965 was the year in which private television became a viable market in its own right. Since then, a great deal of equipment has been designed to satisfy its specific needs and requirements. An example is the ¾-inch videotape recorder. This machine was designed and directed toward the private TV market. It later found acceptance among broadcast news teams and for remote commercial production.

Discovering applications for existing technologies and the development of new technologies to serve specific needs are essential for the continued growth of the television industry and, in particular, for the growth of private television. Representatives from private television must work with equipment manufacturers to identify needs and requirements. The private television community must cooperate with manufacturers so that they will be able to provide equipment and facilities that will contribute to overall professionalism and cost-effectiveness.

RESEARCH AND DEVELOPMENT OF VIDEO EQUIPMENT

The tools for private television are not limited, however, to only those developed specifically for that market. Private TV managers must look beyond their own area of experience to find related technologies that can be either used as they exist or modified to make private TV communications more cost-effective.

The entertainment industry, broadcast news, and sports have historically been lucrative video markets. The cost of extensive research and development can be supported easily by the high price for equipment that this market will bear. Much of this technology contributes to the equipment designed for private television. Color cameras, introduced to bring color television into the home for entertainment, now are used in large numbers in the private TV market. Satellites, once used only to televise events live from Europe, now serve as the technological base for video teleconferencing.

Another growth area that has recently stimulated the development of television technology is the consumer video market. Here, high volume sales support research. A great deal of time, effort, and expense can be devoted to developing video products for this market because the probability of recovering expenses is high once the product is perfected.

Many of the products developed for the home video market have found acceptance in TV applications in the private sector. Contributions to the technology of private television from the consumer video market include the ½-inch videocassette, the videodisc, and, more recently, microcomputers.

NEW TECHNOLOGY FOR PRIVATE TELEVISION

The future of private television is, in part, technology-dependent. To manage effectively, one must recognize the potential technology offers. Included in this section are descriptions of some of the equipment and technology that will influence the direction of private television.

Satellites. Teleconferencing by way of satellite is becoming a viable and frequently chosen alternative to travel for sales calls, training and education, and executive meetings. Skyrocketing costs, plus the time involved, are major problems associated with business travel. Video conferencing offers an attractive solution.

One hour of *up-link* or transmission time to the satellite, on a when-available basis, can be obtained for about $200. The facilities for a complete transmission link cost approximately $2000. This covers transmission time, rental of receiving facilities, and appropriate land-line interconnects to carry the signal from the point of reception to the viewing site. The expense involved for two-way video teleconferencing is now very close to the cost of coast-to-coast travel. Remember that travel expenses must include air fare, automobile rental or other ground transportation, hotel accommodations, meals, and the cost of the time spent for the trip.

With privately owned receive-only stations, the $2000 expense for video teleconferencing will be further reduced, making satellite transmission an efficient means for private TV communications.

One-Half-Inch Videocassette. A second area where technology is increasing the potential of private television is in videotape recording. Large private TV networks (2000 to 3000 units) are now more economically feasible because of the ½-inch videocassette. In many cases, the format is ideal for program distribution to field offices, classrooms, and remote locations. Through inputs from the private TV sector and the manufacturers' realization of a potential market, the units have been modified and redesigned for the more stringent requirements of the corporate, government, and educational communities. Today, ½-inch videocassettes feature slow-motion and stop-action capability, with options available for videotape editing and computer interface.

Videodisc. The development of the videodisc in many ways parallels that of the ½-inch videocassette. Like the videocassette, the videodisc was developed to capture the enormous consumer video market. Like the ½-inch videocassette with two noncompatible formats, there are several formats for videodisc units, all incompatible.

Although compatibility presents a major problem within the consumer market, it offers little obstacle to the acceptance of the videodisc in the private TV sector. A private organization can easily select and use one format as the basis for an entire distribution network. This eliminates any compatibility problems that might exist within the organization. Discs requested from independent suppliers can usually be ordered in the format selected as the standard through the organization.

All videodisc units, regardless of format, can be classified into three categories defining their degree of sophistication.

- Category 1 includes videodisc players designed for the consumer markets. These units permit random access playback only, continuous or single-frame viewing, and little or no processing power or memory.
- Category 2 includes industrial and educational units with random-access playback only, continuous or single-frame viewing, and programmable memory.
- Category 3 covers professional units with random-access playback, continuous or interrupted recording, but not including erase and rerecording capability. Category 3 videodisc units are capable of being interfaced with an auxiliary computer system.

The videodisc is an extremely flexible medium for communications. In addition to its ability to play back standard video images, units are available that can store 54,000 single images on a side and access any image within 5 seconds. This feature of the disc makes it an ideal medium for information storage and retrieval. Imagine 160 books, each of approximately 300 pages, stored on a disc about the size of a record album. This single-frame retrieval capability also makes the disc a valuable tool for keeping records and transcripts.

Videodisc units can also be interfaced with computers (even the smallest computer of the type sold to computer hobbyists), making the videodisc a communication tool of almost unlimited flexibility.

The videodisc will affect the future of private television and communication in several ways:

- The videodisc is an ideal medium for information storage and retrieval.
- Regardless of the format, the disc is durable and light, making it easy to ship and store.
- The cost of the disc is appealing.

Prices for the discs may range from $7 to $12 each. This price range compares favorably with approximately $20 for the average ½-inch videocassette or the still higher cost of other videotape formats, even when considering that currently available discs cannot be erased and reused like videotape.

Computers. An area that will have a pronounced and dramatic effect in the way people communicate will be the microprocessor and microprocessor-based products, in particular the computer.

Throughout the past several years, microprocessor development has made possible accurate, low-cost editing systems, color cameras that will automatically adjust black-and-white balance and even make necessary operational adjustments, and switchers that will remember moves and the positions of fader and special effects handles.

Microprocessor technology was also instrumental in the development of remote start for videotape recorders, random-access controllers, time-code generators and readers developed by the Society of Motion Picture and Television Engineers (SMPTE).

Practical applications of computer technology permit communication possibilities beyond one's wildest dreams and expectations. Videocassettes and videodiscs interfaced with computers are able to display single pages of text for information, branch to motion sequences for demonstration of a particular procedure, administer tests, and evaluate the results. Employing computers linked to appropriate data banks, users are able to access all the information on a topic located anywhere in the world. This information may be available in the form of written text or video programs.

With such data banks and the appropriate hardware, information can be sent from one computer to another by way of land lines or satellites and received by a local terminal (which is nothing more than a computer functioning as a receiving device). Such terminals can be located in corporate branch offices, colleges, universities, high schools, medical institutions, and even private homes.

Computers coupled with video recording equipment can serve as a clipping service, scanning the daily news for all information relating to a specified topic. This information can be stored by the computer and retrieved at the user's convenience.

The merger of computers and video is the single most important technological development to occur within the field of communications. Most of the equipment and software already exists. All that remains is practical and economic justification to implement such technology.

TRENDS

Other Factors

Although technology makes available the tools, many other factors contribute to the future of private television.

Economy. Private TV with all its electronic intricacies is but one tool of many that can be used to communicate. As a medium for communication, television can be the most cost-effective one available. It can also be the most expensive. This may seem obvious, but in reality it's one linchpin in the successful future of private TV.

Much of the communications technology explained earlier in this chapter exists or could be developed within a very short time. The factor that limits the implementation of such elaborate communications networks is the economies of private television as they relate to the economy of the nation or world.

When the cost of travel becomes prohibitive, video teleconferencing may become a common daily practice. On the other hand, another method, like a newsletter, memo, or telephone call, may satisfy the same objectives more cost-effectively. If the purpose of the trip was to communicate a message, and the message wasn't very important, the solution to the problem of high costs might be to stop everything. The point is that the state of the economy, internally and externally, will have a pronounced effect on private TV.

Private television is an extremely powerful tool for communications, but the communications must be truly needed, and television must be the most economical vehicle.

Two questions must be continually asked to determine what role the economy plays in the future of private television.

1. Is the message necessary? If the answer to this question is yes, then . . .
2. Is private television the most cost-effective answer? If the answer is yes a sufficient number of times, the future of private television will be a prosperous one.

Outside Services. Private television within an organization no longer needs to rely exclusively upon its own resources to satisfy certain communications needs. Keep in mind the goal of cost-effective TV communications and look beyond internal capabilities for the solution.

Independently produced materials are becoming more available. Many times, materials produced for distribution treat common problems. These materials may satisfy the objectives of a particular project. If so, they can usually be purchased or rented for far less than the cost of producing the same or similar material.

Cooperation among different groups involved in private television is also important for its continued growth and development. Many problems that exist in one organization are common to others. Know what solutions were used, and to what success, by others in similar situations. Sharing ideas and programs can be beneficial to all involved.

Professional groups like the International Television Association (ITVA), the American Society of Training Directors (ASTD), and the Association for Educational Communications and Technology (AECT) are all excellent organizations. Each exists to serve particular needs, but all function as a focal point for the exchange of information.

Cable Distribution and Broadcasting. Private television can sometimes better achieve its goals by using services offered by broadcasters and cablecasters. This is not beyond the scope of private television. Examples of this type of interaction is common. In areas throughout the country, colleges and universities offer classes to the community through local cable television distribution. Several medical schools are using the same vehicle for continuing education. Government and service organizations use public service broadcast time to convey their messages. Corporations and private industry can many times get an important message relating to their particular field to the public by way of public interest programming supplied to independent program distributors or local cable stations.

CONCLUSION

Many times, the expression "The use of private television is limited only to the imagination" has been used to explain its potential. That expression was never more true than it is today. Managers and anyone else involved in private television must remember, though, that in addition to a dynamic plan for utilization, private television requires proper planning and the ability to take into account the numerous variables involved. Planning carefully and evaluating all the factors related to using private television correctly will assure its continued growth—and, perhaps, the realization of its full potential.

Index

Abbreviation of key terms, **13**-7 to **13**-8
Accelerated depreciation, **8**-7
Accessories, set design, **14**-19
Accountability of video department, **2**-3 to **2**-5
ACE (advanced digital converter and transcoder equipment), **25**-16
Acoustics, **18**-10
Advertising, **1**-7
Advertising department, **2**-4
Agenda:
 for proposal to board of directors, **4**-12
 for task force meeting, **4**-9 to **4**-11
Air conditioning noise, **18**-12
Ambient sound level, **20**-8
Amplifier, limited, **18**-2
Animation, **15**-14
Approval, prior, **9**-1
Approvals, **5**-2 to **5**-3
Art cards, **15**-11 to **15**-12
Artificial light with daylight, **16**-11 to **16**-12
Artwork to videotape, transferring, **15**-7 to **15**-14
Assemble editing, **22**-2
Assistant director (AD), **20**-6
Associations in field, major, **28**-3 to **28**-4
Attitude:
 audience, **3**-2 to **3**-3
 in location shooting, **21**-13 to **21**-15
Audience:
 current knowledge of, **12**-2
 identification of, **11**-4, **12**-1 to **12**-2
 interviewing, **12**-2
 time invested by, attitude toward, **3**-2 to **3**-3
Audio equipment, **18**-1 to **18**-4

Audio in television, **18**-1 to **18**-15
 background for, music as, **18**-14
 conclusions on, **18**-14 to **18**-15
 copyright for sound recordings, **24**-6
 corrective or sweetening process for, **18**-12 to **18**-13
 creative mixing process in, **18**-13
 equipment for, **18**-1 to **18**-4
 fabric noise in, **18**-7
 filters in, **18**-2
 location shooting and, **21**-12 to **21**-13
 loss of high-frequency sound in, **18**-7
 microphones and sound perspective in, **18**-4 to **18**-6
 music sources for, **18**-14
 postproduction, **18**-12 to **18**-13
 problems and solutions for, **18**-6 to **18**-12
 quality of, **20**-8
 sound effects in (see Sound effects)
 terminology for, **18**-1 to **18**-4
 (See also Microphones)
Audio quality, **20**-8
Audiovisual department (see Video department)
Auto assembly, **22**-2
AV department (see Video department)

Back lighting instruments, **16**-3
Backer, **2**-2 to **2**-3
Background noises, **18**-11 to **18**-12
Backup systems, **21**-6 to **21**-7
Bald head, **16**-13, **17**-4, **17**-14
Barn door, **16**-5
Base coat, makeup, **17**-4 to **17**-5
Basic tape (black), **22**-2
Benefit costs, **8**-8
Beta video cassette format, **1**-8, **6**-9

1

Betamax, **25**-3
Blank tape evaluation, **23**-5
Blemish, makeup for, **17**-10
Board of directors, proposal to (*see* Proposal to board of directors)
Boom microphone, **18**-1, **18**-5, **18**-7 to **18**-8, **18**-11
  definition of, **18**-1
  on location, **18**-11
Bottom line, **3**-2, **10**-7
Branch offices:
  informing, **26**-9
  selecting for video department, **26**-3 to **26**-5
Broadcast standards (*see* Standards, international video)
Brush, D/J, Associates, **2**-4, **2**-6
Brush, makeup, **17**-6
Budget, **8**-5 to **8**-11
  capital, **8**-5 to **8**-8
  charge-back systems, **2**-6 to **2**-7, **8**-10 to **8**-11
  classifications within, **5**-5
  conclusions on, **8**-11 to **8**-13
  form for, **8**-12 to **8**-13
  full charge-back system, **2**-6
  major video facility, **6**-7
  operating (*see* Operating budget)
  outside producer and, **7**-6 to **7**-7
  partial charge-back system, **2**-6 to **2**-7
  per program, **2**-6 to **2**-7
  production, form for, **8**-12 to **8**-13
  production values versus, **20**-1 to **20**-2
  records and audit of, **8**-11 to **8**-13
  self-funded system, **8**-2 to **8**-3
  set design, **14**-12 to **14**-13
  for small video facility, **5**-5 to **5**-9
    capital expenditures, **5**-6 to **5**-7
    operating expense, **5**-7 to **5**-9
  support material, form for, **8**-14 to **8**-15
  (*See also* Costs)
Burned-in time code, **22**-1

Cable television, **1**-8, **6**-8
  distribution of, **29**-8
Cake makeup, **17**-5
Camera angle, **13**-9
  graphics and, **15**-11

Camera angle (*Cont.*):
  set design and, **14**-2 to **14**-4
Camera movement, **20**-9 to **20**-10
Capital budget, **8**-5 to **8**-8
Capital expenditures, **9**-8
  of small video facility, **5**-6 to **5**-7
Cardioid microphone, **18**-1
Career path, **27**-8 to **27**-9
Cash, location shooting, **21**-4 to **21**-5
Cassettes, video (*see* Videocassettes)
Casting, **20**-4 to **20**-5
Catalog, program, **6**-7
Ceiling, low, **16**-13
Central processing unit (CPU), **22**-7
Character generator, computer, **15**-13 to **15**-14
Charge-back systems, **2**-6 to **2**-7, **8**-3 to **8**-4, **8**-10 to **8**-11
Checkerboard auto assembly, **22**-3
Chroma key, **16**-10 to **16**-11
Cinematography, five C's of, **20**-9
Cleanliness, makeup, **17**-15
Client involvement, **10**-5
Close-up in sound, **18**-5 to **18**-6
Close-up shot, **13**-8, **13**-9
Closed-circuit network (cable), **6**-8
Clothing, rearrangement of, **17**-3
CMX 600, **22**-7
Color:
  in graphics, **15**-10 to **15**-11
  in set design, **14**-6, **14**-7
Color broadcast standards, **25**-4 to **25**-12
  FCC standards, **25**-4 to **25**-5
  NTSC system, **25**-4 to **25**-5, **25**-10
  PAL system, **25**-4 to **25**-5, **25**-10
  SECAM system, **25**-4 to **25**-5, **25**-10 to **25**-11
  three systems analyzed, **25**-11 to **25**-12
Color temperature, **16**-2 to **16**-3, **16**-7
Communications:
  branch office involvement and, **26**-9
  management, **1**-6, **10**-4
Communications department, **2**-5
Company grapevine, **10**-4
Company news programs, **1**-6
Compatibility, international videotape equipment, **25**-3
Competitive bids, equipment purchase, **8**-12 to **8**-13

Complicated material, graphics for simplifying, **15**-2 to **15**-3
Composition, **20**-9 to **20**-10
Compromise, **10**-6 to **10**-7
Computer character generator, **15**-13 to **15**-14
Computers in future trends, **29**-6 to **29**-7
Concept:
　for program, **11**-4 to **11**-5
　in scriptwriting, **13**-1 to **13**-7, **13**-13
　in set design, **14**-1 to **14**-4
Confidentiality, **9**-6
Contact sheet, **21**-3 to **21**-4
Content:
　of contracts, **9**-3
　determination of, **12**-3 to **12**-4, **13**-1 to **13**-7, **13**-12 to **13**-13
　nonprofessional talent in rehearsal and, **19**-8
Contracts, **9**-1 to **9**-11
　accommodation in, **9**-2
　conclusions on, **9**-11
　content of, **9**-3
　cost and payment terms of, **9**-3 to **9**-5
　equipment purchase (*see* Equipment contract)
　law department and, **9**-2 to **9**-3
　lease of equipment, **9**-8 to **9**-9
　letter of agreement versus, **9**-2 to **9**-3
　liability in, expected exposure to, **9**-1
　for new product or service, **9**-1
　oral, **9**-2
　ownership and use of item in, **9**-3
　prior approval in, **9**-1
　psychological impact of, **9**-1
　reasons for, **9**-1 to **9**-3
　scope of project and, **9**-3
　service (*see* Service contract)
　standard agreement, **9**-2 to **9**-3
Contrast-balancing techniques, **16**-9
Contrast range, **16**-1 to **16**-2
Control track, **22**-1
Cookaloris (cookie), **16**-16
Copy Guard, **23**-5
Copying videocassettes (*see* Videocassette duplication)
Copyright law, **9**-5, **24**-1 to **24**-10
　class D, **24**-3
　class PA (performing arts works), **24**-4 to **24**-5

Copyright law (*Cont.*):
　class RE (renewal registration), **24**-6
　class SR (sound recordings), **24**-6
　class TX (nondramatic literary works), **24**-4
　class VA (visual art works), **24**-5 to **24**-6
　classification system for, **24**-4 to **24**-6
　deposit requirements for, **24**-6 to **24**-9
　duration of, **24**-3
　fee for filing, **24**-8 to **24**-9
　form for registration of videotape, **24**-8 to **24**-10
　for motion pictures, **24**-6 to **24**-7
　notice of, **24**-4
　permission, obtaining, **24**-1 to **24**-2
　persons registering, **24**-4
　procedures for registering, **24**-2 to **24**-3
　works not registrable, **24**-3
Corporate permission, location shooting, **21**-8 to **21**-9
Corporate television:
　advertising efforts and, **1**-7
　data processing training and, **1**-4
　employee news and, **1**-6
　employee training and, **1**-5
　future of (*see* Future of private television)
　management communications and, **1**-6, **10**-4
　marketing and, **1**-5 to **1**-6
　organizational structure (*see* Organizational structure of video department)
　other uses of, **1**-7 to **1**-8
　pitfalls of (*see* Pitfalls of private television)
　placement of in organization, **2**-3 to **2**-5
　public relations and, **1**-7
　regulatory requirement notification and, **1**-6 to **1**-7
　rise of (*see* Rise of private television)
　safety instruction and, **1**-7
　types of programming for, **1**-5 to **1**-8
　usefulness of, **1**-4
　(*See also* Video department)
Corrective process for audio, **18**-12 to **18**-13
Cost center self-funded approach, **8**-2 to **8**-3

4 INDEX

Cost terms, contract, **9**-3, **9**-5
Costs:
 of copies, **26**-7
 in free-lance category, **5**-8
 of international video network, **26**-2
 of location shooting, **21**-1
 nonprofessional talent and, **19**-2
 outside producers and, **7**-4, **7**-6 to **7**-7
 in outside service category, **5**-8
 production, **5**-8 to **5**-9, **7**-7
 of standard converters, **25**-18 to **25**-19
 typical production, **7**-7
 (See also Budget)
Cover letters, **28**-10
Creative mixing process for audio, **18**-13
Credibility, nonprofessional talent and, **19**-1 to **19**-2
Credits, **9**-6
Crew, **20**-6 to **20**-7
 comfort of, on location, **21**-5
Cross-training, **6**-5
Crowd control, **21**-9
Cue cards, **19**-7
Cue track, **22**-1
Current standards, electrical, **25**-6 to **25**-9
Customs officials, **26**-12
Cyclorama, **14**-17

Damage clause, **9**-6 to **9**-7
Data processing, **1**-4, **2**-1, **8**-1
Data processing training, **1**-4
Daylight with artificial light, **16**-11 to **16**-12
Deadline, **9**-7
Decibel, **18**-4
Delivery, equipment contract, **9**-10
Depreciation, **8**-7, **9**-8
Design process, **5**-1 to **5**-3
Diamond-shaped face, **17**-8
DICE (digital intercontinental conversion equipment), **25**-16
Diffuser, **16**-6
Digital effects capabilities, **20**-3 to **20**-4
Digital intercontinental conversion equipment (DICE), **25**-16
Digital standards converter, **25**-16, **25**-17
Dimmers, **16**-7

Directing, **20**-1 to **20**-11
 audio standards in, **20**-8
 composition in, **20**-9 to **20**-10
 conclusions on, **20**-11
 lighting standards in, **20**-8 to **20**-9
 preproduction in (see Preproduction)
 production in, **20**-5 to **20**-7
 talent and, fears of, **20**-10 to **20**-11
 technical standards in, **20**-7 to **20**-9
 techniques of, **20**-7 to **20**-11
 working styles in, **20**-11
 zen and, **20**-1
Director, job description, **27**-6
Directories of operations and production companies, **28**-4 to **28**-5
Discoloration, makeup, **17**-10
Display monitors, **22**-7
Distributed processing systems, **22**-7
Distribution:
 program (see Program distribution)
 videocassette duplication, **23**-9
Distribution points in network, **2**-6
D/J Brush Associates, **2**-4, **2**-6
Documentary style, **13**-2, **13**-4, **13**-9, **13**-14
Dollying, **20**-9 to **20**-10
Downtime, location shooting, **21**-14 to **21**-15
Drapes, **14**-17
Drop-frame time code, **22**-4
DSC 4000, **25**-16
Duplication of videocassettes (see Videocassette duplication)
Dynamic range, **18**-3 to **18**-4

Economy, **29**-7
Edit pulse, **22**-1
Editing, **11**-7, **22**-1 to **22**-11
 decision list for, **22**-2
 definitions for, **22**-1 to **22**-5
 equipment purchase for, **22**-12 to **22**-13
 future of, **22**-10 to **22**-11
 graphics design, **22**-10
 guidelines for procedure of, **22**-9 to **22**-10
 off-line versus on-line, **22**-9
 preparation for, **22**-8 to **22**-9
 process of, **22**-7 to **22**-10
 rough, testing, **5**-2 to **5**-3

Editing (Cont.):
  shooting and thinking ahead to, **22**-7 to **22**-8
  special effects, **22**-10
  systems for, **22**-5 to **22**-7
Education for video employees, **1**-5, **2**-4, **6**-4 to **6**-5, **28**-5 to **28**-6
Electrical current standards, **25**-6 to **25**-9
Electrical voltage and hertz, **26**-3
Electricity, **8**-8 to **8**-9
Electronic color system with memory (*see* SECAM)
Electronic conditions of television system, **16**-1 to **16**-2
Electronic converters, **25**-16
Elevations in set design, **14**-8, **14**-9
Ellipsoidal spotlights, **16**-4 to **16**-5
Employees:
  benefits for, **8**-8
  news for (house organ), **1**-6
  training of, **1**-5, **2**-4, **6**-4 to **6**-5, **28**-5 to **28**-6
  (*See also* Staff; Talent)
Engineer, job description, **27**-7
Engineer, studio, **5**-4 to **5**-5
Enhancement, videocassette duplication and, **23**-3 to **23**-4
Entertainment, graphics for, **15**-4
Equalizers, **18**-2
Equipment:
  competitive bids on, **8**-12 to **8**-13
  for editing, guidelines for, **22**-12 to **22**-13
  floaters, **26**-7
  maintenance of, **23**-5
  multistandard, **25**-14
  protecting, **21**-6
  purchase of, **8**-12 to **8**-13, **22**-12 to **22**-13
  (*See also* Equipment contract)
  repair service for, **9**-9, **26**-7
  in set design, **14**-5 to **14**-6
  vendors of, **26**-8
Equipment contract, **9**-7 to **9**-11
  delivery in, **9**-10
  loss in, risk of, **9**-10
  for major items, **9**-8 to **9**-11
  operational information in, **9**-11
  payment terms in, **9**-9 to **9**-10

Equipment contract (*Cont.*):
  service contract in, **9**-9
  for small items, **9**-7
  specification sheet in, **9**-10
  standard representations in, **9**-11
  substitution in, reasonable, **9**-11
  warranties in, **9**-11
Equipment standards, **25**-12 to **25**-14, **26**-5 to **26**-7
  for international video network, **26**-5 to **26**-7
  multistandard equipment, **25**-14
Erasing tape, **23**-10
Executive steering committee, **6**-2 to **6**-3
Eyebrows, **17**-11
Eyeglasses, **16**-12
Eyes, makeup, **17**-11 to **17**-12

Fabric noise, **18**-7
Face, makeup, **17**-5 to **17**-8
Facial type, **17**-4
Facilities (*see* Video facility)
Fade up, fade down, **18**-4
Federal Communications Commission (FCC) standards, **25**-4 to **25**-5
Fill lighting instruments, **16**-3
Film, television versus, **20**-9
Film-to-tape transfers, **23**-1 to **23**-2
Filters, **16**-12, **18**-2, **18**-11
Financed purchase, **9**-9
Five C's of Cinematography, The (Mascelli), **20**-9
Flats, **14**-14 to **14**-17
Flexibility, **13**-11
Floaters, **26**-7
Floodlights, **16**-5
Floor plan, lighting, **16**-8
Follow up:
  in job hunting, **28**-10
  for location shooting, **21**-16
  nonprofessional talent and, **19**-9
Fome-Cor, **14**-18
Format of program, **13**-1 to **13**-4
Formats, international video network, **26**-6 to **26**-7
Frame rate, **20**-9
Free-lance staff, **5**-5, **5**-8
Freight forwarder, **26**-8 to **26**-9
Frequency response, **18**-4
Fresnel spotlights, **16**-4

Front elevations in set design, **14**-8, **14**-9
Full charge-back system, **2**-6
Funding, **8**-2 to **8**-5
    charge-back approach to, **8**-3 to **8**-4
    conclusions on, **8**-11 to **8**-13
    record keeping and, **8**-11 to **8**-13
    self-funded cost center approach to, **8**-2 to **8**-3
Future of private television, **1**-8, **29**-3 to **29**-8
    cable in, **29**-8
    computers in, **29**-6 to **29**-7
    conclusions on, **29**-8
    economy and, **29**-7
    markets and technological development in, **29**-3 to **29**-4
    new technology in, **29**-4 to **29**-7
    one-half-inch videocassette in, **29**-5
    outside services in, **29**-8
    research and, **29**-4
    satellites in, **29**-4 to **29**-5
    trends affecting, **29**-7
    videodisc in, **29**-5 to **29**-6

Gel, **16**-6 to **16**-7
Gobo, **16**-6
Grapevine, **10**-4
Graphics, **13**-9, **15**-1 to **15**-15
    art cards, **15**-11 to **15**-12
    camera angle and, **15**-11
    character generator and, **15**-13 to **15**-14
    color in, **15**-10 to **15**-11
    conclusions on, **15**-14 to **15**-15
    developing, **15**-4 to **15**-7
    editing, **22**-10
    entertainment use of, **15**-4
    highlighting key points using, **15**-2
    identification use of, **15**-1
    mechanicals, **15**-7
    purpose of, **15**-1 to **15**-4
    reveals, **15**-12 to **15**-13
    safe title area of, **15**-8 to **15**-9
    simplifying complicated material using, **15**-2 to **15**-3
    storyboards, **15**-4 to **15**-7
    transferring artwork to videotape, **15**-7 to **15**-14
    type fonts for, **15**-9 to **15**-10
    type size for, **15**-10

Ground plan, set design, **14**-6 to **14**-9
Guard band, **22**-1
Guidelines for management (*see* Management techniques, guidelines for success)

Hairline, receding, **17**-10 to **17**-11
Hands, makeup, **17**-8
Hardware, emphasis on, **3**-1 to **3**-2
Heart-shaped face, **17**-7
Hertz, **26**-3
Highlighting key points using graphics, **15**-2
Home VCRs, **1**-8, **29**-4
Hook for program, **13**-3 to **13**-4
Host country, technical requirements of, **26**-2 to **26**-3
House organ, **1**-6
Hums, **18**-11
Hypercardioid microphone, **18**-1

Identification, graphics in, **15**-1
Impact desired, **11**-4
Implementation of video department, **4**-3 to **4**-21
    case study in, **4**-16 to **4**-21
    getting company support for, **4**-3 to **4**-4
    ground rules of, **4**-3
    proposal for, **4**-18 to **4**-21
    step one (write a memo), **4**-4 to **4**-5
    step two (needs analysis), **4**-5 to **4**-7
    step three (form a task force), **4**-7 to **4**-9
    step four (proposal to board of directors) (*see* Proposal to board of directors)
In-house studio, **6**-5 to **6**-6
Industry jargon, **10**-3 to **10**-4
Information packets, **10**-4
Input function, staff and, **27**-10
Insert editing, **22**-2
Instruction in video, **1**-5, **2**-4, **6**-4 to **6**-5, **28**-5 to **28**-6
Insurance, location shooting, **21**-10
Interchangeability, international video network, **25**-4, **25**-13 to **25**-14
Internal rate of return, **8**-6

Internal Revenue Code, equipment lease, **9**-8 to **9**-9
International broadcast standards (see Standards, international video)
International color broadcast standards (see Color broadcast standards)
International equipment standards (see Equipment standards)
International Television Association (ITVA):
  salary survey by, **8**-8, **27**-3 to **27**-4, **27**-11 to **27**-12
  training by, **6**-5
International video network, **26**-1 to **26**-13
  branch offices in: informing, **29**-9
  selecting, **26**-2
  survey of, **26**-3 to **26**-5
  conclusions on, **26**-13
  cost of copies for, **26**-7
  costs of, controlling, **26**-2
  customs officials and, **26**-12
  electrical voltage and hertz in, **26**-3
  equipment selection for, **26**-5 to **26**-7
  equipment vendors for, **26**-8
  formats for, **26**-6 to **26**-7
  freight forwarder for, **26**-8 to **26**-9
  host country in, technical requirements of, **26**-2 to **26**-3
  invoices for, **26**-9, **26**-10
  one-way versus two-way systems in, **26**-5 to **26**-6
  overseas mail and, **26**-11
  packaging for, **26**-12
  planning in, **26**-1 to **26**-2
  program distribution in, **26**-11 to **26**-12
  purchasing policy in, **26**-10 to **26**-11
  repair capability and, **26**-7
  research for, **26**-1 to **26**-5
  selecting offices for, **26**-2
  standards for (see Standards, international video)
  translations for, **26**-12 to **26**-13
International video standards (see Standards, international video)
Interpolation, **25**-17
Interview process, job hunting, **28**-10 to **28**-13
Interviewing audience, **12**-2
Inventory control, tape, **23**-10

Inverted triangular face, **17**-7 to **17**-8
Investment tax credit, **8**-7 to **9**-8
Invoices, **26**-9, **26**-10
Iso VTR, **20**-3

Jacketing of videocassettes, **23**-8
Jam sync, **22**-4 to **22**-5
Jargon, industry, **10**-3 to **10**-4
Job descriptions, **5**-3 to **5**-5, **27**-4 to **27**-9
  director, **27**-6
  engineer, **27**-7
  growth within positions, **27**-8 to **27**-9
  manager, **27**-4
  media director, **27**-6
  producer, **7**-2 to **7**-3, **27**-5
  production assistant, **27**-7
  supervisor, **27**-5
  technician, **27**-8
  writer, **27**-5
Job hunting, **28**-1 to **28**-13
  associations in field for, major, **28**-3 to **28**-4
  competitive marketplace of, **28**-1
  correspondence in, size of, **28**-5
  cover letters in, **28**-10
  directories of operations and production companies for, **28**-4 to **28**-5
  first stage of, **28**-1
  follow-up call in, **28**-10
  interview process in, **28**-10 to **28**-13
  personal record summary form for, **28**-2 to **28**-3
  portfolio in, **28**-13
  publications on, **28**-13
  qualifications for job in production, **28**-5
  résumé for, **28**-6 to **28**-10
  self-study program and, **28**-6
  taking stock of yourself in, **28**-3
  videotape training sessions and, **28**-6
  workshop sessions and, **28**-6
Job in production, qualifications for, **28**-5

Key instruments, **16**-3
Key points, graphics for highlighting, **15**-2

Knowledge Industry Publications, Inc., **2**-4, **28**-6
Kodalith, **15**-12 to **15**-13

Labels, **23**-7
Language of industry, **10**-3 to **10**-4
Lavalier microphone, **18**-1, **18**-5, **18**-7, **18**-11
Law department, **9**-2 to **9**-3
Leading, functions of, **10**-2
Lease of equipment contract, **9**-8 to **9**-9
Lease payments, **9**-8 to **9**-9
Leko spotlight, **16**-4 to **16**-5
Letter of agreement, **9**-2 to **9**-5
Lettering, typefaces for, **15**-9 to **15**-10
Leveraged buy-out, **9**-9
Liability, expected exposure to, **9**-1
Library, master, **23**-10 to **23**-11
Light meter, **16**-9
Light sources, **16**-2 to **16**-3
Lighting, **16**-1 to **16**-15
  bald head in, **16**-13
  chroma key and, **16**-10 to **16**-11
  color temperature and, **16**-2 to **16**-3, **16**-7
  contrast-balancing techniques in, **16**-9
  contrast range in, **16**-1 to **16**-2
  control devices for, **16**-5 to **16**-7
  daylight mixed with artificial light, **16**-11 to **16**-12
  equipment for, **16**-4 to **16**-5
    (See also instruments, below)
  eyeglasses in, **16**-12
  floodlights, **16**-5
  floor plan in, **16**-8 to **16**-9
  intensity controls for, **16**-7
  instruments, **16**-3
    (See also equipment for, above)
  of large rooms, **16**-13 to **16**-14
  light meter in, **16**-9
  limitations of television system and, **16**-1 to **16**-3
  for location shooting, **21**-12
    (See also outdoors, below)
  low ceiling and, **16**-13
  for movement, **16**-14
  outdoors, **16**-14 to **16**-15
    (See also for location shooting, above)

Lighting (Cont.):
  panchro glass and, **16**-9
  problem solving for, **16**-9 to **16**-15
  production procedures for, **16**-7 to **16**-9
  for rear projection, **16**-11
  in set design, **14**-5, **20**-4
  for small space, **16**-13
  spotlights, **16**-4 to **16**-5
  standards, **20**-8 to **20**-9
  in studio setting, **16**-2
  survey for, **16**-7
  waveform monitor and, **16**-9
Lighting equipment, **16**-4 to **16**-5
Lighting instruments, **16**-3
Limiter amplifier, **18**-2
Lips, makeup, **17**-12 to **17**-13
List management, of edit-decision lists, **22**-3
Literary works, nondramatic, copyright class for, **24**-4
Local permits, **21**-7
Locale, **14**-2
Location shooting, **21**-1 to **21**-16
  attitude and, **21**-13 to **21**-15
  audio for, **21**-12 to **21**-13
  backup systems for, **21**-6 to **21**-7
  cash and, need for, **21**-4 to **21**-5
  checklists for: backup systems, **21**-6 to **21**-7
    preproduction, **21**-3 to **21**-5
  contact sheet for, **21**-3 to **21**-4
  convenience and, **21**-2
  corporate permission for, **21**-8 to **21**-9
  cost of, **21**-1
  crew comfort and, **21**-5
  crowd control for, **21**-9
  downtime and, **21**-14 to **21**-15
  drivers for, **21**-11
  food for, **21**-4
  insurance for, **21**-10
  lighting for, **21**-12
    (See also Lighting, outdoors)
  local permits for, **21**-7
  lodging and, **21**-4
  logging of, **21**-13
  microphones for, **18**-10 to **18**-11
  outdoor equipment supplies for, **21**-6
  preparations for, **21**-4 to **21**-7
  preproduction for, **20**-1 to **20**-4
  production for, **6**-6, **21**-12

Location shooting (*Cont.*):
  production assistants and, **21**-9
  protecting equipment and, **21**-6
  realism and, **21**-1
  reasons for, **21**-1 to **21**-2
  release forms for, **21**-9 to **21**-10
  safety and, **21**-6
  surveys for, **21**-2 to **21**-3
  talent and, **21**-15 to **21**-16
  technology of, **21**-2
  thank-you letters for, **21**-16
  transportation for, **21**-10 to **21**-11
  travel for, **21**-4
  visual variety and, **21**-2
  weather and, **21**-6
  wrapping up, **21**-16
Location surveys, **21**-2 to **21**-3
Logging, location shooting, **21**-13
Logos, **15**-14
Long shot, **13**-8
Look ahead, in automatic assembly editing systems, **22**-3
Loss of equipment, **9**-10
Low ceiling, **16**-13

Machinery noise, **18**-12
Mailing overseas, **26**-11
Maintenance of equipment, **28**-5
Major video facility (*see* Video facility, major)
Makeup, **17**-1 to **17**-15
  base coat of, **17**-4 to **17**-5
  for blemish or discoloration, **17**-10
  brush for, **17**-6
  camera as final test of, **17**-14
  cleanliness in, importance of, **17**-15
  clothing rearrangement and, **17**-3
  of eyebrows, **17**-11
  of eyes, **17**-11 to **17**-12
  of face, **17**-6 to **17**-8
  face shape and, **17**-6 to **17**-8
  facial type in, **17**-4
  finishing the look in, **17**-13 to **17**-14
  of hands, **17**-8
  individual conditions in, **17**-4
  of lips, **17**-12 to **17**-13
  outstanding feature and, **17**-4
  overview of: for artist, **17**-2
    for producer, **17**-1
  personality of individual and, **17**-4

Makeup (*Cont.*):
  plan of action for, **17**-15
  powder, **17**-5 to **17**-6
  powder puff for, **17**-6
  of problem areas, **17**-4
  psychology of applying, **17**-2 to **17**-3
  for receding hairline, **17**-10 to **17**-11
  removal of, **17**-3, **17**-14 to **17**-15
  skin type and, **17**-4
  style of, variables affecting, **17**-4
  techniques of applying, **17**-6 to **17**-14
  time limit for applying, **17**-15
  touch in, **17**-2
  types of, **17**-4 to **17**-5
  of under-eye areas, **17**-8 to **17**-10
Management communications, **1**-6
Management support for video department, **2**-2 to **2**-3, **10**-5
Management techniques, guidelines for success, **10**-1 to **10**-7
  bottom line in, **10**-7
  client involvement in, **10**-5
  communications in, **10**-4
  compromise in, **10**-6 to **10**-7
  knowledge of product in, **10**-3 to **10**-4
  management involvement in, **10**-5
  selling of video operation in, **10**-3
  staff selection in, **10**-6
  standards in, **10**-6
  video as something special in, **10**-4 to **10**-5
Manager, **6**-9
  job description for, **27**-4
  for small video facility, **5**-3 to **5**-4
Managing, **10**-1 to **10**-2
  components of, **10**-1 to **10**-2
  meaning of, **10**-1
Marketing, **1**-5 to **1**-6
Marketing department, **2**-4
Markets for private television, technological development and, **29**-3 to **29**-4
Master library, **23**-10 to **23**-11
Master tape evaluation, **23**-3
Master VTR, **20**-3
Mastering, **23**-1 to **23**-2
Materials for set design, **14**-9 to **14**-12
Mechanicals, **15**-7
Media director, job description, **27**-6
Media for presentation, **12**-4 to **12**-5
Medium shot, **13**-8

10   INDEX

Meeting services staff, **27**-11
Meetings, microphones for, **18**-9
Memo, implementation of video
    department, **4**-4 to **4**-5
Microphones, **18**-8 to **18**-12
    background noises and, **18**-11 to
        18-12
    balancing, **18**-8 to **18**-9
    boom (see Boom microphone)
    changing, **18**-12
    for large meeting, **18**-9
    lavalier, **18**-1, **18**-5, **18**-7, **18**-11
    for location shooting, **18**-10 to **18**-11
    mixing, **18**-8 to **18**-9
    multiple, phasing problem of, **18**-12
    for panel discussion, **18**-9
    phase problems with, **18**-12
    popping noises and, **18**-11
    sibilance and, **18**-11
    sound patterns and their uses, chart
        of, **18**-6
    sound perspective and, **18**-4 to **18**-6
    various, **18**-1 to **18**-2
    wind noises and, **18**-11
Microprocessor development, **29**-6
Microwave, **6**-8
Mixer, **18**-2
Mixing process in audio, **18**-13
Model, scale, set design, **14**-6, **14**-7
Modest video facility (see Video
    facility, small)
Monitors, display, **22**-7
Montage description within script,
    13-10
Mood, in set design, **14**-2
Motion pictures, copyright deposit
    requirements for, **24**-6 to **24**-7
Motivating staff, **6**-4 to **6**-5, **10**-2, **10**-6
Movement, lighting for, **16**-14
Multicamera, single camera versus, **20**-2
    to **20**-3
Music, sources of, **18**-14
Music library, **18**-14

Narrative, **13**-3 to **13**-4
National Television Systems Committee
    (NTSC) system, **25**-4 to **25**-5, **25**-10
Needs analysis, **4**-5 to **4**-8, **12**-1 to **12**-2
    in implementation of video
        department, **4**-5 to **4**-7

Needs analysis (Cont.):
    user questionnaire for, **4**-8
Net present value, **8**-6
Networks:
    distribution points in, **2**-6
    international video (see International
        video network)
Noise, **18**-7, **18**-10 to **18**-12
    background, **18**-11 to **18**-12
    fabric, **18**-7
    popping, **18**-11
    traffic, **18**-11
    wind, **18**-11
Nondelegation clause, **9**-6
Nondistributed processing systems,
    22-7
Nondramatic literary works, copyright
    class, **24**-4
Nonprofessional talent, **19**-1 to **19**-10
    advantages of, **19**-1 to **19**-3
    attaching faces to memos using, **19**-2
    conclusions on, **19**-10
    controlling cash using, **19**-2
    demonstrating concern using, **19**-2
    demonstrating professional
        competence to management
        using, **19**-3
    disadvantages of, **19**-3 to **19**-4
    enhanced credibility using, **19**-1 to
        19-2
    follow up for, **19**-9
    improving content using, **19**-8
    inappropriate, **19**-4
    inarticulate, **19**-4
    integrating unit into organization
        using, **19**-3
    natural performance using, with Tele-
        Prompter, **19**-7 to **19**-8
    problem of, **3**-3 to **3**-4
    production day and, **19**-9
    psychology of working with, **19**-5 to
        19-6, **20**-10 to **20**-11
    pushing, **19**-6
    rehearsal stage and, **19**-6 to **19**-8
    rehearsal time and, **19**-4 to **19**-6
    during the shoot itself, **19**-9
    uncooperative, **19**-4
    unpresentable, **19**-4
NTSC (National Television Systems
    Committee) system, **25**-4 to **25**-5,
    25-10

Object of project, 7-2
Objectives, determination of, 12-2 to 12-3
Oblong face, 17-7
Off-line editing, 22-2, 22-9
On-camera personnel (see Talent)
On-line editing, 22-2, 22-9
On-location production, 6-6
On-location shooting (see Location shooting)
One-half-inch Video Home System (VHS), 6-9
One-half-inch videocassette, 6-9, 29-5
One-way systems, 26-5 to 26-6
Operating budget, 8-5, 8-8, 8-11
  lease payments in, 9-8 to 9-9
  of small video facility, 5-7 to 5-9
Operational information, equipment contract, 9-11
Optical converters, 25-16, 25-17
Oral agreement, 9-2
Organizational needs, 3-4 to 3-5
Organizational structure of video department, 2-1 to 2-7
  distribution points in network, 2-6
  job descriptions in, 27-9 to 27-10
  parent department for video department, 2-3 to 2-5
  pitfalls of, 3-4 to 3-5
Organizing, functions of, 10-2
Outdoor equipment, 21-6
Outdoor shooting, lighting for, 16-14, 16-15
Output function, staff, 27-11
Outside producer, 7-1 to 7-7
  budget and, 7-6 to 7-7
  concluding thoughts on, 7-7
  cost considerations and, 7-4
  hiring, 7-1 to 7-2
  packaged productions and, 7-5 to 7-7
  payments for, 7-7
  pricing, 7-6 to 7-7
  program development and, 7-1 to 7-2
  reasons for using, 7-3 to 7-4
  rules to follow in choosing, 7-6
  services to buy from, 7-4 to 7-6

Outside production house, 6-7
Outside services, 5-8, 7-4 to 7-6
Oval face, 17-6
Overhead, 2-6
Overprints, 23-11
Overseas mail, 26-11
Ownership of item, service contract, 9-3

Pacing, 13-9
  pitfalls of, 3-3
Packaged productions, 7-5 to 7-7
Packaging, 26-12
  for videocassette duplicates, 23-7 to 23-8
PAL (phase alternate by line) system, 25-4 to 25-5, 25-10
Pancake makeup, 17-5
Panchro glass, 16-9
Panel discussion microphones, 18-9
Partial charge-back systems, 2-6 to 2-7
Payback period, 8-7, 8-9
Payment:
  equipment contract terms of, 9-9
  of outside producer, 7-7
  for script proposal, 13-7
  service contract terms of, 9-3, 9-5
Performing arts, copyright class, 24-4 to 24-5
Permission, obtaining, 24-1 to 24-2
Personal record summary form, 28-2 to 28-3
Personality and makeup procedure, 17-4
Personnel (see Employees; Staff; Talent)
Personnel department, 2-4
Perspective, audio, 18-4 to 18-6
Phase alternate by line (PAL) system, 25-4 to 25-5, 25-10
Phase errors:
  NTSC system, 25-5, 25-10
  PAL system, 25-10
Phasing, 18-12
Picture composition, 20-9 to 20-10
Pitfalls of private television, 3-1 to 3-6
  bottom line in, 3-2
  errors of emphasis in, 3-1
  foresight and, 3-6
  hardware in, emphasis on, 3-1 to 3-2
  organizational, 3-4 to 3-6

Pitfalls of private television (*Cont.*):
  pacing, **3**-3
  process versus product in, **3**-1 to **3**-2
  production, **3**-1 to **3**-4
  professionalism in, **3**-5 to **3**-6
  time management in, **3**-2 to **3**-3
Planning:
  functions of, **10**-2
  for international video network, **26**-1 to **26**-2
  in major video facility, **6**-1 to **6**-4
Platforms, in set design, **14**-17 to **14**-18
Playback locations for videocassettes, **6**-8 to **6**-9
Pop filters, **18**-11
Popping noises, microphones, **18**-11
Portfolio, **28**-13
Postproduction, sound, **18**-12 to **18**-13
Postproduction services, **7**-5
Powder, makeup, **17**-5 to **17**-6
Powder puff, **17**-6
Preproduction, **20**-1 to **20**-5
  for location shooting, **20**-1 to **20**-4
  in production design, **20**-2
  production values versus budget in, **20**-1 to **20**-2
  for sets, **20**-4
  single camera versus multicamera in, **20**-2 to **20**-3
  for special effects, **20**-3 to **20**-4
  studio versus location, **20**-4
  talent and casting in, **20**-4 to **20**-5
  time of, **12**-5
  (*See also* Program development)
Prescreening, **5**-2 to **5**-3
Present value, **8**-6
Presentation methods, **12**-4 to **12**-5
"Principle of technical priority," **10**-2
Prior approval, **9**-1
Private television in corporate environment (*see* Corporate television; Video department)
Pro forma invoices, **26**-9 to **26**-10
Process versus product, **3**-1 to **3**-2
Processing section, staff function, **27**-10
Producer:
  copyright permission and, **9**-5
  definition of, **7**-2 to **7**-4
  job description for, **7**-2 to **7**-4, **27**-5
  makeup overview for, **17**-1
  outside (*see* Outside producer)

Product:
  knowledge of, **10**-3 to **10**-4
  process versus, **3**-1 to **3**-2
Production, **20**-5 to **20**-7
  crew for, **20**-6 to **20**-7
  on-location, **6**-6, **21**-12
  packaged, **7**-5
  parts of, **7**-5 to **7**-6
  pitfalls of, **3**-1 to **3**-4
Production assistants (PA), **20**-6
  job description for, **27**-7
  in location shooting, **21**-9
Production budget per program, **2**-6 to **2**-7
Production coordinator, **5**-4
Production costs, **5**-8 to **5**-9, **7**-7
Production design, **20**-2
Production environment, **6**-5 to **6**-7
Production house, outside, **6**-7
Production schedule, **6**-3, **9**-5, **20**-5 to **20**-6
  major, **6**-3
  service contract and, **9**-5
Production services to buy, **7**-4 to **7**-6
Production staff, **20**-6 to **20**-7
  qualifications for, **28**-5
Production values, **11**-3 to **11**-7
  budget versus, **20**-1 to **20**-2
  creating, **11**-4 to **11**-5
  definition of, **11**-3 to **11**-4
  editing and, **11**-7
  in scriptwriting process, **11**-5 to **11**-6
  during shoot, **11**-6 to **11**-7
  summary of, **11**-7
Professional growth, **27**-8 to **27**-9
Professionalism, guidelines to, **3**-5 to **3**-6
Program budget, **2**-6 to **2**-7
  (*See also* Budget)
Program catalog, **6**-7
Program development, **12**-1 to **12**-5
  content in, determination of, **12**-3 to **12**-4
  needs analysis in, **12**-1 to **12**-2
  objectives in, determination of, **12**-2 to **12**-3
  presentation methods in, **12**-4 to **12**-5
  six-step process of, **12**-1
Program distribution, **6**-7 to **6**-9, **26**-11 to **26**-12
Program evaluation, **5**-3

Program format, **13**-1 to **13**-4
Program guide, **6**-7, **6**-8
Program selection, **5**-2 to **5**-3
Programming, types of, **1**-5 to **1**-8
Programs, number produced, **2**-5
Project, **7**-1 to **7**-2
   hiring a producer for, **7**-1 to **7**-2
   outside producer and, **7**-1 to **7**-2
   scope of, service contract and, **9**-3
Proposal to board of directors, **4**-9 to **4**-15
   agenda for, **4**-12
   sample, **4**-13 to **4**-15, **4**-18 to **4**-21
Props, **14**-19
Psychological factors:
   of contract, **9**-1
   of makeup application, **17**-2 to **17**-3
   nonprofessional talent and, **19**-5 to **19**-6, **20**-10 to **20**-11
Public relations, **1**-7
Purchasing policy, **26**-10 to **26**-11

Quality control, videocassette duplication, **23**-5 to **23**-7
Questionnaire:
   international video network, **26**-4
   needs analysis user, **4**-8

Radio link, **18**-7
Radio microphone, **18**-2
Realism, location shooting, and, **21**-1
Realistic visualizations in scripwriting, **13**-9
Rear projection, **16**-11
Receding hairline, **17**-10 to **17**-11
Receiver locations for videocassettes, **6**-8 to **6**-9
Recycling, videotape, **23**-9 to **23**-10
Red hue, **15**-10 to **15**-11
Reflective materials in set design, **14**-6
Regenerating time code, **22**-4
Regulatory requirements, notification of, **1**-6 to **1**-7
Rehearsal stage, **19**-9 to **19**-8
   improving content at, **19**-8
   prior to shoot, **19**-6 to **19**-7
   TelePrompter use in, **19**-7 to **19**-8
Rehearsal time, **19**-4 to **19**-6
Release forms, **21**-9 to **21**-10

Rent, **8**-8 to **8**-9
Repair service, **9**-9
   (*See also* Service contract)
Research, design idea, **14**-4 to **14**-5
Résumé, **28**-6 to **28**-10
Retrieving program tapes, **23**-10
Reveals, **15**-12 to **15**-13
Rights:
   obtaining necessary, **9**-6
   videocassette duplication and, **23**-4 to **23**-5
   (*See also* Copyright law)
Rise of private television, **1**-3 to **1**-8
   barriers to, **1**-3
   next decade in, **1**-8
      (*See also* Future of private television)
   reasons for, **1**-3
   selling management in, **1**-4
   technical triumph in, **1**-3 to **1**-4
   types of programming in, **1**-5 to **1**-8
   usefulness in corporate environment and (*see* Corporate television)
Roll-off filters, **18**-11
Rough edit, testing, **5**-2 to **5**-3
Round face, **17**-6 to **17**-7

Safe action area, **15**-8
Safe title area, **15**-8 to **15**-9
Safety, **15**-8
   instruction in, **1**-7
   lighting and, **16**-8 to **16**-9, **21**-12
   in location shooting, **21**-6
Salary survey, International Television Association, **8**-8, **27**-3 to **27**-4, **27**-11 to **27**-12
Satellites, **29**-4 to **29**-5
Scale model, set design, **14**-6 to **14**-7
Scene, time of, **13**-9
Scenery:
   reuse of, **14**-14
   supports for, **14**-16
Scenic design (*see* Set design)
Scoops, **16**-5
Scope of project, **7**-2
   service contract and, **9**-3
Scrim, **16**-6
Script approval form, **8**-16
Scriptwriting, **12**-1, **13**-1 to **13**-13
   abbreviation of key terms in, **13**-7 to **13**-8

14 INDEX

Scriptwriting (*Cont.*):
    approach to, **13**-1 to **13**-4
    conclusions on, **13**-12 to **13**-13
    documentary style, **13**-2, **13**-4, **13**-9, **13**-14
    flexibility in, **13**-11
    format of program in, **13**-1 to **13**-4
    montage description within, **13**-10
    narrative style, **13**-3 to **13**-4
    payment for proposal, **13**-7
    production value in, **11**-5 to **11**-6
    realistic visualizations in, **13**-9
    rules of thumb for, **13**-12 to **13**-13
    steps involved in, **13**-1
    talking head and, **13**-1 to **13**-2
    treatment in, **13**-4 to **13**-7
    vignette style, **13**-2 to **13**-3, **13**-6
    visual concept in, **13**-3 to **13**-4
    writing in, **13**-7 to **13**-12
SECAM (Système Électronique Couleur Avec Memoire/electronic color system with memory), **25**-4 to **25**-5, **25**-10 to **25**-11
Self-funded cost center approach, **8**-2 to **8**-3
Self-study program, **28**-6
Selling of video operation, **1**-4, **6**-3 to **6**-4, **10**-3
Service, repair, **9**-9, **26**-7
Service contract, **9**-3 to **9**-7
    confidentiality in, **9**-6
    copy to writer or producer in, **9**-6
    copyright and, **9**-5
    credits in, **9**-6
    damage clause in, **9**-6 to **9**-7
    deadline and, **9**-7
    detailed description in, **9**-5
    nondelegation clause in, **9**-6
    production schedule and, **9**-5
    rights and, obtaining necessary, **9**-6
    standards clause in, **9**-6
Set design, **14**-1 to **14**-20
    accessories in, **14**-19
    budget for, **14**-12 to **14**-13
    camera angles and, **14**-2 to **14**-4
    colors in, **14**-6, **14**-7
    concept and, **14**-1 to **14**-4
    conclusions on, **14**-20
    cyclorama and, **14**-17
    drapes in, **14**-17

Set design (*Cont.*):
    equipment parameters in, **14**-5 to **14**-6
    flats in, **14**-14 to **14**-17
    Fome-Cor in, **14**-18
    ground plan and, **14**-6 to **14**-9
    lighting in, **14**-5, **20**-4
    locale and, **14**-2
    materials for, sources of, **14**-9 to **14**-12
    mood and, **14**-2
    on-camera test of, **14**-7, **14**-9
    platforms in, **14**-17 to **14**-18
    in preproduction stage, **20**-4
    props in, **14**-19
    reflective materials in, **14**-6
    research and, **14**-4 to **14**-5
    reuse of scenery, **14**-14
    scale model in, use of, **14**-6, **14**-7
    scenery supports in, **14**-16
    set dressing in, **14**-19
    set pieces in, **14**-17 to **14**-19
    sketches in, **14**-6 to **14**-9
    space relationships in, **14**-2
    stairs in, **14**-18
    structural elements in, **14**-13 to **14**-14
    studio parameters in, **14**-5 to **14**-6
    style of production in, **14**-2
Set dressing, **14**-19
Set lights, **16**-3
Set pieces, **14**-17 to **14**-19
Shipment of videocassettes, **23**-9
Shooting:
    editing considerations in, **22**-7 to **22**-8
    on location (*see* Location shooting)
    nonprofessional talent and, **19**-9
    production values during, **11**-6 to **11**-7
Shot sheet, **21**-13
Shotgun microphone, **18**-1
Sibilance, **18**-11
Single camera versus multicamera, **20**-2 to **20**-3
Sketches, set design, **14**-6 to **14**-9
Skin type, **17**-4
Small video facility (*see* Video facility, small)
SMPTE (Society of Motion Picture and Television Engineers) time code, **22**-1, **22**-4 to **22**-6, **22**-8, **22**-9
Software changes, editing systems, **22**-7

Sound blanket, **18**-2
Sound effects, **18**-13 to **18**-14
  realistic, **18**-13
  sources of, **18**-13 to **18**-14
Sound level, ambient, **20**-8
Sound perspective, **20**-8
  microphones and, **18**-4 to **18**-6
Sound recordings, copyright class, **24**-6
Sound in television (see Audio in television)
Space relationships, set design, **14**-2
Space size, lighting and, **16**-13 to **16**-14
Special effects, **20**-3 to **20**-4
  editing, **22**-10
Specification sheet, equipment contract, **9**-10
Spotlights, **16**-4 to **16**-5
Square face, **17**-7
Staff, **27**-3 to **27**-12
  career path of, **27**-9 to **27**-10
  free-lance, for small video facility, **5**-5, **5**-8
  guidelines for selecting, **10**-6
  input function of, **27**-10
  job descriptions for (see Job descriptions)
  meeting services, **27**-11
  motivating, **6**-4 to **6**-5
  organizational structure and, **27**-9 to **27**-10
  output function of, **27**-11
  processing section, **27**-10
  production, **20**-6 to **20**-7, **28**-5
  salary of, **8**-8, **27**-3 to **27**-4, **27**-11 to **27**-12
  for small video facility, **5**-3 to **5**-5
  training of, **1**-5, **6**-4 to **6**-5
  (See also Employees; Talent)
Stairs, set design, **14**-18
Standard agreements, **9**-2 to **9**-3
Standard converters, types of, **25**-16
Standard representations, equipment contract, **9**-11
Standards, international video, **26**-6 to **26**-7
  benefits of, **25**-3
  color broadcast (see Color broadcast standards)
  conclusions on, **25**-17
  for conversion, **25**-15 to **25**-19

Standards, international video, for conversion (*Cont.*):
  converters, types of, **25**-16
  cost comparisons, **25**-18 to **25**-19
  optical versus digital, **25**-17
  electrical current, **25**-6 to **25**-9
  equipment and, **25**-12 to **25**-13
  multistandard equipment, **25**-14
  tristandard ½-inch models, **26**-7
  videocassette compatibility and, **25**-3
  videocassette interchangeability and, **25**-4, **25**-13 to **25**-14
Standards of quality clause, **9**-6
Standards of quality:
  director's responsibility for, **20**-7 to **20**-9
  manager's responsibility for, **10**-6
Stop Copy, **23**-5
Storage of tapes, **23**-10 to **23**-11
Storyboards, **15**-4 to **15**-7
Straight-line depreciation, **8**-7
Structural elements, set design, **14**-13 to **14**-14
Studio:
  in-house, **6**-5 to **6**-6
  lighting and, **16**-2
  preproduction and, **20**-4
  in set design, **14**-5 to **14**-6
Studio engineer, **5**-4 to **5**-5
Style of production, **14**-2
Subject of project, **7**-2
Supervisor, job description, **27**-5
Supplies category of budget, **8**-9
Support material budget form, **8**-14 to **8**-15
Surveys:
  branch office, **26**-3 to **26**-5
  for lighting, **16**-7 to **16**-8
  location, **21**-2 to **21**-3
  salary, **8**-8, **27**-3 to **27**-4, **27**-11 to **27**-12
  user questionnaire, **4**-8
"Sweetening" process for audio, **18**-12 to **18**-13
Switcher, effects capabilities of, **20**-3
Symbols, **15**-14

Talent:
  on location, **21**-15 to **21**-16
  nervousness of, **20**-10 to **20**-11
  nonprofessional (see Nonprofessional talent)

INDEX

Talent (*Cont.*):
  in preproduction, **20**-4 to **20**-5
  (*See also* Employees; Staff)
Talking head, **13**-1 to **13**-2
Tape inventory control, **23**-10
Task force, implementation, **4**-7 to **4**-11
  agenda for meeting of, **4**-9 to **4**-11
Tax credit, investment, **8**-7, **9**-8
Tax life, **9**-8
Technical development:
  new, **29**-4 to **29**-7
  research and, **29**-4
Technical director, **20**-6
Technical standards, **20**-7 to **20**-9
Technician, job description, **27**-8
Technological development, markets in, **29**-3 to **29**-4
Telephone, **8**-9
TelePrompter, rehearsal using, **19**-7 to **19**-8
Television versus film, **20**-9
Television system, limitations of, **16**-1 to **16**-3
Terminology, specialized, **10**-3 to **10**-4
Testimonials, **13**-2
Testing, **5**-2 to **5**-3
Thank-you letters:
  for location shooting, **21**-16
  for nonprofessional talent, **19**-9
Three-quarter-inch U-Matic videocassette, **1**-4, **6**-8 to **6**-9, **23**-5
Time:
  of preproduction stage, **12**-5
  rehearsal, **19**-4 to **19**-6
  of scene, **13**-9
Time base corrector, **22**-3
Time code:
  drop-frame, **22**-4
  jam-syncing, **22**-4 to **22**-5
  reading, **22**-5 to **22**-6
  regenerating, **22**-4
  SMPTE, **22**-1, **22**-4 to **22**-6, **22**-8 to **22**-9
  VITC, **22**-3 to **22**-4
Time management, pitfalls of, **3**-2 to **3**-3
Tint information, NTSC system, **25**-5, **25**-10
Top management involvement, **10**-5
Touch, makeup, **17**-2

Trade publications, **10**-4
Traffic noises, **18**-11
Training, employee, **1**-5, **2**-4, **6**-4 to **6**-5, **28**-5 to **28**-6
Training department, **2**-4
Training sessions, **28**-6
Translations, **26**-12 to **26**-13
Transportation, **21**-10 to **21**-11
Travel, location shooting, **21**-4
Treatment, scriptwriting, **12**-1, **13**-4 to **13**-7
Trends, future (*see* Future of private television)
Triangular face, **17**-7
Tristandard one-half-inch models, **2**-7
Two-way systems, **25**-6 to **26**-6
Type fonts, **15**-9 to **15**-10
Type size, **15**-10
Typefaces, **15**-9 to **15**-10

U-Matic videocassette recorder, **1**-3 to **1**-4, **6**-8 to **6**-9, **23**-5
Under-eye area, makeup, **17**-8 to **17**-10
U.S. Federal Communications Commission (FCC) standards, **25**-4 to **25**-5
Use of item, service contract, **9**-3
Useful life, **8**-6
User bits, **22**-3
User questionnaire, **4**-8

VCR (*see* Videocassette recorder)
Vendors, equipment, **26**-8
Vertical interval time code (VITC), **22**-3 to **22**-4
Video department:
  documenting worth of, **3**-4 to **3**-5
  getting company support for, **2**-2 to **2**-3, **4**-3 to **4**-4
    (*See also* Implementation of video department)
  implementation of (*see* Implementation of video department)
  organization of (*see* Organizational structure of video department)
  parent department of, **2**-3 to **2**-5
    (*See also* Corporate television)

Video design process, 5-1 to 5-3
Video facility:
  major, 6-1 to 6-9
    budget of, 6-7
    definition of, 6-1
    executive steering committee of,
      6-2 to 6-3
    in-house studio of, 6-5 to 6-6
    manager of, 6-9
    motivating staff of, 6-4 to 6-5
    on-location production and, 6-6
    outside production house and, 6-7
    planning in, 6-1 to 6-4
    production environment of, 6-5 to
      6-7
    production schedule of, 6-3
    program distribution at, 6-7 to 6-9
    selling your facility, 6-3 to 6-4
  small (modest), 5-1 to 5-9
    budget for (see Budget, for small
      video facility)
    free-lance support for, 5-5, 5-8
    manager for, 5-3 to 5-5, 6-9
    production coordinator for, 5-4
    reassessment and exploration in,
      5-9
    staff for, 5-3 to 5-5
    studio engineer for, 5-4 to 5-5
Video Home System (VHS), 25-3
½-inch, 6-9
Video network:
  definition of 1-4
  international (see International video
    network)
Video programming budget category,
  8-9
Video programs (see entries beginning
  with term: Program)
Video revolution, 1-8
Videocassette duplication, 23-1 to 23-11
  check of incoming cassettes in, 23-10
  conclusions on, 23-11
  controlled environment for, 23-2
  cost of copies in, 26-7
  distribution of duplicates, 23-9
  duplicating process, 23-2 to 23-5
  enhancement in, 23-3 to 23-4
  erasure of incoming cassettes in,
    23-10
  film-to-tape transfers in, 23-1 to 23-2
  fulfillment of orders for, 23-9

Videocassette duplication (Cont.):
  master library in, 23-10 to 23-11
  master tape evaluation in, 23-3
  mastering in, 23-1 to 23-2
  overprints in, 23-11
  packaging duplicates, 23-7 to 23-8
  quality control in, 23-5 to 23-7
  recycling tape in, 23-9 to 23-10
  retrieval of tape in, 23-10
  rights protection and, 23-4 to 23-5
  storage of tapes in, 23-10 to 23-11
  system design for, 23-3
  tape inventory control in, 23-10
Videocassette recorder (VCR):
  Beta, 1-8, 6-9
  home, 1-8
  U-Matic, 1-3 to 1-4, 6-8 to 6-9, 23-5
  Video Home System (VHS), 6-9,
    25-3
Videocassettes:
  compatibility of, 25-3
  duplication of (see Videocassette
    duplication)
  interchangeability of, 25-4, 25-13
    to 25-14
  one-half-inch, 6-9, 29-5
  playback and receiver locations for,
    6-8 to 6-9
  in rise of private television, 1-3 to
    1-4
Videodisc, 26-7, 29-5 to 29-6
Videotape (see Videocassettes)
Videotape recorders (VTR), 22-5 to
  22-7
  master and Iso, 20-3
  multiple, 22-6 to 22-7
Videotape training sessions, 28-6
Viewer attitude, 3-2 to 3-3
Vignette, 13-2 to 13-3, 13-6
Visionary, 2-2 to 2-3
Visual, time on screen, 13-9
Visual artworks, copyright class, 24-5 to
  24-6
Visual concept, 11-6 to 11-7, 13-3 to
  13-4
VITC (vertical interval time code), 22-3
  to 22-4
Vocabulary of market, 10-3 to 10-4
Voice-over, 18-4
Voltage, 26-3
Volume unit, 18-3

18 INDEX

VTR (see Videotape recorders)

Warranties, equipment contract, **9**-11
Waveform monitor, **16**-9
Weather, location shooting, **21**-6
Wide shot, **13**-8, **13**-9
Wind noises, **18**-11
Wind screens, **18**-11
Wipes, **20**-3

Wireless mike (see Radio link)
Workshop sessions, **28**-6
Writer:
  copyright and, **9**-5
  job description for, **27**-5
Writer's block, **13**-7

Zen and directing, **20**-1
Zooming, **20**-9 to **20**-10